Dzierzon's Rational Beekeeping

by Charles Nash Abbott

with an introduction by Jackson Chambers

This work contains material that was originally published in 1882.

This publication is within the Public Domain.

*This edition is reprinted for educational purposes
and in accordance with all applicable Federal Laws.*

Introduction Copyright 2018 by Jackson Chambers

COVER CREDITS

Front Cover
European honey bee extracts nectar by John Severns = Severnjc
(Photo by John Severns.)
[Public Domain],
via Wikimedia Commons

Back Cover
Beekeeper Charlie Brandts works on the South Grounds of the White House
by *The White House* from Washington, DC (P052915PS-0304)
[Public Domain],
via Wikimedia Commons

Research / Resources
English Wikipedia
Johann Dzierzon Page
https://en.wikipedia.org/wiki/Johann_Dzierzon

Wikimedia Commons
www.Commons.Wikimedia.org

Many thanks to all the incredible photographers, artists,
researchers, and archivists who share their great work.

PLEASE NOTE :
As with all reprinted books of this age that are intended to perfectly reproduce the original edition, considerable pains and effort had to be undertaken to correct fading and sometimes outright damage to existing proofs of this title. At times, this task can be quite monumental, requiring an almost total rebuilding of some pages from digital proofs of multiple copies. Despite this, imperfections still sometimes exist in the final proof and may detract slightly from the visual appearance of the text.

DISCLAIMER :
Due to the age of this book, some methods or practices may have been deemed unsafe or unacceptable in the interim years. In utilizing the information herein, you do so at your own risk. We republish antiquarian books without judgment or revisionism, solely for their historical and cultural importance, and for educational purposes.

Self Reliance Books

Get more historic titles on animal and stock breeding, gardening and old fashioned skills by visiting us at:

http://selfreliancebooks.blogspot.com/

introduction

    Here at *Self-Reliance Books* we are dedicated to bringing you the best in *dusty-old-book-knowledge* to help you in your quest for self-sufficiency and food independence. We are pleased to bring you another wonderful old book on Beekeeping.

    This special edition of **Dzierzon's Rational Bee-Keeping** was written by Dr. Johann Dzierzon, a Polish Apiarist, and translated from the original German into English by Charles Nash Abbott. Abbott held the US copyright on the work, and is therefore listed as the author. It was first published in 1882, making it over one-hundred-and-thirty years old.

    Despite being Polish, Dr. Dzierzon wrote his books in German, rather than his native Polish-Silesian dialect.

    The book features chapters on *Theory of Bee-Keeping, The Sex of Bees, Fertilization of the Queen, The Food of Bees, The Dwelling of Bees, The Different Kinds of Bee-Hives, On the Different Methods of Bee-Culture,* and much more.

    A great old book on Apiculture, and the perfect place to begin for all those starting out in, or even considering taking up Beekeeping.

~ *Roger Chambers*
State of Jefferson, March 2018

PREFACE TO GERMAN EDITION.

Since the appearance of the First Edition of 'Rational Bee-keeping,' the Author's latest and most complete work on Bees, great progress has been made in Bee-keeping, and some very important discoveries have taken place in the theory and practice of this science. We have succeeded in so modifying the straw skep, the commonest bee-hive, as to allow of the removal of the combs, and in constructing a machine to empty the honeycombs of their contents without destroying them; and thus to profit to the greatest possible extent by their moveability. We have further been successful in discovering the nature and the cause of foul brood, the greatest terror of Bee-keepers, and, almost simultaneously, have discovered an infallible cure for it in salicylic acid; and, lastly, we have ascertained that sugared milk and eggs form a harmless substitute for the natural food of Bees, and are a powerful means of promoting breeding. These great advances and important discoveries having been duly considered in the New Edition of 'Rational Bee-keeping,' the publication of which had become necessary, I think I am justified in calling this Edition an enlarged one, more especially as the number of pages, which formerly was 314, has been increased to 358.

The tendency of the work has remained unaltered in the New Edition, and is chiefly practical. The theoretical part has therefore been treated but very briefly, and with special regard to practical Bee-culture, the Author's aim being to show

PREFACE.

how to keep Bees rationally and profitably, and at the same time to enable Bee-keepers to satisfy themselves and others as to the reason for the different operations in their apiaries; for he only is entitled to be called a rational Bee-keeper who clearly comprehends why a thing is done in a certain way, and not in any other, and who is able to give the reason for everything he does.

I trust that all, whom the reading of this book may encourage to study the science of Bee-keeping practically, will find it not only a profitable pursuit, but also derive from it the pure and lasting enjoyment which is afforded to every perceptive mind by the contemplation of the works of the Almighty in the wonders of Nature, as strikingly illustrated in a colony of bees; an enjoyment, which during a period of more than forty years has been experienced by

THE AUTHOR.

PREFACE TO ENGLISH EDITION.

SINCE 1844 the Author has been writing articles on Bees and their management for various Journals, but it was not before 1848 that he published his first book, entitled *Theory and Practice of the New Bee-Friend*, which was followed in 1852 by an appendix. From 1854 to 1856 he published a monthly journal, *The Silesian Bee-Friend*, and in 1861 his last work, *Rational Bee-keeping*. The latest edition, 1878, contains in a condensed form all the experience and improvements which are to be found scattered in the works previously mentioned.

Among the most recent advances in Bee-keeping the following three deserve some mention :—

1. The Introduction of Cyprian Bees. These were welcomed with great enthusiasm on account of their beautiful colour, but they have shown themselves very spiteful and difficult to control. They prolong the deposit of brood into the autumn, and, after middling years, are at the end of the honey-season for the most part rich in bees and poor in honey. The Author, therefore, decidedly gives the preference to the Italian bees, which he introduced nearly thirty years ago, and which are characterised alike by beauty, gentleness, and industry, and are now perfectly acclimatised.

2. Recently sections have been introduced for the bees to build comb in and fill with honey. They pack easily in the

Preface to English Edition.

honey-crates, travel well, and more readily find a market. The Americans construct their hives to open at the top, so as to be able to put on supers with these sections arranged in series; the hives must then be set up separately and require much space. In Ständer Hives the sections may quite as conveniently be put in at the side-door, removing the cover-boards and arranging the sections over the brood-room; in Lager Hives they may be placed at the sides. In Germany the sections may be obtained of Messrs. Schulz and Gühler, Buckow.

3. It is well known that recently Salicylic Acid has been recommended as a remedy for foul-brood, the most dangerous of all the diseases of bees, and the method of treatment has been made known by Mr. Hilbert. Formerly he advised spraying with diluted tincture of Salicylic Acid, but more recently he advises fumigating with refined powdered Salicylic Acid strewed on a metal plate and heated over a flame. Since this treatment is much simpler than the repeated spraying of the separate combs, and is applicable even to hives with immoveable combs; the Author thought it ought not to be left unmentioned.

Dr. DZIERZON,

Karlsmarkt, April 29th, 1882.   *Emeritirter Pfarrer.*

TRANSLATOR'S PREFACE.

IT was a matter of some surprise to me to learn that we had in English no translation of recent German works on Bee-keeping. This fact led me to attempt to supply the deficiency by the translation of the work of Dr. Dzierzon, whom we may very well agree with Baron von Berlepsch in entitling 'the Father of the new era of Bee-keeping.'

It was at the recommendation of the late Mr. Schmid, editor of the *Bienenzeitung*, that this work was selected as being the best for giving a general view of the German methods of Bee-keeping.

On writing to Dr. Dzierzon, I learned that he had already given permission for the translation of his book to Mr. A. Neighbour. Mr. Neighbour and Mr. Dieck kindly placed at my disposal the portion (about one-fifth) already translated, and I very gladly availed myself of their offer.

I have to express my great indebtedness to Mr. Abbott for his careful revision and notes, and for all that he has done to assist in publication.

I trust the present work may help to promote those more rational and humane methods which it is the object of the British Bee-keepers' Association to extend.

S. STUTTERD.

Banbury, August 29th, 1882.

EDITOR'S PREFACE.

The great Bee-master, the late T. W. Woodbury, being deeply impressed with the value of German ideas and experiences in bee-culture, glimpses of which he obtained through translations casually published, learned the German language, that he might benefit to the full by the wealth of knowledge contained in the writings of German authorities. Had he lived, it is more than probable his high appreciation of the justly celebrated Dr. Dzierzon would have induced him to translate and publish the masterpiece of that renowned bee-master—a work, by an authority, that contains riches in every page well worthy the study and consideration of all bee-keepers; and which, being now rendered into English, will, we are assured, be duly appreciated wherever that language is spoken.

Our share in the work now offered to the English-speaking public has consisted chiefly in preventing erroneous interpretations of technical terms, and in noting the passages which might otherwise have conveyed doubtful impressions to an amateur in bee-keeping; and occasionally by offering, in all humility, suggestions, where our own practical experiences have not been similar to those of the learned Doctor. The German system of bee-culture, as exemplified in *Rational Bee-keeping*, though

Editor's Preface.

strongly insisting on mobility of combs as a first principle, differs widely from that now taught and practised in the British Isles and in America; and in this very difference will be found the chief value of the work. That its perusal and study may be agreeable and beneficial to others, as they have been to us, who have been honoured by being permitted to share in its production, is the earnest wish and hope of its Editor and Revisor.

THE EDITOR of the *British Bee Journal.*

Fairlawn, Southall, Middlesex.
  Sept. 1882.

CONTENTS.

	PAGE
BEE-KEEPING: AN OCCUPATION AS PLEASANT AS IT IS PROFITABLE	1
THEORY OF BEE-KEEPING; OR, THE NATURAL HISTORY OF BEES	5
The Sex of Bees	8
The Queen	10
Fertilisation of the Queen	13
The Workers	19
The Drones	21
Activity of Bees	24
Construction of Combs and Cells	25
Purposes of the Cells	26
Production of Wax	29
The Food of Bees	30
The Vital Activity of Bees	35
The Aim of their Activity	36
THE PRACTICE OF BEE-KEEPING	41
Bee-stand (Apiary)	41
The Dwellings of Bees	43
The Material of Bee-hives	44
The Shape of Bee-hives	47
The Size of Bee-hives	49
Convenience of Handling	51
Side Door	52
Moveable Combs	53
Covering Boards	56
Guide-Combs	58
Frames	63

Contents.

	PAGE
The Different Kinds of Bee-hives	67
Log-hive (Klotzstock, or Klotzbeute)	67
The Straw-hive	69
Gravenhorst's Bogenstülper	70
The Thorstock (Door-hive)	72
The Twin-stock	76
The Double Lager-hive	95
The Advantages of the Twin-stocks	97
The Lager-hive, with Four Compartments	105
Ständer-hives	108
The Ständer, of One Compartment	109
The Ständer, with Two Compartments	117
The Ständer, with Four Compartments	119
Pavilion	123
The Hive of Three Compartments	129
The Sixfold Press-hive	130
Four and Eightfold Hives	134
Hives for Queen Raising	135
The Observation-hive	136
ON THE DIFFERENT METHODS OF BEE-CULTURE	139
The Swarm Method	139
The Zeidel Method	141
Rational Bee-culture	142
On the Occupation of the New Hives	143
The Furnishing of the Hive that is to be occupied	146
The Swarms	151
The Different Kinds of Swarms, their Origin, and Treatment	152
Pauper Swarms	152
Regular or Proper Swarms	152
The First Swarm	153
After Swarms	154
The Securing of the Swarms	158
Artificial Swarms	165
Driving	171
What can be done by the help of a Fertile Queen	178

Contents.

	PAGE
How to get Fertile Queens	180
Queen-cells	183
The Introduction of Italian Bees	189
The Propagation of Italian Bees	193
The Further Treatment of Swarms	200
The Management of Honey Stocks	203
The Emptying of the Honey-Room	206
Bell-glasses as Supers	210
The so-called 'Travelling Bee-Culture'	213
The Increase of Honey Production	218
Preparation for Winter	223
Signs of the Possession of a Good Queen and of Queenlessness	229
The Requisites of a Stock which is to Stand the Winter	233
Wintering	238
What has to be seen to in Winter	244
How Chilled Bees are to be revived	248
Stocks are to be kept Quiet as long as possible	250
The ordinary Spring Cleaning	252
Feeding	255
Robbing	259
The Spring Comb-cutting	264
The Time for Comb-cutting	266
How and where should we cut?	266
Further Matters that have to be attended to	268
The Ailments of Bees	270
Dysentery	270
Foul Brood	271
Symptoms of Foul Brood	272
Foul Brood is of two kinds	273
The Method and Treatment of Diseased Stocks	274
Vertigo (*Tollkrankheit*)	278
The Enemies of Bees	280
Destructive Influences of Weather	286
Apicultural Implements	287
The Smoking Apparatus	288
Bee-Cap or Bee-Hood	290

Contents.

		PAGE
The Sieve	. . . .	290
Transport Hives	. . . . . .	291
The Comb-horse	. . . . . .	294
The Pliers	. . . .	294
The Fork	. . . .	294
The Knife	. . . .	295
The Feather	. . . .	295
The Hook	. . . .	295
Queen-cages	. . . .	296
The Drone-trap	. . . . . .	297
The Swarm Net	. . . . . .	298
The Scooping Vessel	. . . . .	299
The Feeding-box	. . . . . .	299
The Tin-pan	. . . .	300
The Syringe	. . . .	300
The Screw	. . . .	300
The Honey-Slinger	. . . . . .	301
The Knife for unsealing Comb	. . . .	303
The Refining of Honey and Wax	. . . .	303
The further Use and Manufacture of Wax and Honey	.	305
Bee-Pasture	. . . .	307
BEE-CALENDAR	. . . . . .	313
January	. . . .	313
February	. . . .	316
March	. . . .	319
April	. . . .	321
May	. . . .	324
June	. . . .	325
July	. . . .	328
August	. . . .	331
September	. . . .	332
October	. . . .	334
November	. . . .	336
December	. . . .	337
INDEX	. . . .	339

DZIERZON'S RATIONAL BEE-KEEPING.

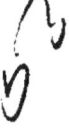

BEE-KEEPING:

AN OCCUPATION AS PLEASANT AS IT IS PROFITABLE.

THE keeping of bees is an occupation as pleasant as it is profitable. Nothing affords such pure and lasting pleasure as the contemplation of the works of the Almighty in the wonders of nature; but nowhere do we find such an exhibition of the wonders of nature as in a colony of bees. It will not be an easy thing to induce a true bee-keeper to exchange bee-keeping for any other occupation; for the more intimately he becomes acquainted with the nature of bees, the more wonders will he discover, and the more enjoyment will he derive from bee-keeping. The genial Baron von Ehrenfels, who has called bee-keeping the 'Poetry of Agriculture,' could not have expressed more beautifully the charm which bee-keeping possesses. But the material advantages to be derived from bee-keeping, if carried on zealously and on rational principles, are also considerable. As a single bee can gather but little, while a large stock may accumulate many pounds, or even hundred-weights, of honey and wax; so bee-keeping, carried on by many, though it be on a limited scale only, may contribute largely to increase the prosperity of a country. We are told in the Scriptures that honey and wax were considered valuable products by the ancients, who used such words figuratively to indicate the fertility of a country. There is still a great demand for these articles of commerce at the present time, and the late rise in the price of wax helps to prove that in spite of the progress made in chemistry, and the application of its discoveries to the arts and manufactures, no substitute for bees' wax has yet been found.

What is realised by bee-keeping is, indeed, all clear gain. The food for domestic animals has to be raised and to be supplied to them; but with bees the case is different. They gather their own food, and in so doing render us an indirect service by causing a more perfect fertilisation of flowering plants, thus increasing the productiveness of our gardens and fields.

It has been asserted that bee-keeping is remunerative only in districts where the soil is but little cultivated, and not in those where agriculture is carried on extensively. The fallacy of this idea is proved by the returns every year, when the season has not been an especially unfavourable one for bees. Even in the most highly cultivated districts the results, in favourable seasons, are such as to cause a considerable and general depression in the market value of honey and wax. What bees are deprived of by extending cultivation is abundantly made up to them in cultivated plants—such as rape, clover, buckwheat, vetches, beans, tobacco, &c., &c., from which they are able to gather honey. The better the soil is tilled and manured, the more luxuriantly also many kinds of weeds grow—such as the wild mustard, which may be seen flowering abundantly in many districts from spring till late in autumn, and which alone affords bees a rich pasturage; thus a farmer may turn to account even the weeds in his fields if he is also a bee-keeper.

Bee-keeping is, moreover, a source of moral improvement, which should not be overlooked. Whoever has once experienced the pleasure to be derived from the study and culture of bees will, I am convinced, spend every leisure hour in his apiary; and in so doing he will escape temptations to immoral and expensive amusements, and while contemplating the indefatigable industry, the cleanly and orderly habits, and the attachment of bees to their queen, he will be encouraged to follow their example; and even should circumstances not permit of any material gain, his moral nature, at least, will be greatly improved.

It is, therefore, very satisfactory to observe that more attention has lately been given to bee-keeping, which formerly was very much misunderstood and neglected. The Author has been awarded a great many certificates of honour, and silver and gold medals; the honorary degree of 'Doctor' has been conferred upon him by the University of Munich, and he has been decorated with the Grand Ducal Hessian Order of Ludwig, and the Imperial Austrian Order of Francis Joseph, besides receiving many other distinctions. He mentions this simply to show that Governments, not less than learned institutions and agricultural societies, now acknowledge the real importance of bee-keeping, and are endeavouring to promote its interests. Bee-keeping may assume still greater importance should the attempts that are being made to check the ravages and further spread of the Phylloxera, the terror of the vine-grower, prove ineffectual; in which case the deficiency in the vintage will probably have to be supplemented by the use of wine made from honey. For wax and honey there will always be a demand, and bee-keeping will maintain its position as an important branch of industry.

But whoever wants to derive profit from breeding or training animals must be thoroughly acquainted with their nature, instincts, and disposition; and must observe the conditions essential to their thriving. In bee-keeping, such knowledge is indispensable, because bees are very delicate little insects, 'fearfully and wonderfully made.' To remedy any disorder in the hive, and thus prevent loss, is often a simple matter; but whoever is not intimately acquainted with the economy of bees will either not know where assistance is wanted, or will take measures altogether unsuited to remedy the evil. In districts where good pasture is plentiful and the season favourable even an inexperienced bee-keeper is pretty sure to realise a profit, for the losses on some hives will be more than covered by the large profits on others. Under less favourable circumstances, however, the effects of bad management will be more evident. The less plentiful the supply of food, the more will bees require judicious management, and the more intimately one is acquainted with their nature, the better can bees be managed. Dexterity in mechanical contrivances is not the only requirement for a bee-keeper, for many unforeseen cases may happen in which a merely superficial knowledge would be useless. But whoever has thoroughly mastered the whole economy of bees will never be at a loss to know how to act, and will always employ the most suitable means to attain the end he has in view. Any man of quick perception and of sound common sense may acquire the requisite knowledge by his own observations; but in order to fully understand the wonderful economy of bees, and to reveal the profound secrets of the hive, a man's whole lifetime would hardly suffice. In order, however, not to waste valuable time, and to save the trouble and cost of expensive experiments, every beginner in bee-keeping would do well to make use of the experience of others for his own instruction. The Author, who has given the most careful attention to bees from his early youth—for more than fifty years,—and who has manipulated hundreds of colonies in all kinds of hives for the last forty years, intends herein to publish, for the use and benefit of the public, the experience he has accumulated, and he hopes that all who follow his directions may derive from bee-keeping as much profit and pleasure as he has done.

THEORY OF BEE-KEEPING;

OR

THE NATURAL HISTORY OF BEES.

THE wonderful little creatures we call bees, of which this book treats, belong to the species of the class of insects called Hymenoptera, which live in large societies, or swarms. It is not, however, by chance, and occasionally only, that they congregate in large numbers, like many other animals—such as grasshoppers, gnats, and many kinds of birds; but they are quite unable to propagate their species, or even to exist, in an isolated state, special functions and special duties being assigned to the different members of such society, which we call a swarm, or a colony, or a stock of bees, if we wish to include the hive also.

Like all insects bees are cold-blooded; but to be able to live and thrive they require a moderate degree of heat, which, according to the season, varies from $10°$ to $30°$ R. ($= 54\frac{1}{2}°$ to $99\frac{1}{2}°$ F.) At a lower temperature they become more and more chilled, and lose the power of moving and of using their limbs, and when the thermometer falls to the freezing point, the cold gradually benumbs them, and they do not recover vitality, or at least do not attain their former strength. In our climate, and in most countries of the temperate zone, where the temperature frequently falls below freezing point, they would be unable to exist at all could they not produce and constantly maintain in their hive the degree of heat necessary to their well-being; and this they are able to do only when united in considerable swarms.

Even when the external temperature is low they are able to produce and maintain sufficient heat, while forming a dense cluster, chemically by breathing, which, as it were, is a slow process of combustion by the combination of the oxygen of atmospheric air with the carbon in the blood, and mechanically by certain movements of the wings and other parts of the body, and by the friction thus obtained; the too rapid escape of heat being prevented by the air confined in the cells of the comb, the hair of their bodies, and even by the hive itself. In order that bees may be able to propagate their species, it is necessary for them to unite in swarms, not only because in congregating together they are able to keep up the temperature necessary to the brood, but also because the different kinds of bees in the hive have different generative functions and perform different duties, all absolutely necessary to the existence of the colony. Regarding these functions and the sexual relations of the different kinds of bees in the hive, most widely divergent opinions prevailed, and have been defended with great obstinacy until quite recently. In order to maintain the views expressed by the Author in his *Theory and Practice*, which appeared in 1848, he was obliged to endure the most violent controversies, which were continued for several years in the *Eichstadt Bee Journal*, the organ of German bee-keepers, but were mostly settled in the *Bienenfreund of Silesia*, which paper was published as a monthly periodical from 1854 to 1856. It was chiefly owing to the introduction and propagation of the Italian bee that the prejudices and errors, which were current for centuries, had at last to give way so quickly and universally.

The Italian bee forms a distinct race of *Apis mellifica*, and is distinguished from the common German bee* by being differently coloured, as well as by other characteristics. While in the German bee all the rings of the abdomen are of an equally grey or black colour, in the Italian bee the first two rings, or rather two and a half, are of an orange colour, looking like a yellow band, which is visible even when on the wing. This beautiful gold-coloured bee was known in early times, and the Roman poet Virgil, in the fourth book of his *Georgics*, which treats of bee-keeping, declared it to possess positive advantages, which still hold good at the present time. It is more industrious than the German bee, more prolific, more capable of defending its hive when attacked by robbers, and of a very gentle disposition, making use of its sting but very rarely, and only when squeezed or very much irritated. The first colony of these bees the Author received in February 1853, since which time they have

---

* And the English bee.—C. N. A.

spread from Carlsmarkt not only all over Germany and the neighbouring countries, but also into Sweden, Denmark, and even into America.* In the following chapters on the relation of the sexes in the hive, it will be explained how this bee contributed to hasten the universal recognition of the long-hidden truth.

Meanwhile, the Egyptian bee has also been introduced into Germany, and has been propagated there. It is somewhat smaller than the Italian bee, which it otherwise very much resembles, but may be distinguished from it by the hairs, which are of a whitish colour, and a small brown plate on the back, at the point of the thorax. The Egyptian bee is of scientific interest only. In the opinion of Mr. Vogel, the well-known bee-keeper and schoolmaster, who was the first to cultivate this bee, it is of no practical value whatever, chiefly because it breeds almost throughout the year, and therefore winters badly.

The Carniolan bee is not considered a separate race, but rather a variety of our German bee; it has been largely introduced into Germany lately, chiefly on account of its gentle disposition and great propensity to swarm. Great difficulty is experienced in keeping it pure on account of its resemblance to the German bee, which renders it difficult to distinguish a hybrid of the two.†

He who is desirous of increasing the number of his stocks largely, will find the Carniolan bees answer his purpose, although he may often be obliged to assist by feeding them; but he who chiefly aims at the largest return of honey, will give the Italian bees the preference.

Even the common black or grey German species may be divided into a honey-bee and a swarming bee. The latter, which is also called the Heath bee, in consequence of the management adopted, swarms continually; and swarming has become quite a second nature with this bee, so much so, that even colonies with young queens of the first year make preparations for swarming and breed drones, which is never done by the honey-bee that is met with in the greatest part of Central and South Germany.

* Parenthetically we may add, that in the British Islands there are now many hundreds of colonies.—C. N. A.
† This applies also to England.—C. N. A.

The Sex of Bees.

On this subject the most erroneous views were held formerly. Even the queen, who, as every beginner in bee-keeping at the present day knows, lays the eggs from which all the bees in the hive originate, was formerly considered a male bee. The poet Virgil, who has just been referred to, calls her by no other name but that of king; and the name of *Weiser*, or *Weisel*, formerly in use in Germany, shows that the same erroneous view prevailed there. Any one who is but slightly acquainted with insects knows that the females of bees, and of humble-bees, wasps, and hornets, which are allied to bees, are provided with a sting, while in male bees it is wanting. From the presence or absence of the sting the sex of bees can at once be determined. The drones, which, as is well known, have no sting, are male bees; the workers and the queen, which are provided with stings, belong to the female sex. Male bees, or drones, are larger than the common worker-bees, and wider cells are therefore required for breeding drones. Drone-cells are about a quarter of an inch in width, and about sixteen of them would occupy a square inch. They are also of somewhat greater depth than worker-cells. But although in some cases—which, however, are quite exceptional—drones are developed in the more narrow worker-cells, which are one-fifth of an inch wide, and of which twenty-five go to the square inch, drones hatched in drone-cells do not differ from those developed in worker-cells, except that in the latter case the drones are smaller. Small drones are also perfectly capable of copulating with a queen-bee.

To make a distinction between large and small drones is quite inadmissible;* small drones are not produced in the hive regularly as a distinct kind of bee—it is occasionally only they make their appearance, and when they do it is generally a sign that there is something wrong in the hive. But the case is different with female bees, which possess a sting. It is true they

---

* We are inclined to dispute this point with our learned Author. If the females come forth undeveloped, as compared with queens, through having been reared in smaller cells, it is but reasonable to suppose that drones (males) reared in small (worker) cells are also not fully developed. They are certainly not a distinct kind of bee, but stand in the same relation to drones that workers do to queens.—C. N. A.

all originate from female eggs, which are all quite similar, but whether they become partially or perfectly developed depends on whether they have been hatched in narrow worker-cells, supplied with less plentiful and less nourishing food, or whether they have been produced in the so-called royal cells, which are wide and long and hang down from the comb, and have been supplied with more plentiful and more nourishing food. Perfectly developed females we call queens, mothers, or *Weisel*, and they alone are capable of propagating the species. The other class, and by far the most numerous in the hive, are imperfectly developed females. They are qualified to do all the necessary work, but are unable to perform the act of copulation, and to propagate their species, on account of their seminal receptacle being undeveloped. They have therefore quite appropriately been termed worker-bees. In any case, it is not correct to call them neuters—a term used in many books on bees and in works of natural history—for they originate from the same eggs as the queens, they possess a sting like all females of this class of insects, and under certain conditions, especially when the stock is deprived of its queen and the regular business in the hive is interrupted, they are even capable of laying eggs, their ovary, the rudiments of which have remained in spite of their imperfect sexual development, in such a case becoming active.

Although there are only two genders of bees, male and female, yet at the time of highest development—the prime of their vigour, which coincides with the time of flowering of most plants—three kinds of bees, differing in shape and design, are met with in the hive, viz.:

1. The perfect female, queen, *Weisel*, or mother;
2. Imperfect females, or worker-bees; and
3. Male bees, or drones.

If we wish to add, as a fourth class, as has been done by many writers, those egg-laying worker-bees which only appear in exceptional cases, it would be necessary to mention as a fifth or sixth class those unfertile queens, or queens that lay drone eggs only, which are also sometimes met with, but this would be out of place and objectless, as such conditions are abnormal.

The Queen.

The queen is the principal bee, on which the prosperity of the colony chiefly depends. The resemblance between the queen and the worker-bee indicates that they are both of the same sex and produced from similar eggs, but the more perfect development of the queen is shown by a stronger frame of body, more especially by a larger thorax, of a rather brownish colour, and a stouter and longer abdomen, which is hardly half covered by the wings at the time when she is most actively engaged in depositing eggs.

Her legs are likewise of a brownish or yellowish colour. In the Italian bee the greater part of the abdomen is of a golden yellow colour, and black only towards each extremity. It is, therefore, much easier to discover an Italian queen among a number of worker-bees, than it is to find a black queen. As a rule there is but one queen in the hive, except when rainy weather sets in at the time of swarming, in which case a young queen often hatches while the old one is still in the hive. Several young queens may also be together in the hive for some time. But at any other time than the swarming season it is the exception for two queens to be in the hive in autumn, winter, and spring, and then one is generally an old queen, frequently mutilated, who remains on sufferance, while the other is vigorous and young and the reigning queen. In such a case the wings of the old queen are generally bitten off entirely, no doubt in consequence of the attacks made on her by the young queen. Jealousy, one of the faults of the queen-bee, appears to be greatest immediately after hatching; it then subsides by degrees, and as soon as the young queen has become fertile her whole attention is devoted to depositing eggs in the brood-cells.

Queens are reared in special cells, called royal or queen-cells. Although there is no essential difference among royal cells, yet in regard to their origin a distinction is made between swarm-cells and supplementary cells. The workers make the foundation of a swarm-cell generally at the edge of a comb, in the shape of a small round cup, slanting downwards, into which an egg is deposited by the queen.* As the royal larva grows and requires

---

* Here we fear there will be differences of opinion; but the Author having divided queen-cells into two classes, the second of which offers the

more space, the cell is carried further downwards, and at last it is sealed, when it appears not unlike an inverted acorn. Supplementary cells occur when bees have accidentally been deprived of their queen but possess suitable brood from which to rear a young queen. Being unable to remove eggs or larvæ from one cell into another, a common worker-bee cell containing an egg or a larva is changed and widened into a royal cell. When cells with small larvæ or eggs are situated near the edge of the comb, or above a passage which might happen to be in the comb, the bees prefer raising the royal cells there, because they meet with no obstacle in continuing the cells downwards, there being no need for destroying any cells beneath them. Yet they will sometimes commence royal cells in the middle of the comb, from which they may then be seen projecting, point downwards.

Many bee-keepers are of opinion that queens hatched in supplementary cells are weaker than those produced in swarm-cells, which, however, is not always the case. The more perfect development of young queens is due to the more strengthening food administered, the efficacy of the food depending on the supply of nitrogenous matter, *i.e.*, pollen, and also on whether there is a sufficient number of young bees in the hive to attend to the care of the larvæ, more especially to those in royal cells.

The more perfect development of a queen is due partly to the greater width of the cell and partly to the larva being supplied with more plentiful food, the latter being throughout of an equally nourishing quality until the cell is sealed, while the food which the larvæ in the small cells receive becomes less and less nourishing, so that sexual development does not take place to the same extent. But as the food of the larvæ for the first few days is alike, it matters not whether the bees select an egg or a larva a few days old from which to rear a queen, the only difference being that the older the larva selected the sooner will the young queen leave the cell. It was formerly supposed that it was impossible to rear a queen from a larva more than three, or at most four, days old,

best opportunities for observation, and few having opportunity for observing the operation described in the first, ourselves amongst the number, we are unprepared to dispute the point. We, however, respectfully beg to differ with the Author as to the inability of bees to remove eggs from cell to cell, having had satisfactory proof that they can and do occasionally remove them.—C. N. A.

the Author, however, has proved in his *Supplement to Theory and Practice*, published in 1852, that perfect queens may be reared from even older larvæ, as long as they remain unsealed. In a swarm made artificially, we often, ten days after, find a young queen. Curiously, however, the development of a queen from the egg to the perfect insect, when the bees commence at once to nurse and attend to it carefully, is completed in sixteen days. The worker-bees, in the former case, must therefore have selected a larva at least six days old; for, supposing the larva to have been at the outside four days old, this would allow fourteen days only for the development of the queen, which is an impossible supposition.* We may also convince ourselves of the correctness of what is stated above if we continue to destroy the royal cells commenced in an artificial swarm, when it will be found that the workers again and again commence new queen-cells, so long as there is unsealed brood in the hive, till about the seventh day. Royal cells are then frequently only just raised above ordinary cells, and scarcely to be distinguished from drone brood-cells; nevertheless, they produce perfect queens. This, at the same time, proves that the sloping position of royal cells is not absolutely necessary for the perfect development of the queen, although the suitability of such a position of the cells is unmistakable. For the cell to be lengthened horizontally the available space (between the combs) would often be insufficient, nor would it be possible to construct the cell durable enough, while an upward direction would not protect the cell from falling *débris*. A royal cell serves but once for rearing a queen, and is generally destroyed soon after the queen has hatched. It is only when a colony has been weakened by swarming, and part of the comb is left unoccupied by bees, that royal cells are occasionally allowed to remain. In all other cases royal cells are only met with in the hive when required as cradles for young queens.

* Our Author has evidently not taken into account the fact that the egg remains unhatched for a period of about three days, after which three days as an ordinary larva, and ten days under transposition, will make up the sixteen required for the evolution of a queen. There is, therefore, nothing novel in finding queen-cells newly raised on the seventh day after a queen has been removed, as many larvæ will then only be from three to four days old.—C. N. A.

Fertilisation of the Queen.

When the young queen has left the cell, she is not yet a perfect mother. She acquires the faculty to lay the eggs, from which all the bees in the hive originate, only by fertilisation, or copulating with a male bee or drone, an act which always takes place outside the hive and high in the air, often at a great distance from the hive. Every queen-bee, therefore, as soon as she has become the acknowledged regent in the hive, and has attained a certain maturity, which, according to the temperature and activity in the hive, occupies from three to eight days, takes one or more so-called nuptial flights on a warm, bright day, during the warmest hour of the day, when it is also usual for the young bees to play before the entrance of the hive. It seems that on her first trip the queen is seldom impregnated; she reconnoitres her hive carefully, and leaves the neighbourhood very slowly. Perhaps she only intends to make herself acquainted with the place in order to make sure that she will be able to find her hive on her return from successive excursions on the same or the following day.

The warmer and clearer the atmosphere, and the greater the number of drones swarming in the air, the sooner the object of the queen's flight will be achieved. While her other excursions are but of short duration, she does not return before a quarter of an hour, or even half an hour, after she has met with a drone and become fertilised; and then she generally comes back with the male organ of the drone adhering to her extremity, as an unmistakable sign that fertilisation has taken place. It is possible that the queen frees herself from the drone, which, in any case, does not survive the act of fertilisation, just as a bee that has stung, when the sting has not penetrated very deeply, tries to free itself by continually turning round. This, however, is only a supposition, and it will hardly ever be observed, as fertilisation takes place far, often very far, from the hive. It has been observed in many cases, that half-bred Italian bees have appeared in hives of common black bees, the young queen having been impregnated by an Italian drone, although the nearest Italian colony was more than a mile* distant. Even supposing the drone and the queen to have met half-way, they both would have had to travel

* A German mile is equal to about five English miles.

more than half a mile (two and a half English) from their hive to meet where they did. It is most probable, if not certain, that the meeting takes place high in the air, where drones, as well as young queens, on their wedding trips at once repair; and it is certainly the sound which both emit in their flight, and which may even be distinguished by the human ear, whereby they communicate with each other, and by which they are attracted. The union most likely takes place there in the same manner as in the case of dragon-flies and may-flies, which we frequently have an opportunity of observing, as these insects do not fly so high. The couple then probably descend and settle on the nearest object, where separation takes place in a violent manner. Once impregnated, the queen remains fertile for life, the average duration of which is four years. If, by accident, she should happen to cease laying eggs, she never becomes fertile again, for when once she has commenced depositing eggs, which in summer is the case two days after fertilisation, she has no longer the capacity of pairing with a drone, nor does she ever fly out again, except she leaves with the whole colony, or with a swarm, or happens to be driven out of the hive forcibly.

It is easy to ascertain, by dissection, whether a queen has been impregnated or not. When the tip of the abdomen of a queen is cut away with a sharp pair of scissors, there will be seen protruding from one side or the other of the section so formed a small vesicle, surrounded by a network, the so-called *receptaculum seminis*, which is intended to receive the male sperm during the act of fecundation. If the contents are a clear fluid, the queen is still unfertilised; but the presence of whitish viscous matter shows that she has been impregnated. When the contents of the small vesicle are viewed under a high power of the microscope a very large number of moving little bodies are seen, of which one at least has to penetrate each egg by a small opening at the upper end, called the micropyle, or the egg cannot be fertilised.

So far, there is nothing uncommon in what has been stated above; but we now have to discuss phenomena in the stock, which are unique of their kind, and which bee-keepers formerly in vain tried to explain. In many hives we meet with queens— young queens, as well as old ones weak from age—which are only able to lay drone-eggs. Worker-bees, also, under certain conditions, deposit eggs, which always produce drones; but a regularly fertilised queen has the marvellous power of always adapting the eggs to the cells, depositing female eggs in the small cells and male eggs in the drone-cells, and thus determining the sex of the eggs at her own will.

To solve this enigma various hypotheses were propounded, which, however, made the matter only more complicated. It was attempted to explain the fact of male and female eggs being deposited in their respective cells by supposing that certain worker-bees were the mothers of the drones, and that the queen, contrary to all experience, only laid the eggs, from which workers are produced; but every attentive observer will often be able to watch the queen depositing eggs in drone-cells. This subject, nevertheless, led to a long and bitter controversy, which was finally decided by the Italian bees, for when an Italian queen was introduced into a hive of common bees, it was found that not only Italian worker-bees appeared, but Italian drones also, proving plainly, that the latter also owed their existence to the queen. The occurrence of females, whether of queens or worker-bees, which only deposit drone eggs, was explained on the assumption of an incomplete fertilisation, because, according to Busch, it had taken place in the hive, or, according to Huber, it had been retarded; but there is no proof in support of such a conjecture, and, when examined more closely, it is apparently quite untenable. Impregnation never takes place in the hive. However great the number of drones, a young queen does not become fertile if bad weather or the time of the year prevent her from flying out. The queen and the drone are not even disposed to pair in the hive; and could the passion of the drones be excited in the hive, there would be no rest for the queen while there were drones about. In general, the young queen is capable of becoming perfectly fertile so long as she continues her nuptial flights, which in the warm summer she does at the very most for a month; in the cool spring and autumn, however, when life and development in the hive are more at rest, for five or six weeks, or even longer. There is, indeed, no reason why impregnation, when retarded, should be less perfect, and why it should only enable the queen to propagate the male sex.

The truth is, mothers that produce drones only have not been impregnated at all, or fertilisation has remained ineffectual or become so, because eggs which produce drones do not require fertilisation; they contain the germ of life on leaving the maternal ovary, and are drone-eggs from the very reason of their being deposited unfertilised. But if the egg be fertilised by one of the spermatozoa from the seminal receptacle of a fertile queen entering it as the egg passes down the oviduct, it becomes transformed into the germ of a worker-bee or a queen. This statement contains the key of all the problems which until now it seemed impossible to solve. The above conclusion forced itself upon the Author while he had an opportunity of watching several young queens, which, on account of having lame wings, or of their having hatched during the cold season, had evidently not been able to take their wedding flights, and which proved unfertilised when dissected afterwards; nevertheless they deposited eggs, from which, however, only drones were produced. As these same eggs, the first developed in the ovary, would certainly have become worker-eggs in case fertilisation had taken place, nothing was more natural than the conclusion, that originally the eggs are all alike, or are without distinction as regards the sex, and that they become male or female according as they are deposited unfertilised or fertilised.

Hence it is easy to understand why unfertilised queens, or worker-bees, which are altogether incapable of being fertilised, can only lay drone-eggs, while fertilised queens are able to deposit worker-eggs or drone-eggs at their own will, it being possible and easy for them to prevent or cause the fertilisation of the egg by the movement of a muscle. Nevertheless, this doctrine, which the Author, in the early years of the *Bienen Zeitung*, did not venture to put forward but as a mere hypothesis, to be brought out more fully and clearly in his later writings—this doctrine was received with the most violent opposition, as being opposed to the, till then, universally recognised theory, that without fertilisation life is impossible. But as this theory explains the phenomena in the hive as completely as Copernicus's hypothesis accounts for the appearances in the heavens, it gradually gained more and more adherents; even professed physiologists began to take interest in it, and it has now become a recognised fact, having passed the ordeal of science under the microscope and the dissecting needles of the great physiologist, Professor Theodor von Siebold, formerly of Breslau, now of Munich. The latter undertook the tedious task of examining a number of worker-eggs, as well as drone-eggs, as to their fertilisation, the result being that the drone-eggs showed no trace of fertilisation either internally or externally; whereas in the worker-eggs one or more spermatozoa, some of which still showed vitality, were distinctly seen. The worker-eggs had been properly prepared previously by being burst between slips of glass, so as to allow of the interior being viewed.*

At the present time the correctness of this theory is no longer disputed, but opinions still differ as to how it happens that the queen deposits unfertilised eggs in the large cells, and fertilised eggs in the narrow cells. It has been explained that fertilisation and non-fertilisation are a purely mechanical result of mere external pressure or change of position on the part of the queen, necessary in order that she may be able to deposit eggs in the narrow cells and in the wider and deeper ones at will; but this explanation is untenable since the queen frequently deposits eggs in wide and in narrow cells scarcely commenced upon, adapting each egg to its corresponding cell, although there cannot well be a question in this case of pressure† or change of position. We must, in any case, assume that the queen possesses the instinctive power to lay fertilised eggs in small cells and unfertilised eggs in large cells; and this is no more inexplicable than the capability of worker-bees to pass from the construction of small cells to that of large cells at the proper time, and of the queen at the right moment to begin depositing eggs in drone-cells and to leave off again, and then only to lay worker-eggs.

* Compare *True Parthenogenesis in Moths and Bees*, by Theodor von Siebold. Van Voorst, London, 1857.

† The Author in this case ignores the internal pressure altogether. The circumstance under which a queen deposits eggs in cells 'scarcely commenced' are when she is at the head of a swarm newly hived, and the weather and the incoming of honey satisfactory. Her desire for oviposition is then stimulated by the bees, and eggs are produced internally in large numbers, and our suggestion is that the accumulation of eggs in the ovaries enlarges them and causes sufficient pressure to secure their fertilisation as they pass the spermatheca.—C. N. A.

As the queen is capable of adapting the sex of the eggs to the cells, so she is also able to adapt the number of eggs to the requirements of the stock, and to circumstances in general. When a colony is weak and the weather cool and unfavourable she only lays a few hundred eggs daily; but in populous colonies, and when pasture is plentiful, she deposits thousands. Under favourable circumstances a fertile queen lays as many as 3000 eggs a-day; of which any one may convince himself by simply putting a swarm into a hive with empty combs, or inserting empty combs in the brood-nest of a stock, and counting the eggs in the cells some days after.

We occasionally read in books on bees, or works on natural history, that the queen in her lifetime lays about 60,000 eggs. Such a statement is simply ridiculous; 600,000 to 1,000,000 would be somewhat nearer the truth; for most queens, in spacious hives and in a favourable season, lay 60,000 eggs in a month. The queen, as a rule, commences laying eggs in February, and continues until September, though not always at the same rate. An especially fertile queen in the four years, which on an average she lives, may thus lay over 1,000,000 eggs. The Author once had a queen fully five years old, which was still remarkably vigorous, and might have lived for another year or two if she had not been destroyed. It is, therefore, quite credible that the age of the queen occasionally extends to seven years, as we are assured by some bee-keepers who have made this observation; yet when we are told that in exceptional cases queens have continued alive for eleven to twelve years, the assertion probably rests on a delusion, or such a case is as rare as that of a man attaining the age of one hundred years or more. There is certainly a great difference among queens as regards fertility; the best mothers are those that lay a great number of eggs and deposit them in the cells regularly, neither laying two eggs in one cell nor missing a cell. With such a queen in the hive the brood is nicely arranged, and much of it hatches simultaneously, thus making it easy for the queen to repeat the operation of depositing eggs when the cells have been emptied. When such is the case the stock will be thriving, its well-being depending chiefly on the queen, who, as it were, is the soul of the hive.

The Workers.

The workers are by far the greatest number of bees in a colony, and are so called because they perform all the work in the hive. They collect the materials for comb-making and for the food, they keep the hive clean, construct the cells, keep the brood warm and feed it, and guard and defend their hive. As everyone has an opportunity of seeing and examining them for himself, either in the hive or on flowers, when the weather is fine, we will not stop to describe them. While in the egg, or as small grubs, they are all capable of becoming queens. In the small cells, however, and because less nourishing food is supplied to them, only the organs intended for work attain perfection, while the sexual organs remain undeveloped; they are, therefore, imperfect females. As, however, they belong to the female sex, and unmistakably possess a rudimentary ovary, it need not surprise us that under certain circumstances, such as when the stock is without a queen, they deposit eggs. But being unfertilised, and not even capable of being fertilised, the eggs deposited by them, of course, produce drones only. They prefer drone-cells to deposit their eggs in if there are any in the brood-nest, but they also deposit them in royal cells, which bees are in the habit of constructing from anxiety to rear a queen. In the absence of large cells they deposit their eggs in the small worker-cells.

For a long time it was the firm opinion of the followers of the old school that worker-bees regularly laid the eggs which produced drones, even when a queen was in the hive; but, thanks to the introduction of the Italian bees, this opinion has had to give way, as being faulty and untenable, as we may convince ourselves by a close examination of the manner in which eggs are deposited both by the queen and the workers. The queen deposits drone-eggs with the same regularity as worker-eggs, but the workers lay their eggs irregularly, often at the edge of the cell, depositing several eggs, and sometimes a whole cluster, in one cell.

Many people supposed, and some still believe, another kind of bees to exist, which have been termed black bees. Their black colour, however, is a purely accidental one, caused through heating, rubbing against each other, biting, smearing, licking, &c.

As a rule the glossy black bees are robbers, which have been pursuing their trade for some considerable time. There is only one circumstance, their age, on which to base a distinction among worker-bees, which otherwise are quite alike. They may be divided into young workers, or brood bees, which attend to the internal affairs, and more especially to the care of the brood, and old workers, carriers or gatherers, which procure the necessary materials from without.

When the young worker-bee has left the cell—which, reckoning from the egg, will be the case at the end of nineteen days, under favourable circumstances, but generally at the end of twenty to twenty-one days—it does not fly out immediately. Even in a warm summer several days pass before it makes its appearance outside the hive, and then only during the warmest hour of the day, in order to cleanse itself, and then to return to the domestic duties in the hive, with which it is chiefly occupied during the first period of its life.

When there is a sufficient number of young bees in the hive to nurse the brood, the older bees take little or no part in home affairs; during the night they quietly hang about under the comb or rest against the walls of the hive or at the sides of the comb, but in favourable weather they are indefatigable in carrying home the necessary materials, especially honey and pollen, which they accumulate in their hive until death at last overtakes them. If not killed by accident, as no doubt they are in most cases, they die from exhaustion, when their wings, worn out by long usage, are no longer able to support the heavily-laden body, especially in windy weather.

The term of life of worker-bees varies, according to circumstances. Of the workers produced in May or June, few will be alive at the end of two months, if the weather allows them to be continually active. However populous a colony of common black bees, if an Italian queen be introduced in spring or summer there will be very few black bees left in the hive at the end of six weeks, and none, perhaps, at the end of two months. The distance which bees have to fly, and the kind of flowers they visit, however, make a difference in this respect. They appear to get very much aged when, for example, they visit the blue corn-flower, the sharp-edged leaves of which, as well as the thick corn among which this plant grows, wearing out their wings very soon. When visiting the buckwheat blossoms they seem to preserve their strength much longer, partly because their visits, although very frequent, are continued for a few hours of the day only, partly because they are able to hover comfortably above the blossoms, their wings not coming in contact with them. But they preserve their strength best, and hardly get aged at all, while remaining in a state of repose. Consequently, the workers produced in September look as young and strong in February and March as if they had left the cells only a few days before. But if they pass the summer in a similar state of repose, as a stock without a queen or otherwise inactive may do, they may then live for twelve months, or even longer.

It is very improbable, however, that a worker-bee, supposing it to escape all danger, will attain the age of the queen, and live for several years, as Baron von Ehrenfels believed. If he had been acquainted with the Italian bees, and had made experiments with them, he would hardly have made such an assertion. As to the number of worker-bees in a hive, it varies very much, according to the difference in the hives and the time of the year. Towards the end of the winter, or in the beginning of spring, when the number of bees in a hive is generally smallest, many weak colonies will be found to contain only a few hundred bees, but the population of a strong colony at the time of its highest development in summer may perhaps be computed at 60,000 worker-bees. On an average, there are 20,000 workers in a large swarm, 12,000 to 15,000 in an ordinary swarm, and 6000 to 8000 in a weak swarm.

The Drones.

The third class of bees in the hive are the drones, the male gender of which nobody doubts who has any knowledge of natural history at all. Their sole purpose is to fertilise the young queens, and their whole activity is therefore limited to taking an excursion high into the air during the warmest time of every fine day, when it is also usual for young queens to take their nuptial flights. During the remainder of the day they are idle in the hive, and nearly motionless. As in the vegetable kingdom, pollen, on the male parts of the flower, is produced in abundance, so does Nature produce an abundance of males in a colony of bees, in order that the queen, upon which the well-being of the whole colony depends, may be fertilised the sooner, and with as little danger as possible.

It is obvious that drones were not also intended to produce heat in the hive, as has often been attributed to them, for when the young queen has been successfully fertilised * and begun to lay eggs, while perhaps at the same time cool weather sets in, at this very time, when the temperature in the hive would require to be raised, the drones are driven out as being no longer of any use. Like the males of the allied humble-bees, hornets, and wasps, the males of hive-bees do not live through the winter. The workers of the former disperse and perish in autumn, because the fertile females hibernate singly, in a state of total torpidity, and therefore do not require warmth; they are also able to construct the first cells in spring, and to feed the first brood themselves.

The queen-bee, however, is unable to do so, and on this account a certain number of workers are indispensable in a hive of bees even in winter, when drones are altogether useless. The queen has stored up in her spermatheca the fertilising material,

* We cannot accede to the Author's assertion that the fertilisation of young queens is 'the sole purpose' of drones' existence. It is well known that when a swarm has left a hive there is often but a handful of worker-bees left at home to care for the huge mass of brood in all stages that the hive contains; and should a cold night follow a swarming day, as is often the case, this handful of workers would find it impossible to maintain the necessary heat in the hive, and there would be great loss of brood and bee life. In this condition of things the drones, the great majority of which are stay-at-homes (few accompanying a swarm), are of immense service, maintaining heat which otherwise the few workers would be compelled to generate for themselves, and setting the latter free to nurse the newly-hatching larvæ. It is true that when the young queen has hatched and been fertilised, and the weather becomes cool, the drones are slain, but at that time there will be little, if any, unsealed brood in the hive, while thousands of young bees will have hatched into life, rendering the hive populous and the drones unnecessary. Nor must it be forgotten that drones are not usually slain *until* 'cool weather sets in,' or, in other words, until the honey harvest has ceased, a fact upon which is hinged a belief in our mind that they are of service in helping to evaporate the honey prior to its being sealed for winter store. Many have noticed the large number of drones often to be found in supers, and though it is generally supposed they are there as consumers only, it by no means follows that

and she herself fertilises the eggs to be laid. When, in the following spring, the time approaches for swarms to issue and young queens to be reared, drone-breeding commences again. This is a remote preparation for swarming for which hibernated drones are not at all necessary. For, according to the theory stated above, drones are produced from unfertilised eggs, which not only fertile queens, but unfertilised queens and even workers, are able to lay.

It seemed that the Italian bees might also afford an excellent means of testing the correctness of this theory, if Italian drones were different from German drones, which certainly they are, although the difference is not so striking as that between Italian and German workers. If the theory be correct, then black queens will only produce black drones, and Italian queens only Italian drones, even if they have been impregnated by a male of the other race. The consequences of the mingling of different races of bees can only be visible in their female progeny, if only the female eggs require fertilisation; and this, indeed, is the case. In one instance a few yellow drones made their appearance in a colony which had a black queen that had been fertilised by an Italian drone. But it was possible for a yellow worker to have laid the eggs, as there were already many Italian workers in the hive.

such is the case. The old saying, 'Give a dog a bad name,' &c., is fully carried out with drones, and no one seems to seek for or believe there are any good qualities in them; yet, as many will doubtless be able to substantiate, some of the best results have been achieved in hives where drones were, at least, numerous. In that case the thought has been, if the bees did so well with that immense number of drones to keep, what would they not have done without them? ignoring the possibility that the drones may have assisted in procuring the good results. And is it not true that under the present system, with drone-traps on during the honey harvest, many have cause to complain that their supers, though filled, are left unsealed by the bees? We have hundreds of times seen bees returned from the fields give their honey to drones, and have as often seen drones with their heads in honey cells. Is it certain that the drones in this case are not, in a sense, honey-carriers? We know they have no honey-sacs, as have the workers, but that will not make our suggestion ridiculous; the bees prepare the food for the nursling bees in their stomachs, may not drones prepare honey for storage in a similar way? They have no honey sac to collect it in, their duty lying at home except on special occasions at certain hours of the day. Bee anatomists who search only for what they hope to find may perhaps overlook truths that have not been suggested. But let the influence be what it may, we cannot believe the drone to be as useless as he is accredited.—C. N. A.

The case of a worker laying eggs when there is a fertile queen in the hive is very rare, yet it sometimes happens, like the exceptional occurrence of two fertile queens in one stock at the same time.

Although the queen meeting with a drone outside the hive in the air renders it somewhat difficult to preserve the Italian race pure, and thus to increase the number of Italian colonies, yet as even those Italian queens, which have been impregnated by a black drone, produce Italian drones exclusively in the following year, this difficulty is not so very great. An Italian queen which has crossed with a black drone, at the head of a populous colony, frequently lays drone-eggs in her first summer, which young queens of the native German race but very rarely do.

The question has been raised, whether drones produced from eggs laid by worker-bees are perfect males, capable of pairing with a queen.* This question must certainly be answered in the affirmative, for as in every respect they are like the drones, originating from a queen, there is no reason why they should not also possess all the capabilities of the latter.

Activity of Bees.

The primary object of the activity of bees, as of all other animals, is self-preservation and increase of their species; in the words of Scripture, 'Increase and multiply,' the direction of their actions is defined for them.

The first object of a young swarm is to secure its own existence. For this purpose its first requirement is a new home, *i.e.*, a cavity with a comparatively small entrance, which the swarm usually reconnoitres before leaving the parent hive by means of so-called scouts. A number of bees, which, as it were, go before the swarm to take possession of the new abode and at

---

* This is a question distinct from one we have frequently been consulted upon, though nearly allied to it, viz., whether drones bred in worker-cells are capable of the fertilising act? and, reasoning from analogy, our opinion has been in the negative. No one can give proof of many things asserted with regard to the reproduction of bees, but seeing that the worker-cell cramps, and almost destroys, the generative organs of the female, it is reasonable to conclude that a similar effect is produced on the male bees.—C. N. A.

once set to work to clean it out, may be seen during the swarming time in hollow trees, cracks of walls, clefts of rocks, and in empty straw and wooden hives, especially if these contain a little comb. Yet many swarms seem to leave the parent hive without having provided a new home, and fly away at random, some of the bees settling in one place or another to rest, others flying away, partly to look for a suitable abode and partly to gather food.

When the swarm has entered its new abode, or been placed in a suitable hive, it settles, in the shape of a bunch of grapes, in the highest place in the hive, whence the bees begin to clear out everything that is unclean, uneven, and not suitable, and to build combs which gradually fill up the interior of the hive.

CONSTRUCTION OF COMBS AND CELLS.

The structure which bees erect in their hive, and which gradually fills up its interior, consists of several combs, or cakes, of about an inch in thickness, which, fixed to the top of the hive, are continued more and more downwards, according as the colony requires it or the room in the hive permits. When constructed regularly the combs run parallel, a distance of about half an inch being left between every two combs, so that the width of one comb, together with the empty space or passage (between it and another), is one and a half inches. If, therefore, the hive were about twelve inches in width and length, the bees would be able to construct eight combs in it, each of which would be twelve inches long. Now, the position and the direction of each new comb being governed by those preceding, the swarm cannot begin work upon all of them at the same time, but is obliged to commence one after another, except there be guide-combs in the hive, which the bees would then at once set about lengthening. In order that the bees may be able to get from one passage to another and round the different combs, they do not cement them to the hive-wall throughout, especially on the entrance side, but leave about half an inch free, and fasten the combs only here and there to give firmness to the structure. Occasionally they also leave a passage here and there in the middle of the comb. With a continuance of good pasture the bees work the combs down towards the floor of the hive, to which, however, they do not fix them, but leave an interval, in order that the wax-moth, which is frequently met with on the floor, may not be able to get into the comb so easily.*

Each comb consists of a double row of regular hexagonal cells, which are separated from each other by very thin walls of wax. The ends of these two rows of cells opposite each other open on the two surfaces of the comb, and they all appear as regular hexagonal figures. Between the two rows in the centre of each comb runs a partition wall, forming the common basis of the cells. As the combs are built from the top downwards, the cells are, of course, arranged horizontally, but towards the open end of the cells, especially of those intended exclusively for the storage of honey, they are slightly turned upwards.

Purposes of the Cells.

In addition to their general use as affording bees a resting place and also greater protection against cold, the combs in the hive chiefly serve the double purpose of holding the brood and of receiving the necessary provisions of honey and pollen. Now, as male bees, or drones, are sensibly larger than workers, the so-called drone-cells are also perceptibly wider than worker-cells, as shown in the illustrations below.

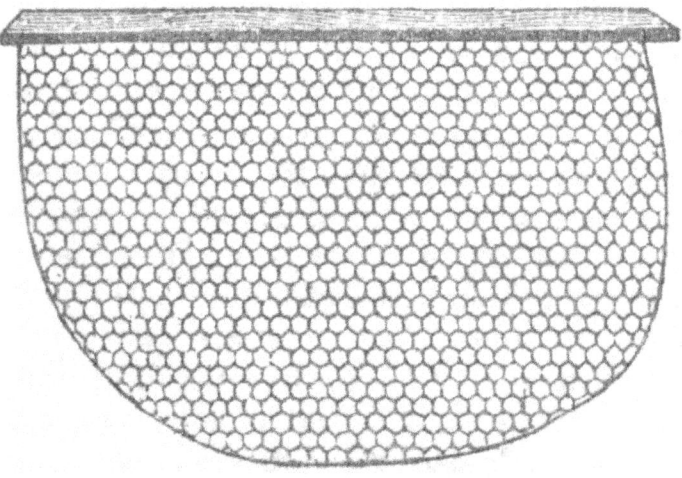

WORKER-CELLS.

We have already given the proportion between worker-cells and drone-cells, in stating that in an inch in length there are about five worker-cells or four drone-cells. But all the cells of these two kinds perfectly resemble each other in shape. Where mathematical precision is not required they might, therefore, be

---

* And obviously that the bees themselves may be able to pass easily under them.—C. N. A.

used as a measure, which would be understood at all times and in all countries where our honey bee is to be found. Writers on bees, at least when stating measures, should reduce them to cell widths. In order that there may be no doubt as to the scale adopted by the Author when, for example, giving the dimensions of bee-hives to be described hereafter, he thinks it advisable to mention that he has taken the width of five worker-cells to represent one inch, and that of sixty worker-cells as twelve inches, or one foot.

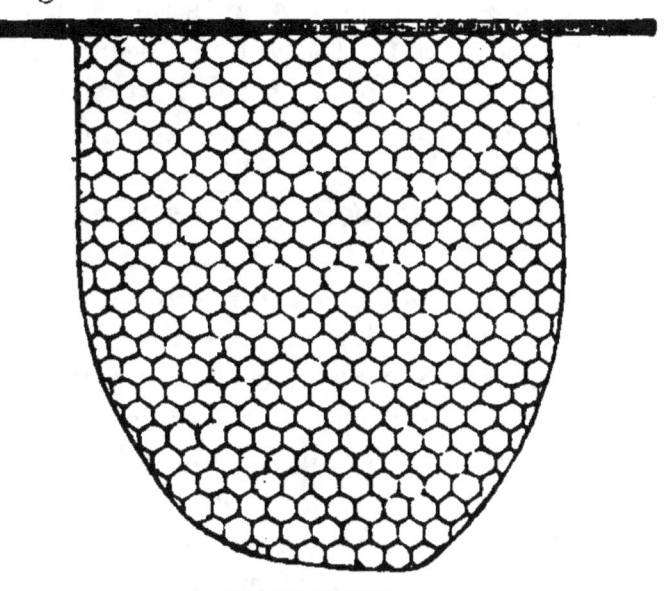

DRONE-CELLS.

Worker-cells are much more suitable than drone-cells for fixing a measure, because there is a much greater number of them in every hive, and it is more easy to find a regularly-constructed comb with small cells than one with large ones. Many young colonies do not even build a single drone-cell in the first year, as they do not yet require them.* Self-preservation being their sole consideration, they only rear worker-brood at first, and not until later—either in the same summer, if good pasturage continues plentiful, or only in the following spring, when the hive is partly filled with comb and the cells are full of brood and honey, when the number of bees has increased and the temperature in the hive has reached a high degree—does the stock, sensible of its strength and as a more remote preparation for swarming, commence drone-breeding, in order that there may be no want of males to fertilise the young queens to be reared. Bees, in this case, pass from the construction of small cells to that of drone-cells more towards the lower

* We do not agree that drone-cells are built by the bees only when they require drones, experience having convinced us that they are (except in queenless colonies) primarily built during a glut of honey, for storage purposes. Many stocks build drone-cells in their first year and fill them with honey, not a drone being produced.—C. N. A.

parts of the combs, or else they build combs at the sides of the hive, consisting of drone-cells only. Moreover, queenless stocks, if comb-making has not ceased in the hive altogether, generally build drone-cells only,* because they instinctively feel the necessity of having drones for the young queen that is to be reared.

It has already been stated that the third kind of cells—the so-called royal or queen-cells—are only made in small numbers, and are generally destroyed as soon as the young queens have hatched, and that these, therefore, are not always to be met with in the hive.

As every young bee on hatching leaves its casing (cocoon) behind, a cell that has once been used for breeding is easily recognised, being of a dark or brownish colour. The more frequently a comb has been used for breeding, the darker will be its colour and the thicker the walls of the cells, the latter becoming more and more narrow and less and less fit for use, so that in time it becomes necessary for the combs to be renewed, although in case of need the bees themselves partly remove the casings, or even pull down the cells entirely.

The cells further serve as vessels to hold the honey and pollen. Under certain circumstances bees make use of all the cells in the hive, large and small, formerly used for breeding, to store their honey, yet there are also cells which are exclusively used to keep the honey in, and which cannot even be used as brood-cells, because they are situated in the cooler parts of the hive, and are of such depth that the queen would not be able to deposit her eggs regularly at the bottom of these cells.

It is remarkable, however, that pollen is only deposited in worker-cells,† and as near as possible to the brood, the food for which is chiefly prepared from it. If, on account of the loss of the queen in a hive, breeding ceases at a time when pasturage is plentiful, the bees fill most of the cells of the brood-nest with pollen, but consume it again by degrees, using it as food for the brood when there is again a fertile queen in the hive. Many cells, also, which are half full of pollen, are filled up to the top

---

* It is nevertheless true, that if worker comb-foundation be given to queenless stocks, they will build worker-cells almost certainly, a fact which gives immense advantages to those who use that material.—C. N. A.

† We admit this to be the rule, but occasionally we have found drone-cells packed with pollen.—C. N. A.

with honey and sealed by the bees, so that frequently much pollen is found even in sealed honey-combs, especially in those from the neighbourhood of the brood-nest, where it is particularly well preserved, being excluded from the air, and is of great use to the bees in the following spring, when they have commenced breeding, but are yet unable to procure fresh pollen, while that which remains in open cells often becomes mouldy and dried up, and is thereby rendered partially or entirely useless.*

Production of Wax.

The material of which bees build their combs and cells is known to us as wax, that valuable substance whose production is one of the principal aims of bee-keeping.† But where do bees get this wax from? Do they find it already prepared in Nature, and is it only necessary for them to collect it, like propolis, with which they stop up crevices in their hive? Or are the little pellets which they carry into their hive the wax? By no means. Wax is a peculiar product of the organism of bees—their fat, as it were. Like the spider, which produces the substance of its web out of its own body, bees also elaborate the material for their combs and cells out of themselves. It makes its appearance—when plenty of food has been taken by the bees and the temperature is suitable—outside the rings of their abdomens, in the shape of thin, mica-like, oval scales. In summer some bees may be seen, on the outside of whose bodies small lumps of wax have accumulated in the manner described. Some little time after a swarm has been put into an empty hive a number of such scales may be noticed on the floor of the hive, looking like whitish specks of froth. The bees having only just commenced making comb are unable to work up as much wax as they produce, and, consequently, many of these little plates fall to the floor. The bees withdraw with their hind feet these little plates of wax, and

---

* Pollen, when dried or mouldy, as described, is removed by the bees in the spring, when the cells are required for other uses. The casting out of the pellets is a healthy sign of progress within the hive.—C. N. A.

† This statement will grate on the ears of British bee-keepers, who have been taught to prefer the production of honey to that of wax, and to give the latter to the bees in the form of comb-foundation, that they may produce the former more abundantly.—C. N. A.

place them between their mandibles, with which they mould the wax and deposit it where they intend to continue the structure. They generally attach the plates to form the somewhat thicker edge (or border) which encompasses all cells, even those only just commenced, and by which they acquire greater firmness. By thinning this edge more and more the cells are lengthened until they have acquired their proper depth. But the border is allowed to remain when the cells are finished, and the wax employed to give this border thickness is used, partly at least, to seal the cells when full of honey, or when the larvæ have attained their proper size. In continuing their structure the bees do not, however, only use wax recently made, but often bite off old wax, which happens to be in the neighbourhood, and use it in making cells. When, therefore, bees build new cells in the hive in continuation of old ones, these new cells are of a darkish colour at their commencement, and only gradually assume the white colour which new wax possesses. Queen-cells are generally of the colour of the comb on which they appear, because bees take the wax for royal cells chiefly from cells in the neighbourhood.

It is well known to all experienced bee-keepers that wax is a secretion of the bees themselves, and it was a great error to suppose, as was done formerly, that the small leg pellets and wax are similar substances; yet bee-keepers are not quite agreed yet as to what wax is prepared or secreted from.

Wax, of course, can only be prepared from something which bees are able to eat and to digest, or which serves them for food. We, therefore, now pass to the section on

The Food of Bees.

The food of animals may in general be divided into two classes. The food-stuffs of the first class serve to maintain respiration and to generate heat, and have been called 'means of respiration.' They are composed of the elements, carbon, hydrogen, and oxygen, and such foods are starch, sugar, fat, alcohol, and all digestible substances which do not contain nitrogen. Foods that come under the second division serve to form flesh and blood, to build up and to more fully develope the body, and they have, therefore, been called flesh, or blood-forming, or plastic foods, of which albumen is the most important. These are distinguished from the former class by containing nitrogen in addition to carbon, hydrogen, and oxygen. Honey, being a non-nitrogenous food, serves to maintain the process of respiration and to generate heat; and bees are able to subsist on it alone while in a state of repose in autumn and winter, when no brood is reared, and all other occupations which waste the strength of the body are entirely discontinued. But as soon as the instinct of multiplying awakes in the hive in spring, or as early as in February, if not before, the necessity for another kind of food also becomes apparent. The bees do not only consume far more honey than hitherto, because a higher degree of heat has to be generated and to be maintained in the hive, but they also show a great desire for nitrogenous or albuminous food, and such a food is the pollen stored away in the cells, which is now eagerly consumed, even if partly covered with mould. When the days are warm the bees are also busy in gathering pollen. It is well known that when pollen is wanting, common flour, put into a comb and placed in a sunny and quiet spot, is eagerly loaded by bees on their hind-legs and carried into their hives as greyish* little pellets, chiefly on account of the nitrogen it contains. But in order that bees may be able to prepare in their bodies from dry flour the nutritious milky food for the brood, they also require water, which they carry into the hive eagerly in spring, and even in summer, if the honey they gather does not already contain it in sufficient quantity. Water is quite indispensable to bees in preparing food for the brood. They can better do without pollen for a time, and it is certain that without a single cell of pollen a colony is able to live through the winter, and even to rear some brood before their first flight in spring. To explain this strange phenomenon it was hitherto supposed that bees are able to keep a certain quantity of nitrogenous food stored up in their stomachs for a considerable time, which supposition, however, is probably erroneous, for there can be no doubt that honey and pollen—the food of bees—are soon further digested, their nourishing constituents being conveyed to the blood in a short time. Bee-keepers

---

* When pea flour is used the pellets are of a bright yellow colour, and appear almost like split peas attached to the hindmost legs of the bee.—C. N. A.

in general had not previously formed a correct idea as to how the chyle is prepared in the body of the bee and conveyed into the cells.

It was supposed that pollen and diluted honey are consumed by the bees which have the care of the brood, the nutritious part being extracted by them from the chyme and returned into the cell as chyle. This view, however, is contrary to experience; for if a nurse-bee be crushed, we always see honey appearing but never any brood food. The great nutritive property of chyle, which passes into the blood of the young bee entirely, without leaving any particular residue, is further evidence of its being secreted from the blood of the nurse-bee, in the same way as milk is secreted from the blood of mammals. Quite recently, naturalists have also discovered the organs by which the secretion of chyle from the blood is effected; they open in the mouth of the bee and ramify in the head and the thorax. Professor von Siebold gave an interesting lecture on this subject, which was illustrated by drawings, at the annual meeting of bee-keepers held in Salzburg. It has certainly not yet been settled whether these organs only assist in the preparation of chyle, by causing a more complete assimilation of its component parts, or whether they produce the entire mass of the food for the brood; yet the latter is more probably the case, and those organs might very properly be called chyle-glands (*Futtersaftdrüsen*), or more briefly, milk-glands (mesenteric glands) (*Milchdrüsen*). It is now easy to explain how bees are able for a time to breed, and consequently to prepare the chyle for the nourishment of the brood, without nitrogenous food. The elements in which the food is deficient are for a time supplied by the blood, yet exhaustion gradually ensues, the blood becomes thinner and poorer; and as experience teaches, bees which breed out of season die in large numbers—doubtless from exhaustion.*

But however large their stores of honey and pollen may be, bees are unable to prepare food for the brood if they are in want of water. Water is not stored up in the hive, but the bees fetch it when needed, either from flowers, simultaneously with the fresh honey they collect, which is generally very watery, from damp

* This is a most important fact and should be remembered by all who are over anxious for early and late breeding in their hives.—C. N. A.

sunny spots near river banks and pools; or if the weather does not permit of their flying out, from damp places in the hive where the warm vapours condense on cool surfaces, or they carry it into their hive as dew or rain-water.

But bees not only require water to prepare the food for the brood, they also frequently require it when honey has become too thick, or has solidified entirely, to dissolve or dilute it, and to render it fit for their own use; and if they are unable either to carry it into their hive, or to procure it inside, the colony may perish from what has been called water-dearth (*Durstnoth*). How much bees sometimes suffer from thirst may be seen from the eagerness with which they suck up every drop of the water, which is squirted into the entrance of the hive to keep them back from flying out at too early a period, and even lick the snow thrown into the entrance for the same purpose.* But they require far more water when breeding is going on to any extent. Damp weather, therefore, during which flowers generally yield more pollen, is favourable to breeding and swarming; while in a dry season generally more honey is gathered, but then there are few or no swarms.

We know from experience that breeding and comb-making progress most favourably when there is a certain amount of moisture in the air, these two occupations always being dependent on, and accompanying, each other. Extension of the area of the brood chamber requires also an extension of the combs, and both demand a higher degree of temperature in the hive. As soon as breeding commences, the bees generally produce also a little wax, required for sealing the brood-cells; whereas, if breeding be interrupted, it may be by the removal of the queen, wax-making is also discontinued immediately, even during the most favourable season.†

---

* We have come to the conclusion that bees cannot resist the temptation to take up water whenever it appears within their hive, and hence, in damp hives, in cold weather, they often get so overcharged with it as to become dysenteric.—C. N. A.

† These statements, though true in principle, are not in all cases strictly correct. It is evident that when bees are decreasing in numbers through having been long queenless there will be no temptation to comb-building; but it has been admitted, in a former page, that queenless stocks build drone-comb. It is also well known that after swarming, when sufficient bees have been hatched, and all the brood is sealed, comb-building in supers will go on rapidly in favourable weather; and as many

From this alone it follows, that they are in error who are of opinion that the food for the brood is prepared from pollen and that wax is prepared from honey alone. Such a distinction is not admissible at all. The one as much as the other is a product of the organism of the bee, when its activity is raised to the highest degree. At that time, however, bees do not live upon honey alone, but also consume pollen, and only when taking both do they keep up their proper strength; wherefore, also, both foods are necessary for the continued production of wax, though they are possibly not taken in equal proportions. Even if wax, as a fat, is a non-nitrogenous substance, and even if feeding upon honey or sugar alone is sufficient to enable the bees to prepare it, it does not therefore follow that pollen is not necessary for its continued production. For, as already remarked, bees are able to prepare food for their young for a considerable time without pollen; yet no one would assert that pollen is unnecessary for the nourishment of the brood. In the one case, as in the other, the bees are sustained by a certain store taken into their bodies, but which by degrees becomes exhausted.

The question has often been asked, and frequently been discussed, How many pounds of honey the bees consume to produce one pound of wax? Opinions on this subject differ very much. While many bee-keepers consider 20 lbs. of honey necessary to yield 1 lb. of wax, others reduce the quantity to 10 lbs. An agreement on this point will hardly ever be possible, because the quantity of honey consumed by the wax-producing bees no doubt very much depends on how much pollen they consume at the same time, which would increase the quantity of wax secreted, on account of pollen containing carbon in addition to nitrogen. It is, therefore, quite natural that where good pasture is plentiful, the bees are able to gather far more honey if the cells in which it is to be stored are ready in the hive to receive it, than if they are obliged, first of all, to build them, supposing the activity of bees to be the same in both cases. But bees which are obliged to construct new cells and combs are, as a rule, much more active than those for which this necessity does not exist.

bee-keepers have experienced, if a comb be removed from a hive newly rendered queenless, the bees will immediately fill the vacant space with drone-comb.—C. N. A.

The Vital Activity of Bees

varies a good deal according to circumstances and the time of the year, and the quantity as well as the quality of the food necessary are dependent thereon. The quantity of honey which a bee is able to hold in its stomach may, under certain circumstances, afford it ample food for more than a week; and, under different circumstances, may be insufficient to prevent death from starvation within twenty-four hours. If we compare life to a process of combustion, then a bee's life is at one time like a spark glimmering under the ashes, and at another like a bright flame, which in a few minutes consumes the combustible matter that would have fed the but glimmering fire for a much longer time.

The vitality of bees is certainly never lowered to the extent of rendering them insensible and totally torpid, which is the case with many insects, and even with the kindred species of the humble-bees, wasps, and hornets; yet late in autumn and at the beginning of winter—say, from October till January in our climate,—when vegetation almost entirely rests, it is reduced to the lowest possible degree. The bees then keep perfectly quiet, and not the least humming is heard in the hive, especially when the temperature of the air is mild.* They do not at that time in the least exert themselves to raise the temperature, the air immediately surrounding the cluster of bees being about $8°$ R. ($= 50°$ F.), at which bees are only just able to move their limbs, and are momentarily incapable of flying. Breeding will, of course, have ceased now, nor would the brood thrive, although the temperature in the interior of the cluster of bees might be $12°$ to $15°$ R. ($= 59°$ to $65\frac{3}{4}°$ F.). Even the young bees in sealed cells—if any are left in the hive—mostly die off, and are afterwards thrown out of the cells. The consumption of honey is very small now, amounting to about half an ounce a-day in a tolerably populous colony, or about 1 lb. a-month. The bees, especially if they possess in their cells, mostly sealed, a supply of

---

* In very cold weather bees use their legs and wings to promote circulation, and often create a humming or buzzing. Men, under similar circumstances, buffet their arms across their breasts and stamp their feet. Remembrance of this fact, and that bees only require as much heat as will sustain life, will explain why they do not make a humming within the hive with the temperature at 50° F.—C. N. A.

healthy honey gathered from flowers, are then able to dispense with a cleansing flight for three months and more, although it is very desirable for them to have a sport before the hive more frequently—perhaps once a-month.

When, towards the end of winter and the beginning of spring, the sun rises higher and higher in the heavens, and the days are getting longer and longer, and all nature awakens to new life, the colony is roused to increased vitality, and especially so when the sun is shining warm and the temperature has risen to a degree at which bees are able to leave their hive. While, during their sleep in winter, slow movements of their bodies and a certain vibratory motion of their wings were the only signs of their being alive; while, further, they were unable to separate from the cluster, and could not even prevent a mouse which had intruded into the hive, from gnawing at the comb and eating their honey, they now rush through the air as swiftly as an arrow, removing the dead bees from their hive, fetching water, putting to flight any enemy approaching their hive, and having the full use of their limbs, and strength to perform all their duties. On account of their increased vitality the consumption of honey is also greater now, and still increases; the more so, as breeding, which had been discontinued for four or five months, commences again—the brood extending more and more as the temperature rises, especially when the first spring blossoms yield pollen again, and the weather allows the bees to carry it into their hive. It is pollen chiefly for which bees at this time have a great longing, and which they are eager to gather and carry home every time the sun breaks through the clouds, because they require it to prepare the food for their brood, to the care and increase of which their whole attention is now directed. As the degree of the vital activity varies, so varies also

The Aim of their Activity

at different times of the year. It is true the bees breed from February till September; they gather honey and pollen and carry water into their hive to dilute solidifying honey, and propolis to stop up crevices in the hive and to fasten the combs more firmly, if the weather allow it; but according to the difference in the time of the year and change of the weather they at one time chiefly—though not exclusively—pursue one aim, and at another time another aim.

In spring, when all nature awakens to new life, the activity of the stock is chiefly directed to the increase of the population by the depositing of eggs. At first, indeed, eggs are deposited in worker-cells at the rate of a couple of hundred a-day; later, however, the number is increased to thousands—every colony in the first place seeking to make its own continuance secure. When the number of bees by daily augmentation has gradually been increased considerably, when the pasture has become more fully developed, and the temperature in the hive has risen, then, in the confidence of its strength, and a sort of maturity, drone-breeding also commences, as a more remote provision for increasing their species. When the hive is not yet quite full of comb, the bees pass from the construction of small worker-cells to that of the larger drone-cells, on one or even several combs, and the queen is not long in occupying them with eggs. When, at length, the hive is tolerably full of brood and the colony in its greatest strength, when the temperature—for bees, the measure of their strength—has risen to such a high degree, say $30°$ R. ($=99\frac{1}{2}°$ F.), that the heat is beginning to get uncomfortable for them, royal cells are commenced here and there, especially at the edges of the combs, mostly in the neighbourhood of drone-comb, if not on these combs themselves—and eggs are deposited in them by the queen.* Such royal cells are called swarm-cells, to distinguish them from supplementary queen-cells, which are constructed over worker-cells already occupied by an egg or larva when the colony is unexpectedly deprived of its queen. As soon as one or the other of the royal cells is sealed over, and the royal larva has entered into the chrysalis state, the old queen feels no longer safe in the hive, and leaves it on a fine day about noon with the so-called first swarm. The same thing is done nine or more days later by the first-hatched young queen, which by her calls of 'Tüt!'

---

* We have never seen a queen deposit an egg in a royal cell, as stated in a previous note; and, seeing the antagonism displayed by queens to the cells' existence in a hive, have always doubted their willingness to aid in their production. And if queens laid eggs in royal cells built upon drone-comb, as stated by our Author, they would surely often hatch into queens, which, in our experience, has never happened, though we have had hundreds of such cells under our notice.—C. N. A.

'Tüt!' which she utters in a clear and shrill manner, expresses her jealousy and fear, and thereby announces the second, or after-swarm. This is sometimes followed by a third and even several later swarms from similar causes and announced in a similar manner, mostly on the day following the announcement, if the weather remains favourable and the colony has not yet been weakened too much, and when there are still royal cells in the hive, which, because all the eggs are not generally laid at the same time only attain maturity successively.

As the voluntary departure of the old queen with the first swarm is generally followed by after-swarms with young queens; similar events happen when the queen has been removed either alone or with an artificial swarm, or when the colony has accidentally been deprived of its queen. But as in such a case royal cells are only commenced after the queen's removal, it is at least thirteen,* but generally fourteen days, before the appearance of the next swarm with a young queen.

Such swarms are called 'Tüt' swarms (pipers) because they announce themselves by the piping of the anxious and jealous young queen, and may appear at an unusual time of the year, either very early or late, if the weather be in some degree favourable, and the colony not too weak—otherwise the remaining royal cells are destroyed after a young queen has hatched—the young queen commencing the work of destruction, and the workers completing it. From the time the old fertile queen leaves the hive till a young queen reigns in her place and is fertilised, egg-laying is discontinued entirely in the parent-hive; thus—in case the queen's nuptial flight is not delayed by unfavourable weather—there is an interval of about three weeks, during which time no eggs are deposited. When the queen has been fertilised breeding commences again, and parent-hives, which in the meantime have accumulated a quantity of pollen, continue breeding for a longer time than colonies in which no change of queen has taken place.

---

* In our experience the so-called supplementary queens have hatched within ten days. From the newly-laid egg it takes sixteen days on the average, and as it takes three for the egg to hatch out a larva, and a larva of three to four days old is eligible for queen-raising, it will be evident that young queens may often be due on the ninth or tenth day after a stock has been rendered queenless; and swarms may issue almost immediately. —C. N. A.

With a uniform continuance of pasture, even swarms, and especially first swarms, make preparations for swarming again by making drone-cells and depositing eggs in them, when to a great extent they have filled their hive with comb, and a great many of the worker-cells contain brood. In most years and in most districts the bee-pasture, however, reaches its climax before the end of the swarming season, and then diminishes. The activity of bees now shows itself in another direction. In order to store as much honey as possible for the approaching time when food is not procurable, the bees now adopt a system of saving. To make up for the loss of bees which is unavoidable while they are out gathering, worker-bees still continue to be reared, but in diminished numbers, while not only does the queen give up laying drone-eggs, but the drone-brood already in the hive is generally thrown out of the cells, and the drones themselves are driven out, as being no longer of any use. Wax-making also, which was carried on so eagerly by the bees in spring, now ceases entirely, because it would be at the expense of honey. When the autumn pasture is very favourable the bees, indeed, fill all the cells with honey, the majority of which had previously been used for breeding, but they very rarely construct new cells. It is now the aim of their activity to secure their future by accumulating as large a quantity of honey as possible, to protect themselves as much as possible against draughts and cold, to fill up any crevices in their hive with propolis, and to narrow the entrance of the hive, and these they do as long as the air is warm enough for them to be able to fly out, for which a temperature of about $13°$ R. ($= 61\frac{1}{4}°$ F.) is required. The bees' eager desire after honey is never satisfied, and now when the pasture is drawing to a close, as well as in spring before it becomes developed, they are very much inclined to rob each other. First of all, they look about for queenless colonies and plunder them, and then they attack very weak stocks, which are unable to defend themselves properly. Robbing is a bad habit of bees, for which those bee-keepers are to blame who give them the opportunity of robbing by allowing faulty stocks to remain in the apiary, and by not being careful in feeding and in cutting honey-combs out of the hive; it is the robbers which are the best and most industrious bees. As the Italian bees are particularly industrious, they are also very much disposed to rob.

But, however fine the weather, when there is nothing more to be gathered, the bees, in order to save their strength and honey, fly out no longer; they keep at perfect rest, and only now and then, after an interval of several days, do they fly out for a sport before the hive about noon-time of a fine day, in order to cleanse themselves once more before the winter.

As the cold increases the bees leave the side-combs and draw together in as dense a cluster as possible in some of the middle passages, as near the entrance of the hive as possible, and below their stores, which they continually follow slowly.* In this manner, while the whole vegetable world is dormant, bees also pass the winter in a state similar to sleep, until the warmer rays of the sun, which now rises higher again in the heavens, awaken them to renewed vitality to resume anew the occupations briefly referred to above.

In what has been stated above we have by no means exhausted the natural history of bees, but have only given a most general outline of it. A more intimate knowledge of their wonderful instincts and capabilities can readily be obtained by handling them and from experience. We will, therefore, hasten to that part of our work which treats of the practice of bee-keeping, where we shall be able to discuss at the proper moment many theoretical points more clearly and comprehensively than has been done in the preceding pages.

* This is perhaps true of bees in tall German hives, where they are well stored with honey, but in shallow hives, with the honey disposed around the brood-nest and not above it, the bee-cluster will be found reaching to the top of the combs, and in a manner gradually revolving about the hive centre as they consume their stores. We mention this that amateurs may not be under a misapprehension.—C. N. A.

THE PRACTICE OF BEE-KEEPING.

Bees thrive in most countries of the temperate zone, and also in most parts of the tropics. In the wild state their natural abode is the forest, where a hollow tree will afford them a domicile. They also live in crevices of rocks and holes in the earth. But on account of the great benefit to be derived from bees, since they furnish the two valuable products, honey and wax, and because the contemplation of their industry, their wonderful economy, and their skill, is a source of great enjoyment, particular attention has been paid to them from very early times. They have, as it were, been made domestic animals, and, in order to facilitate watching them, have been brought into the neighbourhood of the abode of man, where they are kept in artificial dwellings of various shapes, sizes, and arrangement, and constructed of different materials. The spot where the hives are placed, and which is generally in the garden attached to the house, is called a bee-garden, bee-stand (apiary), or bee-house, the latter name being given to it if the hives are placed in a separate building, which, however, is not necessary, and not even desirable.* The whole management, including all that is necessary to be done, is called Bee-keeping.

Bee-stand (Apiary).

Success in bee-keeping is to a great extent influenced by the position of the hives. It is by no means necessary that these should be placed in a special hut—a so-called bee-house. Although such a house offers certain conveniences and affords shelter to the hives, yet it has also its inconveniences and disadvantages. When the hives are placed closely together a great many bees from different colonies playing before the hives at the same time lose their way, and even queens get into wrong hives; the bees are frequently disturbed in their winter sleep by the noise transmitted through the bee-house in case of any manipulations, even

* We fully endorse the Author's opinion, for the reasons given in the ensuing paragraph.—C. N. A.

if one hive only is interfered with—they are more troubled by mice, and cannot be handled so conveniently as on an open stand. The expenses of building a special bee-house are also very considerable, which may be saved if the hives are placed quite separately in suitable spots in the garden. The chief consideration ought to be to place the hives in a position which affords bees the greatest possible protection against prevailing storms. It makes very little difference whether the hives are placed more or less in the sun, or whether the bees fly out towards the east, south, west, or north. However agreeable to bees the warm rays of the sun sometimes are, at other times, again, they are just as troublesome and ruinous to them, so that the advantages and disadvantages are pretty evenly balanced. It is a great advantage to have close by, and to the north* of the apiary, either a high hedge, a wall, a house, or other high object, on the warm surfaces of which the bees are able to rest and bask in the sun during their flights to cleanse themselves, which they often undertake when the air is pretty sharp. Such an object would also save bees during gusts of wind from the necessity of descending upon the still cold ground, where they are frequently benumbed. Any tumbling noise is injurious to bees, especially in winter, because it disturbs their rest, yet they gradually get accustomed to it, and it is only when the thing which causes the noise is directly in contact with the hives, or when the ground is shaken too violently, that injurious consequences result.†

Large surfaces of water, lakes, ponds, or wide rivers in the immediate neighbourhood of the apiary are injurious, because many bees are drowned in them during a storm; but small brooks, shallow ditches, or water puddles are very desirable. Where they are wanting, or when perhaps they have become quite dry, it will be well always to keep a supply of water in shallow troughs or other vessels, in which a little moss might be

---

* The recommendation of a high hedge or wall to the north of the apiary clearly points to a fact often alluded to by us in the *British Bee Journal*, viz., that it is highly undesirable that bees should be placed on the north, or sunless side, of such shelter. Sunshine on the hive may sometimes be a disadvantage in summer, but in winter its total absence should never be contemplated.—C. N. A.

† This points to the fact, often pointed out in our *Bee Journal*, that proximity to railways should be avoided.—C. N. A.

placed, in a sunny spot, sheltered from the wind. This has a favourable influence on the undisturbed progress of breeding, the bees requiring much water in spring, and in summer also, during a continuance of dry weather, to dissolve or dilute the honey which has crystallised or become too thick, and to prepare the food for the brood. Pure rain-water is preferred by bees, but they readily take also water from other sources, especially if it has been warmed a little by the rays of the sun. The sweetish juice which flows abundantly from birch-trees in spring, when tapped, they suck up greedily, and as it contains sugar, although the quantity be insignificant, it is to be preferred to water. It might, therefore, be worth while to plant a few of these trees in the neighbourhood of the apiary. The tree furnishes plenty of pollen for the bees, and its thick crown affords them shelter against storms. It is true a row of pines, which are always green, offer such shelter still more completely, and the planting of such a hedge around the apiary, but more especially towards the weather side, is very much to be recommended.

If a road happen to pass by the apiary, either such a row of trees, a high wall, or a high hedge, is quite indispensable in order to compel the bees to fly higher, and to make it impossible for them to molest man and cattle, or to do any mischief.

After the stand, it is

The Dwellings of Bees

which chiefly require to be taken into consideration in bee-keeping. The allied species of wasps and humble-bees at any rate are able to build an abode for themselves, the former from the well-known substance similar to blotting-paper, the latter from moss. Bees, on the other hand, are indeed able to clean their dwelling, to widen it by gnawing off rotten parts, and to render it more habitable by stopping up crevices and narrowing openings which are too large, but they are unable to make it entirely. It must be provided for them by Nature or be prepared by man, and consists of a cavity of about one, or sometimes several, cubic feet, limited in all directions, and having at least one opening. This dwelling must afford them shelter against storm and rain, cold and heat, as well as against their numerous enemies, which prey, partly on their stores of honey, partly on their combs, and partly on the bees themselves, and the more completely it affords them this shelter in every respect, the better will bees thrive in it and the better will it answer the purpose. The suitability of the dwelling chiefly depends on

The Material

of which its walls are made. Warmth is the chief necessity of bees and their brood, it is essential to their life. This indispensable warmth, which they are able to produce only by a certain amount of exertion and at the expense of their provisions, the hive must keep together as much as possible, which will be the case if the walls are made of a substance which conducts heat very slowly.

Different substances, or bodies, as is well known, show a very great difference in this respect. Metals conduct heat very quickly —they are good conductors of heat. If one end of an iron bar be put into the fire the other end will in a short time become so hot that it cannot be held in the naked hand. A metal plate, if heated on one side, soon gets hot on the other side, while a board may be burning on one side and be quite cold on the other side, because wood is a much worse conductor of heat than iron, although there is, again, a very great difference as regards conductivity of heat among the different kinds of wood. While many kinds of wood by their density and weight approach to the metals, and, like them, sink to the bottom at once when put into water, others are porous and light as straw. Generally, it may be taken as a rule that the lighter the wood in the dried state, the less woody matter it contains in a given space, or the more porous and spongy it is, the less does it conduct heat, and the better does it retain heat, for the pores or empty intermediate spaces are filled with air, which, being enclosed in definite spaces, where it is motionless, prevents the heat escaping. A bee-hive will retain heat the better the thicker the walls are and the more air they contain, and the more completely this air is enclosed in them and isolated from the external air, which is in motion.

If the walls of bee-hives are made of a material which is a good conductor of heat, or which transmits heat readily, there is this twofold disadvantage—1st. That the bees suffer from cold, they are obliged to consume more food to raise the temperature, and are not able to breed so early and so extensively, &c. 2nd. That moisture is formed on the walls of the hives, which causes mildew, mould, an unhealthy atmosphere, dysentery, &c. As water on being heated, or when it absorbs heat, is converted into steam, so by reversing the process and withdrawing heat from steam by means of a cold body, or a cold surface, it is reconverted into water, and the cold surface gets 'misty,' as it is customary to call it. Like the window-panes of a warm room, the walls of bee-hives get damp if they allow heat to be withdrawn rapidly. When there is a difference between the temperature inside and outside the hive, and when their interior surfaces have cooled down below the temperature of the air in the hive, moisture is being formed, which, if the external cold continues and increases, is converted into hoar-frost and ice. If the hive have several compartments,* this will never be the case on the common partition wall. There being the same degree of heat in all the compartments, no withdrawal of heat, no cooling, and consequently no dampness, occurs. Even if one of the compartments were unoccupied, the layer of air enclosed in it forms a warmer wall than could be made in any other way. Altogether, it would be impossible to make any walls of bee-hives in contact with the external air retain heat better than by making a double wall and filling up the space between of about two inches in width with dry moss, flax, refuse, sawdust, thin shavings of wood, straw, hay, &c., in order to prevent the enclosed air from circulating.

But everything has its fixed limits, which cannot be exceeded with impunity, and even in regard to the retention of heat in bee-hives it is possible to go too far. Too much is as bad as too little. Direct injury cannot certainly result from the hive being too warm, for the bees will not produce more heat than they want; and it would be easy for them to expel the heat produced in excess through the entrance of the hive by ventilation. Indirectly, however, too warm a hive becomes injurious by possibly causing the bees to suffer from want of water in winter, and in spring before they commence to fly out, there being nowhere a cool surface in the hive on which moisture might be precipitated. It will, therefore, be well for us to keep to the happy medium, and to

---

* This is in allusion to twin or triple hives, with bees on both sides of the division boards.—C. N. A.

make our bee-hives in the manner approved of by long experience, in order that we may not, after all, be losers after having incurred considerable expense.

It was formerly much disputed among bee-keepers whether the preference should be given to wood or straw as the material for bee-dwellings. Straw recommends itself by being cheap, light, and warm; wood, on the other hand, by being more durable and cleanly. Bee-hives of any possible shape, especially the angular one, can more easily be made of wood, and they retain the exact shape given to them; combs are loosened more easily from the smooth wooden walls, the bees are brushed off more conveniently, &c.

If one happens to have some boards or planks, one and a half to two inches in thickness, of a kind of wood which naturally is soft, light, and warm, such as the wood of the poplar, willow, lime-tree, aspen-tree, or fir-tree, or wood which has become spongy, light, and warm through a kind of rottenness, walls of wood for hives are to be preferred, if only for the reason that bee-hives of any desired shape may be made from it in a convenient and easy manner, while it is troublesome and takes much time to make straw hives. It is possible, however, to employ wood and straw together, thin boards being used to make the walls of the hives, and a layer of straw fixed on the outside to give them the necessary warmth.

Bee-hives made in this way combine the advantages of straw hives and wooden hives, and leave nothing to be desired. The Author has all his bee-hives made in this manner, and he finds them insurpassable.

Hives of clay were formerly recommended in the *Bienenzeitung* by the Director Stöhr of Würzburg, and special attention has recently been called to them by Pastor Scholz of Hertwigswalden. If colonies of bees have been able to exist for a long time in crevices of rocks and walls, formed accidentally, as bees are known to have done, it is certain they will get on better in hives carefully made of burnt or unburnt clay. Yet as clay readily absorbs moisture, it is necessary that such hives should rest on a proper foundation and have a wooden floor-board; the clay should be mixed with as much straw as possible, or hollow or very porous bricks should be used, which may be obtained if plenty of finely chopped straw or turf is mixed with the clay. But the chief fault of these bee-dwellings is that they are not moveable, for even bee-keepers, who are not in the habit of removing their bees from place to place for the sake of pasture, may unexpectedly be compelled to remove their hives, as, for example, in the case of an inundation or a fire, or when a new house is being built.

The means of transport being frequent and cheap, it will be possible to procure at a moderate cost the few small boards which are required to make a hive, even in districts where wood is scarce; the wood of old boxes will possibly answer every purpose. There is no doubt that most bee-keepers will stick to wood and straw for the present, and we shall therefore only refer to bee-hives made of these two materials, which we will now proceed to describe more fully.

The Shape

of bee-hives varies. There are round hives and angular hives, also perpendicular and horizontal hives. Angular hives are certainly to be preferred to round ones, because they have a firm position; the bees are able to construct their fabric of combs in them in a regular manner, panes of glass to allow of observations being made can be fixed in them more conveniently, but chiefly because the moveable combs, about which we shall have something to say presently, can be more conveniently arranged in them.

It makes very little difference whether the hives are made so as to occupy a perpendicular or a horizontal position, *i.e.*, whether their height be greater than their length, or whether their length be greater than their height. The bees know how to accommodate themselves to circumstances. They enter and inhabit a perpendicular hollow trunk just as well as a horizontal hollow branch of a tree. Yet Lager hives possess this advantage, that, if we wish to remove them we are able to place them in the cart in the same position in which they stood in the apiary, and that, on account of the bees building a greater number of combs in them, filling the upper cells of each comb with honey as quickly as possible, they generally contain more honey,* and as the individual combs are shorter they are better secured against breaking off when the weather is very hot. The chief objection to Lager stocks is this, that it becomes necessary for the bees, on account of their pro-

* Lager hives are such as have all their frames on one floor; Ständer hives have two or more sets of frames, one above another.—C. N. A.

visions for the winter being stored in many combs,* not only above, but for the greatest part on both sides of their winter quarters, in course of time to move from side to side, which sometimes, when the cold is severe and the walls of the hive are covered with hoar-frost, they are unable to do, and thus often the whole colony, or a great number, of bees die of starvation, even though the hive contain provisions enough out of their reach. It is, therefore, not advisable to make hives too low, but to keep pretty well the medium between Ständer and Lager hives, making them about just as high as long.†

As to their width, nine to ten inches is perhaps most suitable. When the hives are not too wide, even colonies which are somewhat weak are able to build several full combs, at one end at least, and to occupy them properly, while the space left empty is easily partitioned off by inserting a small board. But if a swarm in a hive of equal width and length‡ has one corner only filled with comb, it is exposed on all sides to the influence of the cold, and it can recover in it§ only with difficulty and slowly. In a wide square hive, or even in a round one, a colony which has once become strong will generally keep up its strength, because the

* The Lager hives of the Author have their entrances about the middle of the long side, and the brood nest being about the centre of the hive, as is usual in what are in that case 'collateral' hives, the honey is stored on both sides of the brood-nest, with the possibility, which too frequently happens in winter, that the bees consume the honey on one side and are unable to get to the other.—C. N. A.

† The Lager hive described has moveable ends, and it is by these that the combs are removeable. An operator standing at one end of a hive would consider the distance from left to right, the width; the perpendicular, the

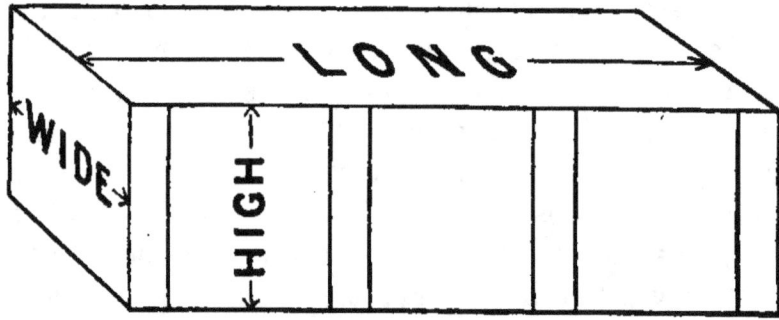

height; and the distance from one end to the other, the length, as in the wood-cut annexed. British bee-keepers must therefore understand the Author's recommendation to be that Lager hives should be as high as they are wide across the front, and from nine to ten inches from front to rear.—C. N. A.

‡ *i.e.*, as large from front to rear as it is across the whole front.—C. N. A.

§ After a winter's cold.—C. N. A.

warmth spreads equally in all directions, and breeding therefore takes place in all parts of the hive; but it is precisely on account of the excessive extension of the brood, and especially of drone-brood, which, because the heat is so great, is raised in large numbers, that the returns from bees in such wide or round hives are less than expected.*

In Log-hives, which may be considered the original hives, this can be observed very well. The narrower and the longer they are the more honey is generally accumulated in them. Bees inhabiting such hives limit the raising of brood to a certain space, for which a few of the combs in the middle of the hive are selected, and the combs which are situated sideways in the cooler parts of the hive the bees are sure to fill with honey during the first plentiful honey harvest. The shape of bee-hives should, therefore, be such as to insure their being twice as long as broad.†

The Size

or the interior space of bee-hives, which in the case of square boxes is obtained by multiplying together the interior length, width, and height expressed in the same measure, generally in inches, depends partly on the abundance of bee-pasture, and partly on the method of management. Whoever desires early swarms will have to keep his bees in comparatively small hives, but he who prefers large returns in honey will have to use more spacious hives. The hives need not, however, be made very spacious if they are constructed in such a manner that any superabundance of honey gathered by the bees may at any time be taken away conveniently and without injury to the colony, by removing one or more combs filled exclusively with honey, either altogether or returning the combs to be filled again by the bees after having been emptied of their contents by means of the honey extractor to be described hereafter.

As it is possible with plenty of good pasture and a continuance of favourable weather to repeat the operation referred to

* The Author's opinion is evidently antagonistic to the use of the Pettigrew hive, and in no way favours its principles.—C. N. A.

† That is to say, if nine inches from front to rear as the combs run, they should be eighteen inches wide along the front, and, as per note on page 48, they should be eighteen inches high.—C. N. A.

in a few days, a considerable quantity of honey may be obtained from a colony in a less spacious hive in the course of a good summer for bees. A space of about 3000 cubic inches will be sufficient in most districts; nevertheless, a space of 5000 to 6000 cubic inches will do no harm, if there be no difficulty in partitioning off the superfluous space.*

In order to give bees always as much room as they require—to be able to assign to a large swarm a spacious hive and to a small swarm a more limited one—divisible bee-dwellings have been invented, which are made by joining together single hives, and these combinations have been called Magazine hives. The Author has also given these hives a trial, but long ago discontinued using them after having commenced the construction of bee-hives, which are far more convenient for the management of bees, and which possess all the advantages of the Magazine hives, but none of their disadvantages.

The more complicated a bee-hive is the more expensive it is, as it gives pretty much the same trouble and takes about the same time to make a small hive as it does to make a large one. But time is money—it is capital. Is it not more suitable to make each bee-hive at once of the normal size, and to raise the swarm which is to inhabit it to the normal strength at once, thus adapting the swarm to the hive, instead of the hive to the swarm, as this can be done so easily? And if the room in the hive is to be diminished, is it not better to effect this by a partition board, than to have to add to the hive afterwards? If the proprietor of an estate has a thousand sheep, but wishes to increase their number to three thousand in a short time, he will certainly build a stable (fold) for the latter number at once, and not add to his premises as his flock increases. All the arguments which have been brought forward in favour of separable boxes† for bee-hives are, therefore, only seeming reasons, which are not likely to deceive an experienced bee-keeper. The separableness of bee-hives must, therefore, be looked upon, not as an advantage, but as a very great drawback, for the parts of the hive may tumble to pieces in

* Exactly the principle of Abbott's Combination Longitudinal, or Irish Hive.—C. N. A.

† As, for instance, the 'Stewarton' or 'Nutt's Collateral' Hive.—C. N. A.

our hands, the colony be destroyed, and harm be done to man and animals. A bee-hive constructed so as not to take apart is very much simpler, cheaper, and more solid, and yet by a simple arrangement the object of divisibility may be completely attained. Even if a hive of normal size should become too small for a colony in a particularly favourable year, then it may be enlarged temporarily by a collateral adjunct, or a super, but so far as it is intended to contain the brood-room and provisions for the winter, the hive should be non-separable in order to be warm, inexpensive, easily moved, and otherwise handled.

But it is possible for a bee-hive to be constructed of the most approved material and to be of the most suitable form and size, so that the bees get on well in it, and nevertheless the hive may be quite useless to the practical bee-keeper. We have examples of escaped swarms having found habitations quite suitable to them in hollow lime-trees, poplars, oak, and other trees, or in buildings, where they remained for many years in excellent condition; notwithstanding this, their proprietor did not derive the least profit from them, because he was unable to take the superfluous honey without destroying, or at least greatly damaging, the tree, and at the same time the colony. A stock of bees, however, is in the position of a colony in such a tree if their hive is so arranged as to allow of the removal of the superfluous honey only with much trouble and to the great injury of the bees, for the brood-room of the bees is sure to be destroyed during the operation, or an empty space formed, which cools it and thus causes injury.

Many a bee-keeper is scarcely compensated for the hard work and the stings he receives during the tedious and troublesome operation, by the small harvest he obtains and which costs him dear, considering the injury inflicted on his stocks.

A bee-hive, in order to be of value to the practical bee-keeper must, therefore, afford

Convenience of Handling,

i.e., it must be possible to do all that is necessary to be done with ease, without giving trouble to the bee-keeper and without injuring or molesting the bees. If a bee-hive requires to be raised merely for the purpose of looking into it, or if it be necessary to place a nadir under it in order to give the bees more room, it can no longer be called a convenient hive. When the stocks are getting heavy these and similar operations can no longer be performed without an assistant, who is frequently not to be met with just at the moment he is most wanted, or he takes to his heels if he happens to get stung. One must at any time be able to do everything quite alone. It must be possible to open and to close a bee-hive like a clothes-press; like the latter it must, therefore, have a

Side Door,*

like the so-called Log-hives, which are closed by a side door. In hives made of logs this, however, is somewhat inconvenient, as the hollow in the log is smaller at one end than at the other, and irregular all through. But the boxes, which we shall describe more fully presently, have the same width through their entire length, and the moveable door is of the same width, or, properly speaking, half an inch wider, as, though not merely leaning against but let in between the walls, it rests both on the right side and on the left against a quarter of an inch rabbet, in order that it may not get farther into the hive than its own thickness. No rabbet is necessary on the cover (crown-board) of the hive above or on the floor-board below; those on the two side-walls are sufficient, and render a rabbet on the door itself quite superfluous. The door may, therefore, be made of a smoothly-planed piece of plank of some soft kind of wood, of the proper length and breadth, and about one and a half inches thick and kept in position, to prevent it from slipping down, by buttons or wire tacks, which, being driven in above, either on the right-hand side or on the left, and bent round, can be turned conveniently.

If the hive has been made perfectly rectangular, it will also be possible to reverse the door, putting the inner side, which perhaps may have become damp, temporarily on the outside to get dry, although a coat of varnish renders it pretty well proof against moisture. How to make very neat and suitable doors for bee-hives from thin laths and straw will be explained further on.

On opening the side (or end) door, we shall have the fabric of combs before us and be able to cleanse the hive from any

* For side door, read 'door or opening at the end.'—C. N. A.

dirt on the floor, or a vessel with food may be inserted, &c., but nothing much is gained from this. We also wish to have a view of the interior of the fabric, and to convince ourselves of the presence of the queen, as well as to examine the brood and to see whether the bees have sufficient stores of honey, or we may wish to deprive the colony of the one or the other, all this is rendered possible by having

MOVEABLE COMBS,

or the arrangement in the hives which permits us to remove every comb from the hive and insert it again as conveniently as the wooden door can be removed and put in its place again. But how is this possible, a person unacquainted with this arrangement will ask, as the bees cement their combs to the cover of the hive, loosened from which it will hardly be possible to fasten them again at once, especially if filled with honey and brood, which increases their weight. The loosening of the combs from their top attachment is just what does not take place. Under the covering are separate small, thin, and narrow pieces of board, from which the comb which is to be removed is suspended, which are taken out and the combs attached to them of course accompany them, they having previously been loosened from the side-walls.

These are not, however, the real cover of the bee-hive, which is generally nailed down and immoveable, but they form a cover below the cover, being inserted in grooves provided on the right and left sides of the hive, in the same manner in which sometimes, when a plaster ceiling is made, boards are inserted in opposite grooves in the rafters, instead of being nailed to them.

Formerly the Author used to place these moveable pieces of board* on two ledges nailed to the inner sides of the hive, one to the right and the other to the left, and fixed equally high; but for some time past he has been using quarter inch rabbets, which he decidedly prefers. It need hardly be mentioned that the small pieces of board which carry the combs must be half an inch longer than the width of the hive, as they not only reach from wall to wall, but each end enters the wall to the depth of a quarter of an inch. It is best to make the rabbets wedge-shaped, or of an

* Really, *comb bars.*—C. N. A.

acute angle, by making one cut straight into the wall and another above it obliquely, or by sawing into the wall to the depth of a quarter of an inch, and then cutting away with a chisel the upper edge obliquely.* When the rabbets are made of this shape it is easy to loosen any combs fastened by propolis—two cuts with a knife will be sufficient; it is also easier to cut down to the proper size any pieces of wood intended to carry comb, which perhaps are found to be too long, if the section, corresponding to the rabbet, be made obliquely (see woodcut on page 26), than if it be made crossways against the fibres of the wood (see woodcut on page 27).

Although the comb-bars and the combs suspended from them are pretty well parallel to the door, yet it is a good plan if that end which we generally take hold of and draw towards ourselves when taking out the bar—which as a rule is the end on the right hand side—is somewhat nearer the door—be it ever so little—the comb-bars being shaped accordingly by giving them the form of an oblong with slightly oblique angles (parallelogram), rounding off the corners a little. Comb-bars of such a shape, if pulled towards the operator but very slightly, at once come out of the grooves, and can be taken out of the hive conveniently.

A quarter of an inch thickness will give comb-bars the necessary strength to carry even the heaviest honeycomb, as the weight presses not only on the middle of the bar, but is uniformly distributed over its entire length. The width of a comb-bar must, however, be one and a half inches, as every comb with the space between two combs takes up this room. But in order to be able to take out more conveniently the separate bars carrying the combs — which if pushed closely against one another and cemented together by the bees it would be difficult to separate —and also to enable the bees to get above this set of bars, as well as below a second set fixed lower down in the hive, and to continue the combs there, it is necessary that passages should be left between the separate bars, which may be done by cutting away

---

* The hive here described is in the nature of a collateral hive with the crown-board fixed, and mobility of combs secured by having doors at each end of it. Instead of frames, bars only are used, which are slid into grooves formed on the side walls under the crown-board, and the combs attached to the bars are also attached to the hive walls, and have to be separated with a knife to make them moveable, and they are then drawn out by the bars being slid along the rabbets.—C. N. A.

part of the wood in some places, leaving, however, the full width of one and a half inches here and there that they may rest against one another, and thus retain a firm position. Now, we may either reduce the width of the bars in the middle to one inch, leaving small quarter inch projections at both ends and on both sides, or we may leave them of the full width in the middle, reducing it towards both ends, or leave projections in the middle as well as at both ends, or at once make the comb-bars only one inch wide, and secure the necessary distance of half an inch to make up one and a half inches, by driving in towards each end a pin, projecting half an inch; finally, even the two pins may be omitted, as a little practice will enable any one to judge with his eyes and fingers of the half-inch distance between the comb-bars, and these, being soon propolised by the bees, remain at the proper distance from each other, even if not kept separate by projections or pins. The shape of the comb-bars might therefore be as shown in the accompanying illustrations.

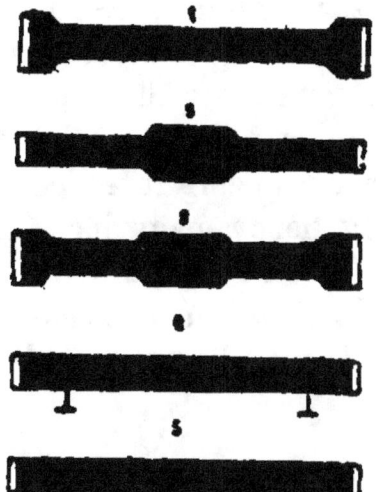

Though No. 3 is the most difficult comb-bar to make, it is certainly the most suitable one, because one is able to hold even a thick comb by the projections in the middle of the bar, without pressing against and damaging any cells.

As the bees fill up the empty spaces in the grooves with propolis, and also fasten the combs to the walls, here and there at least, no marking the spots for the separate comb-bars and combs is needed, all that is required being a little care in first inserting the comb-bars. If as many comb-bars are inserted as there ought to be in a hive of a certain length, and if all the bars are placed equally distant from each other, everything will be in order. In a hive fifteen inches long there ought to be ten comb-bars. If the hive were sixteen inches long, the bars might either be separated from each other a little more—especially in the place where we know that only honey will be stored, and by doing which thicker combs would be obtained—or an additional comb-bar might be inserted and the others moved a little more closely together. The latter is perhaps to be preferred, as it is easier to operate with thin and light combs than with very thick and heavy ones, which are more liable to break off. In the brood-room particularly the comb-bars must not be separated from each other too far, for if the distance between two combs be more than half an inch, the bees will possibly build a wedge between them, commencing from the wall, and depositing eggs in the cells, thus causing impediments to future operations.

In Lager stocks, the length of the combs in which are not more than about twelve inches, one pair of grooves and one set of comb-bars will be sufficient; in Ständer-hives of twenty-four to thirty inches in height, a second and a third, perhaps even a fourth pair of grooves must be made at a distance of 8, 9 to 10 inches lower down, and a set of comb-bars be inserted—though not necessarily all at once—in each groove, because combs more than ten inches deep are liable to break off when the operator loosens them from the walls and attempts to lift them out of the hive. In Ständers it is best to have three divisions,* or storeys, the two lower ones being intended for the brood-room, and the upper one for the honey magazine. But in order that the bees may arrange their brood-room and winter quarters in the middle and lower division, they must be entirely shut off from the upper part. This is done by a layer of small thin boards, the

Covering Boards,

which are placed over the set of comb-bars, and may be obtained from cigar boxes, or be made from logs of wood which splits well. They can be made three to eight inches in width, but their length depends on the width of the boxes. They may be placed either lengthwise or crosswise. The broader ones it is better to place at the back of the hive, the narrow ones more in front, in order that there may be no need for removing more boards than absolutely necessary, when only one or two combs are removed. With quite plain comb-bars without projections and pins, the covering boards may partly be dispensed with, and at the same time the proper distance between the comb-bars be determined by half-inch pieces

---

* These would be somewhat in the nature of Stewarton hives when in actual use, but without the facilities for subdivision the Stewarton enjoys.—C. N. A.

of wood inserted between every two bars, which exactly fill up the intermediate spaces.*

Near the door this arrangement has the advantage, that it is not necessary to break away any covering boards when a few combs are removed, no disturbance is caused in the hive, and there is less difficulty in catching the queen if it be desired to remove her. Altogether there is no need for being anxious about the careful placing of the covering boards. When the colony has once arranged its brood-nest in the lower division, the covering boards may partly remain displaced without any harm to the stock.

The uppermost pair of grooves (in the hive walls) must be one to two inches below the cover of the hive, in order to be able to measure the distance between the comb-bars with the finger, and to more conveniently get hold of the bars and remove them. Covering boards may be omitted at the top, but if it be desired to prevent the bees from filling with comb the small space between the cover and the frames, and in order to facilitate the removal of the honey-combs, they may also be placed there. Comb-bars, even when placed close under the cover, may indeed be removed by means of a small hook or a pair of pincers; yet it is more difficult to manipulate when there is not sufficient playroom for the fingers.† It need hardly be mentioned that if the combs are not intended to be inserted into light hives, but are to be broken up in the honey-pot, comb-bars may be omitted in the honey-room and the bees be left to fill it with comb as they like, which, however, is generally not done until the second year. But it is preferable to have the combs attached to bars, even if they are to be broken up a short time after, because they can be removed from the hive far more easily and quickly, and in a more cleanly manner, even in hot weather.

In order, however, that it may be possible not only to take the combs out of the hive and to replace them, but also to put them into any other hive and to make them fit there exactly, it is

---

* The Author does not seem to think much of the English bugbear, propolis, but makes his hive comfortable for the bees, apparently accepting propolising as a natural condition of things.—C. N. A.

† It seems impossible to account for the Author retaining the crown-board a fixture, when, if it were moveable, all the operations described could be so much more easily performed.—C. N. A.

necessary that all wooden hives should be of the same internal width, in order that any two combs may be placed by the side of each other without coming too closely together in some places, and being at too great a distance from each other at others; the combs must be made as even and straight as possible, following the bars as nearly as possible. This does not only give them a neat appearance, but is also a suitable arrangement. The bees, if left to themselves, would but seldom work thus. Instead of commencing the combs from the middle of the bar, they would rather begin at the edge, and perhaps make them from one bar to another, making a future removal very difficult, if not quite impossible. In order to prevent this it is necessary that the bees should be

Guided to Construct their Combs Regularly,

and this is done by fixing to every new bar about to be used a piece of comb, which should, if possible, be of the entire length of the bar. In the absence of large pieces of comb, two smaller pieces may be fixed to the bar and joined, so as to form a larger guide-comb. This may be done with the well-known curd-cement, but it is more according to nature to employ wax, which in this case may be mixed with resin, which increases its tenacity, and is also used by bees to fix their combs more firmly. If the comb, or at least the side to be fastened to the bar, is of pure wax, all that is necessary to fix it is to heat the bar before the fire or on a plate of metal, placing the comb on the bar and putting a piece of wood to lean against it, which keeps the comb in a perpendicular position and at the same time presses it gently down against the bar. Combs of a brownish colour that have been used for breeding once or several times, but which on account of their greater firmness are particularly suitable to be fixed as guide-combs, must have their edge (side) dipped into wax which has been melted in a longish pan of tinplate or earthenware over a plate (of metal), and must then be quickly pressed against the bars. The cells, it is true, are slightly inclined upwards; with the small guide-combs, however, it is not necessary to be particular in placing them in the exact position they formerly occupied. That part of the comb which was previously situated sideways may now, without hesitation, be fixed so as to come against the bar. It will then be possible to get straight strips of comb more easily, even from bent and twisted combs, as the combs are continued downwards in a perpendicular direction. Large combs, however, it will be better to fix in their former position. Even if somewhat bent, they can easily be made straight, and if warmed a little—which makes them flexible—and then laid on a table, any protuberances may be pressed down with the palm of the hand, a piece of board, or a book. This can be done while the comb is being fixed to the bar, the comb being pressed flatly against the table with the right hand, while the comb-bar is held in the left and pressed against the comb until the wax into which the comb had previously been dipped becomes hard. Combs properly fastened do not easily fall to the ground unless they are rotten, in which case they are gnawed off or entirely reduced to powder. The fixing of honey-combs is attended with more difficulty. It will be necessary, in the first place, to let the cut surface drain off completely, then to fix the comb in the manner described, and further to handle it carefully, placing it perhaps somewhere in the hive in an inverted position—bar downward—until it has been more thoroughly fixed by the bees themselves. Brood-combs are readily fitted upon bars which have already been in use, and on which a good deal of wax has been left; the bees, being always more active around the brood and in the neighbourhood of the brood-nest, will soon fix the combs to the bars, after which they may be inserted properly. Combs which contain much pollen, and which it is difficult to fix to bars, may be left to be fastened by the bees themselves, in the same manner as brood-combs. A bar need only be provided with a guide-comb once. When the comb is being broken up for the sake of the wax or honey, only a small strip need be left as a guide for the next new comb to be made. If the comb to be broken up is a honey-comb, the best way is to cut it off at about an inch from the bar, and if the cells on both sides are then partly cut away and the comb cut down to a point so that hardly anything remains except the middle wall, almost all the honey runs out, and what remains we may leave the bees to lick up, and keep the bar with the clean strip of comb for future use, in case we do not return it to the bees at once. The larger the piece of comb left on the bar the more labour we save the bees, and the more quickly will they be able to rebuild the comb and to fill it with honey, when good pasture is plentiful. Much would already be gained if only the middle wall of the comb could be preserved undamaged as nearly as possible, which is quite possible if the cells be cut down carefully with a very thin and sharp knife.* A greater number of bees would then be able to take part in the work of rebuilding the comb, and it would be impossible for drone-cells to be made on it if it had been a comb with worker-cells only.† We may even empty a honey-comb by merely carefully cutting away the material with which the cells have been sealed, placing the comb obliquely in a warm place and allowing the honey to trickle out slowly. It may be done still better if the side of the comb to be emptied, which should be cut as evenly as possible, is placed obliquely on a wide-meshed texture of bast (rush) or yarn, or on a piece of board with a great many small grooves cut into it. The fibres of the texture, or of the wood, when the honey has once penetrated them, attract it and conduct it downwards. Recently a machine has been invented, by means of which it is possible to empty quickly, and completely, combs filled with honey, when not yet sealed over or after they have been unsealed, as long as the honey in them has not yet crystallised or otherwise become very thick. The great advantages of moveable combs have been considerably increased by the invention of this machine. A full description of it will be given when we come to speak of the different implements which are either necessary, useful, or convenient in bee-keeping.

It is not difficult to understand that in consequence of this invention it is possible to increase the returns from bee-keeping very much during a continuance of abundant pasture, as the bees are able to fill a comb much more quickly than to build it.

The attempts which have lately been made, and which have been partially successful, to make wax-combs, or at least the middle walls, artificially, will hardly obtain much practical importance, because it requires a good deal of wax to manufacture them, for which reason they will remain a pretty expensive article, and because it is difficult to fix the artificial combs or middle walls

* This is practically a very strong recommendation of comb-foundation, the Author's objection to which is most mysterious.—C. N. A.

† This is not strictly correct, for bees will occasionally disregard the size of cell-bases.—C. N. A.

to the bars.* The bee-keeper is to produce but not to consume wax, and if he only be careful in keeping and preserving from wax-moths all the pieces of comb fit for use which he obtains when uniting colonies in autumn, or when trimming the combs in spring, he will not be in want of guide-combs.† Even in summer fresh combs may be procured by cutting away part of the fabric in parent hives which have given off swarms or been divided, as all the worker brood will have hatched three weeks after the fertile queen has left the hive, when comb may be cut away anywhere in the hive without causing any damage.‡

In order to preserve spare combs from destruction by moths, they must not be piled one upon another in the warm season, but should be kept in a place which is dry and as cool as possible, or where draughts of air are allowed to circulate freely. If the moths have already commenced breeding in the comb their brood may be destroyed or expelled by a temperature of about 40° R. (= 122° F.) For this purpose the combs may be exposed to the hot rays of the sun, but the heat must not be allowed to melt the combs; for, the air contained in the cells being stationary, the heat in them rises quickly to a high degree. For this reason brood-combs should not, during operations, be placed in the sun for any length of time in such a position as to expose the cells to the direct rays of the sun, because the brood might easily be killed. Combs made soft by the heat of the sun, which renders them very flexible, may then be made straight if placed on an even piece of

---

* The Author is evidently wrong here, as every bee-keeper who has used comb-foundation will know. Foundation is now recognised as the mainspring to swarms, since with it judiciously supplied a hive will be furnished with combs in three or four days, when, instead of the bees remaining at home as wax-makers, they go abroad as honey-gatherers.—C. N. A.

† We regard the practice of utilising scraps of old comb for guides as one of the most effectual means of propagating what the Author himself describes in his preface as 'the greatest terror of bee-keepers,' viz., the disease 'foul brood,' and we miss no opportunity of discouraging the practice. With comb-foundation, new comb is at command, and every bit of old or any that has been once used should go straightway to the melting-pot.—C. N. A.

‡ In England we should advise, if the combs in a hive needed renewal, that they be taken away wholly and frames filled with foundation put in their place; but in England the craving is for honey, while in Germany, as shown by the astute Author, the *production* of wax is held to be of very great importance.—C. N. A.

board and gently pressed down. When cool they retain a straight shape. To employ water to remove moths' brood is less advisable, because there is a possibility of the combs becoming mouldy and decaying if the moisture is not soon removed from the cells. Water is to be used only to clean combs which have been made dirty by bees suffering from dysentery. Such combs are soaked in water, the loosened dirt is then removed by means of a soft brush or a (washing) bath sponge, and the comb rinsed in water until it is clean, the water being removed from the cells as much as possible by letting the comb fall several times heavily against the palm of the hand or against a piece of board, after which it is dried in the air or near the stove.

Another effective means of preserving comb, or entire fabrics of comb, from destruction by moths, and of killing moths' brood already in the cells, is the vapour from sulphur, sulphur being lighted from time to time in the straw or wooden hive, or other vessel containing the comb, and care taken to prevent the rapid escape of the sulphurous vapour by carefully closing all the openings.

To fill the combs with dry sand has likewise been suggested as a means of preserving spare combs; and the use of sand recommends itself by its simplicity.

Combs slightly damaged by moths are not by any means unfit for further use. The bees know how to repair small damages very skilfully, and to remove the moths' brood. Damaged combs may, therefore, be inserted in populous stocks either at the top or sideways, and the cleansing and further preservation of them may safely be left to the bees. In the cool season, however, it is better to remove any empty combs and guide-combs from the honey-room, because they easily get mouldy there, and decay.

It is more difficult to preserve combs filled with pollen than combs entirely empty, because pollen turns easily mouldy, and it then becomes hard and unfit for food. Its removal gives the bees a great deal of hard labour, and the pollen is entirely useless, while in spring it would be of great value to the bees in the preparation of the food for their brood. It is well known that pollen is excellently preserved under honey, and it is therefore only necessary to cover the cells with honey in order to protect pollen from getting spoilt. When a stock dies of starvation in winter, the dead bees hanging between the combs or sticking in the cells must be removed soon, otherwise the bees get mouldy, and even the combs will partly be spoilt. The bees in the cells may partly be removed by beating against the combs, and those remaining are got out with a pointed instrument. But if a colony has died of foul brood—of which we shall speak hereafter—care should be taken that the combs of such a stock are not used again, as the disease might easily be communicated to healthy stocks.

Although moveable comb-bars allow of the removal of every comb, it has been attempted to render this still more convenient. The bees fix their combs not only at the top, but also in different places at the sides, which necessitates our loosening every comb we wish to remove. This can be done easily and quickly with a very thin common knife, especially if it be heated a little before the fire. But people wished to save themselves even this trouble, and not only were bars made for each separate comb, but entire

FRAMES

were prepared in which the bees were obliged to build their combs. But the side pieces attached to the comb-bar, as represented by figure No. 1, or better still by No. 3, on page 55, must not rest against the wall of the hive, else they would be cemented to it, and the removal of the combs would not be facilitated, but only rendered more difficult.  The distance between the side pieces of the frames and the wall must be about a quarter of an inch, or sufficiently large to enable the bees to pass through conveniently, because passages of less width are immediately filled up by them with propolis. But these passages, and especially the one at the wall opposite the entrance, are unnatural, and they carry off the necessary heat and moisture from the brood-nest and winter quarters of the bees, so that colonies generally winter rather badly in frame-hives.* It is, therefore, not advisable to insert frames in the

* We are simply delighted with the support the Author here gives to our views, expressed on so many occasions, in favour of close-ended

brood-nest; in places, however, where generally only honey is stored—near the door where one is most frequently obliged to remove the combs—they are very convenient and less injurious, especially if made of greater length, say about the same length as the height of the whole brood-room.* The lower cross-piece must not, of course, rest against the bottom of the hive; it must be about half an inch from it, and even a space of one inch may be left for the sake of being able to clean the floor more conveniently. The distance of the frames from the walls, as well as the distance between the separate frames, may be regulated by wire-pins driven into the frames and projecting quarter of an inch and half an inch respectively. They may be put together of four pieces or of two, and nailed together with wire-pins. In the latter case we may take a piece of a tolerably tough kind of wood, such as aspen wood, and cut strips one inch in width and quarter of an inch thick, and of the length of the lower three pieces of the frame taken together, making two notches more than half way through the wood at the proper places, where the piece is then bent so as to form right angles, as in the figure. They are then united to the upper part—the real comb-bar, so as to form a closed frame, either by tenons, or, what is more simple, by two wire-pins each.† When frames are made of four separate pieces, the lower cross-piece is fixed to the two side pieces with wire-pins, like the upper one. If the length of the lower cross-piece be made of the width of the bee-hive, it will project a quarter of an inch on each side beyond the side-pieces, and these will by that means be kept at the proper distance from the sides of the hives. In order to regulate the mutual distance of the frames

frames. There is nothing more unnatural in hive arrangement than the absurd practice of making or leaving spaces round the frame ends, either as conveniences for the bees or for the prevention of propolisation. The Author is truly a 'Rational' bee-keeper; he considers first the comfort of the bees, and accepts their own evidences of what is best for them; hence he prefers the simple bar to the open frame for the brood-nest, studying the comfort of the bees rather than conveniences for the bee-keeper.—C. N. A.

* This is in relation to hives that have two or more stories of comb-bars and combs. Instead of making separate frames for each story, the Author recommends that the honey-frames should be of the depth of all of them.—C. N. A.

† The intention is that the under portion of a frame shall be grooved and folded, as are the American one-piece mitred sections.—C. N. A.

from one another—as they are easily moved out of position—the lower cross-bars may have projections at both ends. But because wax *débris* and other rubbish are liable to accumulate on the horizontal parts of the frame, especially in winter, it might be more suitable to make them only an inch broad throughout the entire length, and to provide the perpendicular sides with appendages, or ears, as shown in fig. 1, page 55. With this arrangement the upper comb-bars never touch one another, not even at their ends, but stand everywhere half an inch apart, so that there is nothing to hinder the introduction of a hook behind them by which they may be drawn out.

But with all their convenience the frames have their disadvantages. If they reach nearly down to the floor they make it difficult to clean the hives, to feed, &c.

The lower cross-bar of the frame is especially inconvenient at these and other operations, and just this part is of least service, as it is well known the bees never extend the comb downwards as far as the floor, nor yet build it to this cross-bar. Many bee-keepers have, therefore, in adopting the frame, left it open below, which certainly deserves the preference.* A firm position is given to the side pieces, and the proper quarter-inch distance is easily kept by means of small wire nails lightly driven into the sides of the hive, or the side pieces may be kept apart by little pieces of wood, at the same time inserting quarter-inch strips between the side pieces and the hive walls, until the frames are, at any rate, partly filled with comb.

There is another way, however, by which a firm position may be given to the sides of the frame, viz., by inserting between the sides a second cross-bar, at a distance of three to five inches, beneath the upper one, and fastening it with tacks. In the figure on this page, the lower cross-bar is represented as inserted and fastened at more than half of the entire height, showing a frame enclosed in the upper part of it but open below.

* Here, again, the Author strongly fortifies the opinion we have often expressed, that the bottom rails of frames are worse than useless, and he would take very considerable pains to dispense with them.—C. N. A.

The advantages of frames made in this way must be evident to every thoughtful bee-keeper. Honey may be cut out of the upper enclosed part, and empty comb inserted in its place, without endangering brood that may be in the lower part. By enclosing the upper part with wire gauze on both sides, a cage may be made that will be a grand prison for the queen. If this is further divided into several compartments, like pigeon-holes for deeds, and closed on one side with a pane of glass, and on the other side provided with little doors of tin, it will form a protection for superfluous queen-cells, where the young queens may be kept in reserve and used as required.

As the principal weight of a heavy honey-comb is mostly in the upper part, the danger of such a heavy comb breaking out of the frame is greatly lessened by the inserted cross-bar mentioned above. Most combs have generally toward their lower part brood-cells that are often quite empty; on account of the consequent light weight the bees either do not build them in that part to the walls, or only attach them here and there. It is therefore not at all necessary to make the side-pieces of the frame as long as the brood-room is high. They may, indeed, have only such length as may allow the frames to be inserted into a far lower honey-room. But if the bees in the higher brood-room should, notwithstanding, extend the combs far downwards, that causes no inconvenience in taking them out. It will further afford the advantage that the combs can be cut away at pleasure, just as with simple bars or unmoveable combs. And in this way a supply of beautiful comb is obtained, of which bee-keepers using frames usually stand in such need that they are obliged to have recourse to artificial comb.

Instead of side-pieces of wood, strips of tin may be used, which have the advantages of taking up little or no room, and of surpassing wood in durability. A Danish bee-keeper has hit upon the happy idea of making frames of not very thick wire, and he finds them answer well. An ordinary wooden comb-bar forms the upper part, and the other three sides are made of three pieces of wire, connected like the links of a chain. In order to be able to give the sides a firm position by means of wire nails, it is advisable to provide them not only with two loops at the ends, but also in the middle. In order to make such parts of the frame in quantity, drive four nails into a board at the requisite distance and position, wind the wire round the first nail, then carry it on and wind round the second, third, and fourth, then again from the fourth to the first in a similar way, till the nails are full, when the single parts may be separated with a pair of cutting pliers.

Since the wire is entirely built into the combs by the bees, it in no way hinders their being brushed off the combs, but gives to these considerable firmness, so that there is no fear of their breaking off, though of great length and weight. Even with a considerable jolt, which may occur in moving hives from one place to another, honey-combs would be more likely to break out of wooden frames than out of wire ones, because the force of the motion is mitigated by the yielding, elastic character of the wire.

The Different Kinds of Bee-hives.

As we have treated of bee-hives in general, of how they must be made to be beneficial for the bees, and of how the interior must be arranged to be convenient for the bee-keeper and suited for a rational method of culture, we will now describe the different kinds of hives more in detail, and consider them more with regard to their exterior architecture.

One of the oldest hives, which may be considered as the primitive form, is the already mentioned

Log-hive (Klotzstock, or Klotzbeute).

Log-hives may either be set up perpendicularly or laid horizontally. In the former case they are called Ständer hives, and in the latter Lager hives. In logs of an unusual circumference several hives are hewn out near to or opposite to one another, so there are arranged in them two, three, or four hives, according to circumstances. These altogether are called 'Manifold hives,' a term of which we perhaps stand in need when we wish to speak of a hive with many compartments. If the log is of such thickness that after the hollowing out of a sufficiently roomy hive the walls retain a thickness of two or three inches and more, and if it is of a kind of wood that is suitable, viz., a wood that is soft or that has become so through decay, then the Log-hive is one that is not to be despised.

The entrances may be situated either in the doors, in the walls opposite to them, or in the sides, according as the hive may be set up. They may be most suitably placed by pairs, isolated in the garden. The log-hive may be easily adapted for moveable combs by fixing strips of wood on the right and left sides. This,

though, is only practicable when the side walls are made fairly smooth and parallel, that is, are of about the same distance from one another in front and at back. With old hives, which generally are narrower in front and wider behind, this can only be done when there is in front such a thickness of wood that the necessary breadth can be given to them, or if they can be forced with wedges to a greater width without splitting them. Those which are hollowed out more roundly may be used to most advantage in the usual way. Single combs on bars may be introduced to young stocks by hanging them on two pegs or on nails driven into the hinder wall, or by nailing them on to the cover. This is often done when it is desirable to give a swarm brood as a means of settling it in a new hive, or for the purpose of making swarms, or provisioning a stock poor in honey.

Upright log-hives, especially if somewhat large, will be considerably improved by the introduction of a horizontal dividing-board. The size of the hive may then be diminished for swarms, strong stocks may be forced to swarm earlier, too extensive depositing of eggs may be limited and the more honey be gained, which the bees will store below the dividing-board if passages are left giving access to this space. And since the third or fourth part of the space has been shut off below, the driving of a swarm or capture of the queen is always possible. All that is required is to withdraw the dividing-board, turn the hive upside down, and drive the bees by smoke and drumming to the empty room now to be found above.

If, therefore, no one can be advised to set up new log-hives, since a log cut up into boards will make a larger number of hives of a more convenient kind, yet every one who possesses good log-hives already stocked with bees, may still continue to make good use of them. He will gain from them natural and artificial swarms, as well as fine comb for his boxes; and it will be time enough to do away with the log-hives when he has a large number of boxes containing vigorous stocks.

The Straw Hive.

The bell-shaped straw-hive, though possibly not of the same antiquity as the preceding, is yet very generally used, especially in countries poor in wood. On account of the heat-retaining properties of its material, its round shape, which gives an equal distribution of heat to all its parts, and its generally limited size, this hive is peculiarly adapted for raising brood and swarms, but not for winning honey.* A larger harvest of honey can only be obtained from it by putting on a super at the right time. This may consist of a small straw-hive, a little box, a flower-pot,

* What will the Pettigrew school say to this?—C. N. A.

a bell-glass, or other vessel. For this purpose an opening must be made in the top of the hive, and the larger it is the sooner will the bees take to the super, and when pasture is abundant fill it with beautiful comb and honey.

The ordinary straw-hive, considered by itself, is more of a hive for brood and swarms. On account of its round and bell-shaped form, its heat-retaining material and its limited size, a generally higher temperature obtains all over it, which causes the bees to extend the depositing of eggs in all directions, and with increasing heat to swarm with a regularity that does not often fail. The bees use up again what they carry in, for the most part, in producing comb and brood, and if the late pasture fails, both the weakened parent stocks and the swarms will be entirely without honey. But if the stocks collect some provision in the late summer and autumn, it is mostly heather honey, of dark colour and inferior quality, while the finest nectar from the spring and summer flowers has been used up by the bees,* and a larger provision of it could not be stored up on account of want of room in the limited space of a hive already filled up with brood.

Gravenhorst's Bogenstülper.

Gravenhorst's Bogenstülper is an incomparably better hive than the ordinary straw-hive (Korb, Stülpkorb, Stülper), and though only brought out but a few years ago has already, by reason of its suitability, been widely adopted. It is a straw-hive, extended in length—as it were, a double hive—about as long again as

---

* There cannot possibly be a stronger argument in favour of giving comb-foundation to swarms.—C. N. A.

wide, with corners as rectangular as possible, only rounded off above, and so, in shape, not unlike a high-arched baker's oven. It is not accessible from the side, but below, and is, therefore, quite suitably called Stülper (that which may be tilted). It is called Bogenstülper (Bogen, a bow or arch) from the form of the frames with which it is fitted. These are of the shape represented in the figure—usually fourteen in number—in which the bees, assisted by guide-comb, build the single-combs. In the crown, under the arch, a kind of rack 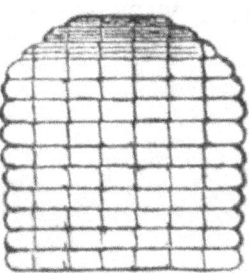 is placed, and the bow-frames are let into its notches. The frames are further made secure in this way; two short nails are driven into the top of the frame, coming one on each side of the rack, and below, the sidepieces are firmly fixed to the hive-wall by longer wire-nails, but these can be drawn out after the hive has been turned up; and when the two contiguous frames have necessarily 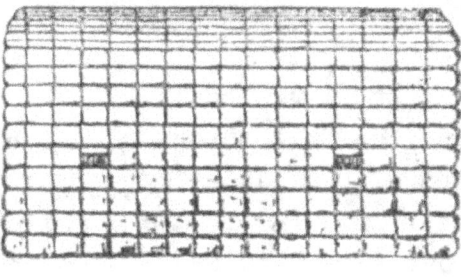 been somewhat pushed aside, the frames can be taken out without difficulty, and be either replaced and made firm in the same or in any other similar hive. That no advantages may be lost, all Bogenstülpers must have a similar width, and are therefore best made upon a machine. Division-boards are necessary for limiting the room, that may be too great for a moderate-sized swarm, or for setting up a special honey-room, as well as finally for placing two or three different stocks in the same hive for the winter. These division-  boards have the same size and shape as the frames carrying comb, and are similarly fitted in and fastened. Perhaps the entrances are most suitably situated at half the height of the hive, and every hive may have two—in one of the long sides, and at some distance from one another. If a third should for a short time be necessary it had better be cut on the level of the floor.

The Thorstock (Door Hive).

The Bogenstülper is as much like the Thorstock, constructed and described by the Author long ago, as one egg is like another.

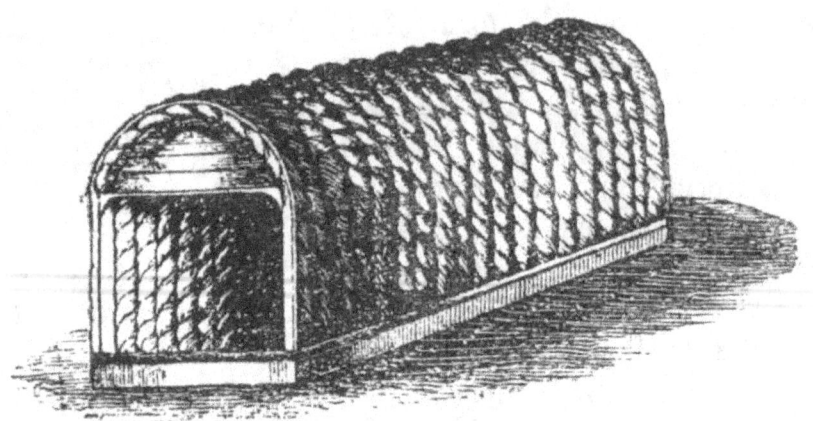

Although the two hives are essentially distinguished by this, that the Bogenstülper is only accessible from below, and must be turned over at every operation, whereas the Thorstock has a fixed floor-board, and is provided at the two ends with moveable gate-shaped doors. Strips certainly are fixed at the two sides where the arching begins, on which the ordinary comb-bars may be laid or the ordinary frames be hung. Further bow-frames of similar height and breadth as in the Bogenstülper might be adapted to it. And it would be so much the easier to make them firm in it, since they would have immediately a firm position on the floor, and would only need to be made safe against displacement by nails driven in to right and left of them. The bow-frames may, however, also be suspended if they are provided at the requisite height with ears to rest on rails fixed to the hive-sides. But it will always be more suitable to fit up the Thorstock only with ordinary frames or simple bars, and for a while to shut off the room under the arching by covering-boards, then, later on, to make it accessible to the bees, for them to fill with comb if they like. They will then have so much the more provision of honey over them, and will certainly winter well—let the winter be hard and long as it may. But young stocks that have not been able to collect the necessary food for winter can be provided there either with sealed honey-comb or pieces of barley-sugar. In order to obtain comb that would be suitable there, if the stocks are strong, the honey-room might be furnished with small bow-frames. These may be made of thin splints bent into the form of a bow, or, rather, of a half-circle, and the two ends connected by a wire brace. Pasteboard soaked in wax, as well as tin cut into strips an inch wide, would be somewhat firmer, and might very well be used for such frames.

With respect to the most suitable dimensions for the hive, the width may be nine to ten inches, or forty-five to fifty worker cells, one and a half to one and-three-quarters of this dimension for the height up to the under side of the arch, and two and a half to three times as long for the length.

The hive may be constructed in different ways. For the purpose of keeping it of an equal width, it is best sewn—as is every other straw hive—over a shape of the given height, width, and length. It is not begun at the crown like the Bogenstülper, but at one of the open ends, and finishes at the other open end. A wooden floor-board is always most suitable, for the sake of cleanliness, and the hive keeps the rectangular form better below, whereas, if entirely of straw it has a greater tendency to assume the round shape, and with the former method the construction is very much easier. For two Thorstocks can be got out of a wide cylindrical hive if this has had the shape of a longish box given to it. Two floor-boards are inserted, then the walls are fastened to the floor-boards by wooden strips nailed on from the outside, and the whole is divided into two parts by cutting down each side between the two floor-boards.

The Thorstock might be made with smoother walls if prepared on a machine similar to Oettle's.* The mould for the straw—one and a half to two inches wide—must have the shape of a horse-shoe, or arched door, and not be closed at the side, but open where the floor-board is inserted. The little pillars between which the straw is laid and pressed firm are not required to be of the length of the whole hive, since the hive can be prepared in two or three sections, and need not be fitted together

---

* A woodcut would be explanatory here, but that assistance is denied the reader; nevertheless, it is not difficult to imagine a miniature railway-arch that requires to be thatched, so that when completed the thatch shall be removable.—C. N. A.

before the floor-board and strips for carrying the frames are put in and fastened. By this the whole acquires a certain firmness, and it would then only need some three seams down the whole length of the upper arched part and two down the length of each side. It would be best, and most easily completed, by split cane or wire fastened inside and out, and mutually sewed up, or clamped. The doors would be most easily made of a board of some soft kind of wood. Two thin boards connected by wooden pegs, with a space between filled with newspaper, felt, straw, or the like, would also furnish a warm door. The outer board might project a little on each side, and would then form a rabbet, by which all chinks would be closed. The doors might also be made of straw without any difficulty. Their middle part might consist of a piece of board, rounded off above; and the entrance, if situated here, might be cut in it. The first straw-band passing up one side and down the other side of the board, after being bound with cane or wire, might be secured with nails, and have the succeeding bands sewed to it. The outside straw-band, both of the doors and of the hive, must, on account of the wear and tear of repeated opening, be well bound with wire or split cane, or might also have a frame of bent wood. If we wish, or are obliged to let the bees fly out from one of the long sides, we could at once—as in the Bogenstülper—make two flight-holes, so that by inserting a closely-fitting dividing-board, two stocks might be wintered, or at swarming-time a stock ready to swarm might be divided. Whether the flight-hole is placed a little above the floor or higher up is a matter of indifference. If it is wished to provide for inspection of the hive from below, all that is needed is to make in the floor-board a longish opening, that can be easily closed with a suitable board, and if bow-frames are wanted to be drawn out of the middle—as in the Bogenstülper—it is only necessary to make the middle, or about a third part of the floor, moveable. But this is really not advisable, since it is easier to take out at the side-doors even half the frames than to reverse the hive and draw a comb out of the middle.

The Thorstock described is certainly an excellent hive, and has—like the Bogenstülper—only one fault, that it must be set up singly, and a large number of such hives require much roofing, and considerable room, which every one is not able to afford. The hive would be more suitable, and easier to make, if it were angular above as well as below. Moveable combs might then be applied in the upper space close under the top, and six or eight of such hives, if they were twice as long as wide, could be piled up by pairs across one another on a couple of sills, and under a small roof covering the top pair. They must then, of course, have strength enough to carry weight, and for this purpose sticks should be inserted at intervals in the straw-walls.

A hive for setting up in pile would be made most easily by preparing from thin boards a box without ends—twenty-four to thirty inches long, nine to ten inches wide, and about sixteen to eighteen inches in height, and covered spirally by a continuous straw-band. The ends might be finished with a border of lath. By means of wires stretched from one end to the other and then nailed, the straw-covering would be made more compact, and the rough surface might be finished and made smooth by a coat of varnish. A couple of grooves or strips are required for the moveable bars eleven to twelve inches above the floor, and another pair close under the cover. The doors should be made with a rabbet on the right and left at least as wide as the grooves are deep, *i.e.*, a quarter of an inch, or the width of a drone-cell. It would not be well to make it broader, because in quickly shutting the doors bees might be crushed.

But if several hives are piled as described, the sides adjoining to and covering one another do not require to be made very retentive of heat. They may be made of thin board, which is the more suitable, for this reason, that the hives by being pushed close together may be made warm, or by moving apart, or inserting wedges, may be made cool, which in certain circumstances is very beneficial for the bees. The whole pile also gains greater firmness, because the hives fit on one another better, and at the same time the smooth surfaces of floor and cover give greater facility for withdrawing a hive from the middle of the pile.*

This brings us to the hive which not only the Author considers the most suitable one of all his bee-hives of various shapes,

---

* We can scarcely reconcile the recommendation of a pile of hives resting one on another, and covered by a single roof, with the direct objection taken by the Author to bee-houses (page 41—Bee-stand).—C. N. A.

but which was also declared to be the most superior hive by the German bee-keepers at their meetings in Dresden and Stuttgart, and was awarded the first prize. This hive, which combines the greatest possible advantages with the utmost simplicity and cheapness, and leaves nothing more to be desired, is

The Twin-stock,

so called on account of these hives always being placed in twos, back to back. It will be advisable to give a full description of these hives, illustrated by numerous woodcuts, in order that every bee-keeper may be able to manufacture them. They are already very extensively used not only in Germany, but also in Sweden, Russia, Hungary, and Italy, and even in America.

In order to be able to place the hives referred to side by side (and across each other), in the way mentioned above, they must, of course, be of the same height externally, and the upper part of the two boxes, placed side by side, must form an even surface, like the top of a table. If the hives, placed side by side, always remained in this position, they alone would require to be of exactly equal height; but for many purposes it may perhaps become necessary to remove one or the other box into another pile, replacing it by another hive. To prevent inconvenience, all hives should be made of the same height, in order that each hive may fit into the place of any other hive.

But what is the most suitable height?

The Author possesses Twin-stocks fifteen inches high, and also some which are nineteen inches high, inside measurement, giving the preference, after long experience, to the former, which are fifteen inches high inside, and rather more than sixteen inches high outside. In both these hives the brood-room is twelve inches high. The space of only three inches above the set of comb-bars, which is quite sufficient to allow of all operations being performed in a convenient manner, will be filled with comb by the bees even in the first year, for which reason they winter well in such a hive, because they have above them sufficient food and also the moisture they require, a copious condensation of which takes place in the hive, especially on the crown-board.* But in the

* We cannot understand how the condensation of moisture within a hive can be considered beneficial, seeing that it will always be greatest in

honey-room of double this height in hives of greater length, comb-making will generally not be commenced by the bees before the second year, and then perhaps they will store more honey there than they can spare from the compartment below, where they are wintering. If, as a measure of precaution, the cutting away the combs in the honey-room be delayed until the following spring, the whole colony will then perhaps be found to have moved into the upper compartment and arranged its brood-nest there.* In hives which have but a low honey-room, this will less frequently be the case, and would be of no consequence, as the honey-combs there are generally allowed to remain untouched, at least those just above the place where the bees are clustered together, the actual spare honey in such hives being found more at the sides,† whence it may be removed much more conveniently and without hesitation. The small upper compartment, when completely filled with comb and when all the cells are full of honey, contains sufficient stores for the winter, and these are always accessible to the bees even when the temperature is very low.‡

If the Twin-stocks be made fifteen to sixteen inches high, it is then possible to place four pairs of them one above another, which is a further advantage. The saving in the cost of covers will be considerable, and it looks prettier and more symmetrical when pairs of hives face north, south, east, and west, than if the pile contains only three pairs of hives, in which case two aspects would have the entrances of two hives each facing them, while each of the other two aspects would be fronted by one hive only. Even when

---

very cold weather, when it cannot possibly be necessary for breeding purposes.—C. N. A.

* This and the preceding sentence plainly suggest that, irrespective of the chamber above the brood-nest affording cool space for the condensation of moisture, it is undesirable in other respects, and when increased becomes positively injurious; for if a colony takes possession of its honey-room, it may be taken for granted that a great deal of its strength will be exhausted in breeding abnormally early drones, because the cells in the honey-room will be principally drone-cells.—C. N. A.

† Or, in other words (as we prefer to put it), on the ground-floor.— C. N. A.

‡ We can only think that the system, being inconvenient as a whole, the Author endeavours to make the best of it. With an open top and quilt, rendering the frames removable from above, the Lager-hive would practically be an English Collateral, the principle of which has been greatly abandoned in favour of the Combination.—C. N. A.

the height is only sixteen inches internally, the two entrances towards the same aspect will still be as much as thirty-two inches apart, so that it will be impossible* for bees or queens to mistake their hive. But a bee-keeper who might wish to insert a set of moveable bars in the upper compartment also—which it would hardly be worth while to do in a space of three inches in height—and who prefers more spacious hives altogether, because he harvests his honey but once a-year, and is therefore obliged to provide hives of sufficient dimensions, need not hesitate to make them eighteen to twenty inches high inside; for the room in the hive cannot well be enlarged by increasing its length and width, because these dimensions have to be in a definite proportion to each other, and the one cannot be altered without the other. As to the width of the hives, whatever width has once been fixed upon must be adhered to, in order to be able to change combs from one hive into another. Whoever is not tied in this respect should make his hives of the width of 8, 8$\frac{1}{3}$, 9, 9$\frac{1}{2}$ to 10 inches; but in case any one should happen to have hives of considerably greater width, he had better not be guided by them, as otherwise the Twin-stocks would have to be made proportionately larger, and would become too spacious.

Viewed externally, the length of the Twin-stocks should therefore be double their width, in order that two hives placed against each other may form a square, or nearly so. It is better still that the length of the hive be three to four inches more than double their width, so as to form a square and make the corners of the different hives coincide when the Twin-stocks are separated by a space of three to four inches.

Supposing the width of the hives to be nine and a half inches inside, and taking the thickness of the back part which is placed against that of another hive as half an inch, but the thickness of the front side as two and a half inches, then the hives will be twelve and a half inches wide outside. The width of two hives would be twenty-five inches, and consequently they ought to be twenty-eight to twenty-nine inches in length. It is not absolutely

---

* Why impossible? There appears to be nothing more likely than that the bees or queen of an upper hive, coming home laden or heavy in a breeze, might miss their own alighting-board, and find themselves on that directly below it—with the usual consequences.—C. N. A.

necessary that one hive should be accurately as long as the other, but it would offend the eye if one Twin-stock were to project beyond another. They should, therefore, be made to fit against one another in such a manner that when two are placed side by side they will appear exactly like a double Lager-hive.

But in order that the hives, when placed in position, may fit perfectly, it is not only necessary that they should be of equal dimensions, but also that they should have a strictly rectangular form, *i.e.*, their long sides—especially the one forming the back—and their two shorter sides should form perfect rectangles. To see at a glance whether the front and back* which contain the doors form right angles, one need only turn the door round, so as to get the outer surface inside and the inner surface on the outside. This has the further advantage of enabling us, when the inner side of the door has become a little damp, to place it occasionally on the outside to dry. Even if the whole hive should turn out to be somewhat warped, it can do no harm. Considering the elasticity of the thin wooden sides, the hive will soon become straight in the pile from the pressure of the hives above and from the doors being made rectangular, and where there is still a considerable space between one hive and the adjoining one, they can easily be drawn together by means of two (French) nails and some string. A small space, however, will do no harm, and is easily filled up with folded paper, or stopped with tow or clay, if it be required to preserve the heat in the hive as much as possible.

Wood, it is well known, has the disagreeable property of swelling when it becomes damp, and shrinking again when it becomes dry. This swelling and shrinking do not, however, affect the length of the boards, but their width only. It is, therefore, necessary to arrange the boards in such a manner as to render their swelling and shrinking quite harmless. It would be injurious, however, if the hive were to become wider at one time and narrower at another, for if the hive increased in width considerably the comb-bars would be loosened, and perhaps even slip out of their grooves; the combs, on account of their being firmly attached to both sides of the hive, would necessarily be broken, and on the hive shrinking again the comb-bars would be squeezed in so tightly that it would only be possible to get them out with

* Or, as we understand them, the ends.—C. N. A.

very great difficulty. A change in the height of the hive might also damage the combs, as they cannot, of course, accommodate themselves to such a change; but if the length of the hive were liable to a trifling change it would not be of the slightest consequence. From this it will be seen that the boards which form the crown-board and the floor-board have to be placed crosswise, and those forming the sides, as well as the door which closes the

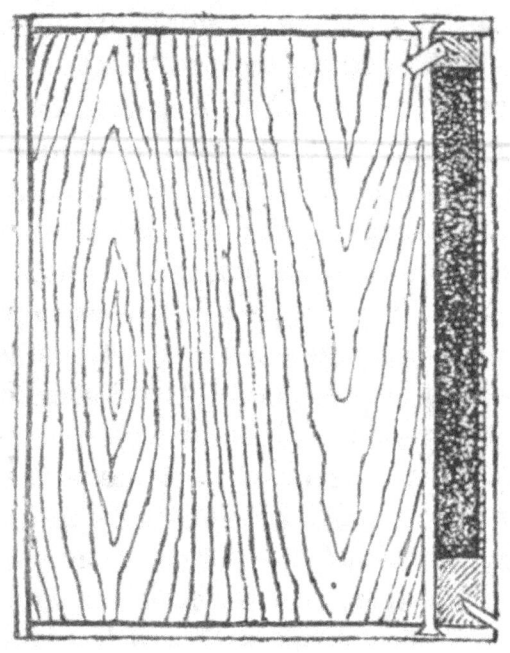

hive, should be placed in an upright position. It will therefore be necessary to join or glue together either two pieces of board fourteen inches wide, or three pieces nine to ten inches wide, or four pieces seven inches wide, for the floor-board and crown-board, as well as for the two sides, of a hive twenty-eight inches in length. It would be advisable to bear in mind not to make the joints in the crown-board and floor-board coincide with those in the boards forming the sides, in order that the hive may not fall to pieces but still keep together, if by chance the glue were to give way.

The two long sides, the front and the back of the hive, are perfectly alike as regards height and length, but they differ in thickness.

The front side of the Twin-stock (compare the illustration opposite, which at the same time shows the alighting-board) is, in addition to the two side doors, the only side which is continually exposed to the full influence of the external air, as well as to the burning heat of the sun and the most severe cold. This side must, therefore, be made as non-conductive as possible, in order that it may neither allow excessive heat to penetrate into the hive nor the internal heat to escape from the hive. While planed boards of half an inch thickness are quite sufficient for the floor-board, crown-board, and the back part of the hive, boards about two inches thick must be used for the front side, if it be intended to construct it entirely of wood. The two, three, or four pieces of

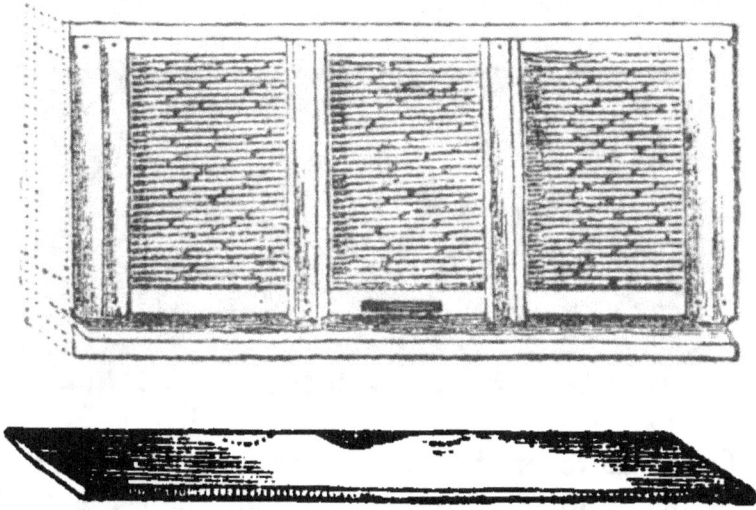

board, which must be joined to obtain a length of twenty-eight inches, should be united as firmly as possible in order to prevent the hive from falling to pieces after it has been in use for a time. If the hive is to be put together by simply nailing the floor-board and crown-board upon the two sides, the latter need only be made of the height the hive is intended to be internally, *i.e.*, fifteen inches, but otherwise it will be necessary to add more than an inch for the over-lap.

Before the hive is put together grooves are cut in the inner sides of the two walls at a distance of twelve inches from the bottom or three inches from the top, and, according to the thickness of the doors to be fitted in afterwards, a quarter-inch rabbet is made for them, one and a half to two inches from each end, by cutting away sufficient wood to make the quarter-inch groove vanish there. The entrance may be made before or after the sides have been put together; it should be three inches wide by half an inch high, and the most suitable place for it is the middle of the front side, one inch above the floor-board.* A piece of wood, three inches long by one and a half inches high, is also cut out of the thin board, which forms the back of the hive, exactly

* We cannot imagine why it is best to have the entrance an inch above the floor-board.—C. N. A.

opposite the flight-hole, and consequently at half length near the bottom, the hole made being exactly as large as the entrance would be if the wood, which is allowed to remain above the floor-board, were cut away. The woodcut below will illustrate this.

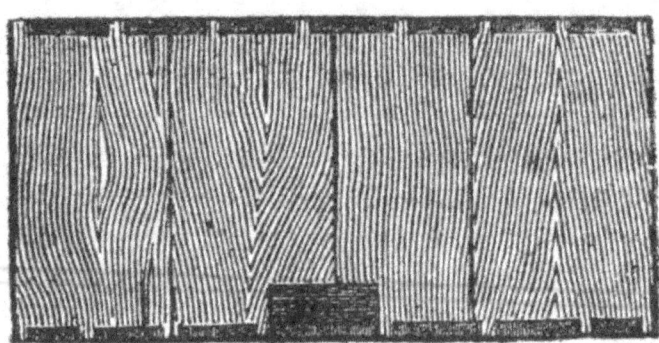

This hole, however, is closed by means of a small block of wood, three inches long by one and a half inches wide, fitting in exactly, and is only opened when two adjoining Twin-stocks are to be brought into communication with each other, which may be desirable in certain cases, to be referred to hereafter. If now the crown-board and floor-board are firmly nailed down upon the sides—the latter having previously been prepared in the manner we have stated—so that a box is formed nine to ten inches wide, fifteen inches high, and twenty-eight inches long, open at both ends, and if these ends be closed by well-fitting doors, then the hive will be complete in its most essential parts. The only thing still wanting is

A Small Moveable Door,

which may consist either of a single thin piece of board of the height and width of the hive, or of two panes of glass fixed in small wooden frames; it should have two apertures in the middle of the door capable of being closed easily by means of a moveable slip of wood.

This door is intended to contract the space in the hive which is too large for a swarm at the commencement, and to compel the bees to arrange their brood-nest is the centre of the hive near the entrance-hole.*

* This, in England, we should call a moveable dummy, or divider, which the Author prefers to have of glass.—C. N. A.

If a swarm were put into a Twin-stock, and the bees allowed to enter at one side,* they would certainly rush to the opposite end, where they would cluster together and commence comb-building; they would also arrange their brood-nest there, and consequently have their winter quarters in a cool part of the hive. But if the little door be pushed inward five to six inches from one end of the hive and fixed there, and if, after the hive has been closed the swarm be made to enter from the opposite end, the bees will settle against the little door, where they will begin comb-making and arrange their brood-nest, as well as their winter residence, and be the more comfortable, just as the centre rooms of a house are always found to be warmer than the corner ones.

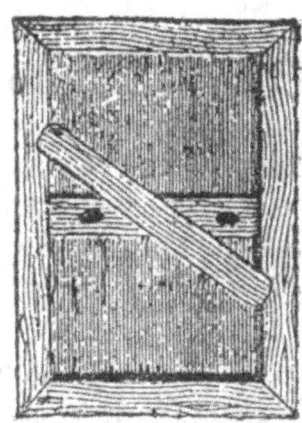

MOVEABLE DOOR.

We also gain a space which, when the bees have filled the room beyond the door, they will likewise furnish with comb, filling the cells exclusively with pure honey, as the queen would show herself here but very rarely. Although quite a thin piece of plain board would perfectly answer every purpose, yet, it is evident, that a glass door is much more agreeable, as it allows a frequent inspection of the colony, and enables us, even in winter, without in any way disturbing the bees, to ascertain whether any provisions are left in the hive, or, at least, in the comb nearest the door, which can be viewed through the glass entirely. The two apertures in the wooden door, made primarily for the purpose of enabling us to take hold of it with the fingers more easily and slide it forward and backward, also to afford passages for the bees into the room beyond, give us some insight into the interior of the hive, but not very much. It is possible, however, with very little difficulty, and without greatly disturbing the bees, to withdraw this moveable door altogether, if its height be about one-eighth of an inch less than that of the hive, and if it be inserted somewhat obliquely— in the manner already recommended as regards comb-bars—so that its right side, by which it is usually taken hold of first, may already be somewhat nearer the outer door.

We might also make this little door considerably shorter—

* *i.e.*, one end.—C. N. A.

perhaps one inch less than the height of the hive—in order to be able, if necessary, to afford the bees larger passages at the bottom as well as at the top, and to partly clean the hive, and feed the bees; and it would be easy to close these passages again by small blocks placed before, or inserted into, the passage-way. But this is not at all necessary, as the whole door may be withdrawn with the greatest ease. Some years, when breeding is not carried on very extensively, on account of prevailing dry weather, or honey being unusually plentiful, this moveable door had better be removed entirely, for it is quite plain that when this is done the bees will fill the space sooner than if access to it had been given them in ever so convenient a manner. In general, when the colony has arranged its brood-nest in the centre, and constructed a good fabric of comb, the little door is quite superfluous, especially when the outer-doors prevent the escape of heat from the hive, and it need not be inserted, except in case a considerable empty space is left in the hive by the removal at one end or the other of several full combs in autumn or spring, when it might be advisable to replace it, in order to shut off the inhabited part of the hive from the empty portion.

A small straw mat will also answer every purpose of the moveable door; and such a mat any bee-keeper may make himself, all that he requires being some straw cut of the requisite length, two needles, which should be long and strong, and some string. The needles being threaded at each end of the string, are passed alternately through the layers of straw, piercing the middle of the first layer and drawing the next one to it; and this process is continued until the mat has attained the width of the hive. Such mats may be put into the hive for the winter, one next each door, and it will be well to air them occasionally, or replace them by dry mats, should they have absorbed much moisture. By so doing we should render the bees a good service, and be able to take the outer-doors away temporarily, for the purpose of drying them before the fire, if we were afraid of their becoming too much warped by moisture. It has already been stated that the doors may also be reversed now and then.

All this can easily be done with the moveable doors; but with the immoveable front of the hive, which, of course, is also liable to swell, the case is different. Lining the inside with a thin board, the front and back edges of which in that case might take the place of the rabbets for the doors, and the upper edge that of the groove of the comb-bars—would certainly have the effect of preventing the wooden front wall becoming damp; on the other hand, it would make the hive more expensive, and heavier also. It would, likewise, be a good thing to encase the outside of the front wall, which would enable it to better resist the effect of moisture and the sun's rays, and lessen the risk of the pieces of board, which form the front wall, coming asunder. It would be well to carry this exterior cover downward only as far as the alighting-board, which runs along the entire front, and is made about three inches wide, and fixed to the hive, below the entrance, in a sloping direction, in order to carry off any water running down the front wall of the hive.

In many districts, however, boards of a suitable kind of wood are not to be had. Hives made entirely of wood are also of considerable weight, which is an inconvenience, not only when the bee-keeper is in the habit of sending his hives to some distance for the sake of a second harvest, but also in other respects. We get a much lighter, and otherwise more suitable hive, if we use thin boards also for the front wall, covering it on the outside with a layer of straw, to make it sufficiently warm. (Compare the illustration on page 81.)

In making this front wall we may use the worst kind of boards, and even outside boards, as long as there is an even surface on that side which is placed towards the interior of the hive, the outer side being entirely covered with straw, in such a manner as to make the surface appear quite even. In order to be able to fix the straw conveniently, the crown-board and floor-board should be made to project beyond the thin board of the front wall just as much as the thickness of the layer of straw is intended to be—say, one inch and a half to two inches. In that case the thin boards of the front wall cannot, of course, be dovetailed into the crown-board and floor-board like the back wall, but must be inserted in a groove, or mortised together; and, it is hardly necessary to say, that, like the back wall, the inner side of the board of the front wall must contain the necessary grooves for the insertion of the comb-bars, as well as a rabbet at each end for the doors to fall against.

The space between the projections of the crown-board and floor-board is filled up with straw, which is laid on horizontally, and fastened down to the board by four perpendicular bars of wood at equal distances from each other. A lath about one and a half to two inches thick, and two inches wide, and of the entire length of the hive, is placed over the floor-board, and fixed to it as well as to the wall by wire-pins, partly for the sake of enabling us to easily arrange the flight-holes (the most suitable place for which is the middle of the front wall, one inch above the floor-board), partly in order that we may be able to fix the four perpendicular bars more conveniently, as well as to give the hive greater firmness. The lath must be fixed rather firmly to prevent its breaking away if the straw covering with the bars be drawn in more tightly. These bars do not extend down to the floor-board, but only as far as the slanting-board fixed below the entrance, which serves as a convenient alighting-board for the bees, and to protect the lower hives in the pile from heavy rain—three inches is a sufficient width, except on the weather-side, where a somewhat wider board may be fixed. Its length corresponds with that of the hive, being adapted to the lower lath in its entire length. A small groove—forming an acute angle, and slightly inclined upwards—may be made in the lath along its entire length, just below the entrance, and the alighting-board inserted into it, in order that this board may take up, and carry away all the water that may collect on the side of the hive during a heavy shower of rain. The alighting-board might, however, be dispensed with if the floor-board of the hive were made to project about three inches—the part projecting being somewhat bevelled to carry off any rain-water driven against the front wall. The groove, just referred to, would then, of course, not be required.

It is not absolutely necessary to have a lath just below the projection of the crown-board, as the bars, which fix the straw, may be fastened to the crown-board projection itself; still it is more suitable, and the hive looks handsomer, if a lath of the same width, but less in thickness, be fixed at the top in the same way as the lath over the floor-board. The crown-board is made to project a quarter of an inch beyond it, so that the bars, which are of about this thickness, may be flush with the edge of the crown-board when nailed against the lath. Compare the annexed illustration, showing two Twin-stocks placed one against another, the doors being open. (Compare with woodcut on page 80, representing a single hive with closed door.)

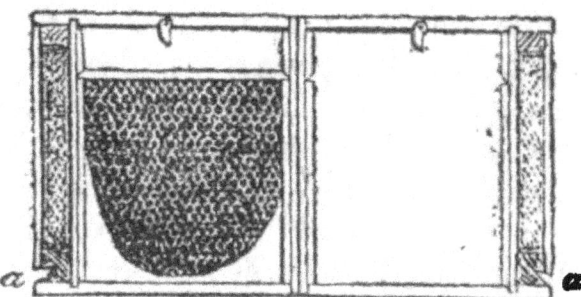

If a bee-keeper were desirous of constructing Twin-stocks by merely joining his boards by wire-nails, and were to employ very thin boards for the front wall, in which the nails would not well hold, he would not be able to dispense with the top lath, as it forms the means of connecting the crown-board with the wall. He would, first of all, make his front wall by nailing boards upon two laths placed crosswise—a hole for the entrance having previously been made in the middle of the thicker one—arranging the groove for the comb-bars and the rabbets for the doors on the smooth side of the boards, and then nail down upon it the crown-board and the floor-board. In order to be able to do so the board forming the back wall must not, of course, be too thin. For the back wall thicker and wider boards ought therefore to be chosen, as they have either to be glued together or firmly joined to one another in some other manner, which is not absolutely necessary in constructing the front wall. The crown-board and floor-board need not be nailed down very firmly, as they are pressing against the side walls when piled one upon another and cannot possibly part. Whether the walls be joined according to the strict rules of mechanics or simply put together by wire nails, which are preferable to ordinary nails, because their heads do not protrude, makes very little difference, and does not alter the appearance of the hives, as long as the outside of the front wall has a neat covering. This external covering, which renders the hive capable of retaining the requisite degree of heat, may easily be made in the following manner:—First of all, the hive is laid on the ground, the side to be covered being placed upwards, and then the wooden bars are fitted. The two corner bars may be made of the width of two inches, and the two inner ones of one and a half inches and a quarter of an inch thick. The inner side of the corner bars and both sides of the inner bars might be nicely bevelled and chamfered, which would improve their appearance. Their length is the same as the distance of the crown-board from the floor-board, *i.e.*, fifteen inches, but they are made somewhat shorter if a special alighting-board be provided, as, in that case, they need only extend downward as far as this board. They are placed at equal distances from each other, and holes are bored in them for the nails. One nail at the top and one at the bottom will be sufficient for the narrow bars, but for the wider ones two nails are necessary at each end. Then the bars are removed for a while, care being taken not to disarrange the order in which they are to be placed, and the trough-like space between the two laths is filled up with well-shaken straw, the thick, or stubble end, of which is placed to the right and left alternately, until the layer, when pressed down without any very great exertion, has attained the thickness of the width of the laths, as intended. Care should be taken, however, not to put in too much straw, as this would only render the task more troublesome; the lath would bend in the middle, and, in proportion as the air was expelled from the individual stalks of straw by excessive pressure, the straw covering would lose in warmth. It should also be considered that the straw is to receive a facing of some material, which gives it a smooth surface, thereby improving the appearance of the hive and rendering it more durable. Reeds — such as grow in ponds — may be specially recommended for this purpose. They are cut into pieces of the length of the hive, and after the straw has been pressed down slightly by the two corner bars, and a sheet of paper spread over it, the pieces of reed are inserted one after another, being first pushed under one of the laths at either end, and then drawn back a little and worked under the lath at the other end. Osiers—either peeled or with the bark on—may also be used; very thick reed may be split or cut open and then pressed flat and inserted. In that condition it is certainly not so stiff, but it makes a very beautifully smooth surface, and covers three times as much as it would if inserted in the round state. Small pieces of thin board, or splits, placed one above another, like Venetian blinds, also look very well, and are still more durable than reed. The whole side having been thus covered, the two middle bars are next put into their places, and all the four bars are then strongly fixed by nails, while the straw, especially in the corners, is being pressed down firmly. A better finish is given by placing the reed so as to hide the black knots under the bars. The straw sticking out at the ends is cut off smoothly with a knife, which should have a very sharp edge, and if the bars are very much bent, as generally happens with the narrower ones in the middle, they may be drawn in towards the wooden part of the wall by means of nails driven through the bars and the straw. These nails should be of such a length as not only to penetrate the wooden wall, but to allow of their being clenched inside the hive. Thus the hive would be complete, and present a very neat and pleasing appearance, as shown in the annexed illustration.

A coat of varnish gives the reed a still more beautiful appearance than it already possesses in its natural state, and besides renders the hive still more durable. Of course, in time it will suffer, especially when facing the weather side, but it will be easy, even if the hive be occupied by bees, either to renew the facing or to replace it by thin splits of wood, which, in order to be able to insert them, need only be long enough to reach from bar to bar. Any one who might wish to cover the entire front side at once with thin pieces of board, which in that case would have to be fastened in the same way as the four bars, and had better not reach down farther than the alighting-board, need certainly not hesitate to do so. The external covering is entirely a matter of taste, and does not at all affect the hive in its essential points. The Author has lately commenced covering his hives externally with paste-board, the use of which simplifies this work very much, and if the outside be varnished over, a paste-board covering will be as beautiful as it is lasting.

In addition to the front wall, there are still the two doors which, when the Twin-stocks are placed in the pile, come into contact with the external air, and are exposed. It is necessary, therefore, that they should be made somewhat warmer and also more neatly. One and a half inch boards of poplar, willow, aspen, lime-wood, or pine, will sufficiently answer the purpose, and if varnished on both sides they are but little affected by moisture and do not easily get warped; and even if they did, they are easily taken out by means of a screw fastened to a large ring, through which, after the screw has been driven into the door, a bar might be put and the door thus be pulled out forcibly. But this can only happen in Schrank (press)* hives, which will be described hereafter, and will never occur in Twinstocks. The thin walls of the latter will always yield sufficiently to allow of the doors being opened easily by means of a pocket-knife, especially if a little play be allowed for the thin wooden wall of the front side, so that it may 'give' slightly towards the straw covering, and return again to its former position when the pressure is removed, thus enabling us to draw the wooden wall towards the door by means of small wedges inserted between the door and the two laths, in case the door shrinks to an unusually small size in a hot summer.

Moreover, no harm would be done if the door were to shrink a quarter of an inch, as there is a quarter-inch rabbet inside the hive for the door to fall against to the right as well as to the left; it would also be easy to close any cracks by fitting in small bits of wood.

But to entirely obviate any inconvenience arising from the wooden doors swelling and shrinking, or if thick boards of a warm

---

* This screwing in of a ring and wrenching out the end doors with a bar, we take it is a little worse than the necessity that formerly arose with the Woodbury Hive, when the crown-board had to be prised off with a garden spade.—C. N. A.

and light kind of wood are not to be obtained, very suitable doors may be made of thin laths, bars, straw, and reed (compare the illustration below). Looking at the reed-covered front wall of a Twin-stock, and supposing it to be placed in such a position as to give the bars a cross direction, we have before us such a door on an enlarged scale, the only difference being that both sides of the door may be covered with reed and be provided with bars. The two laths to the right and to the left are, of course, made of the width of the intended thickness of the straw door, say about one and a half inches, but they may be made only half an inch thick. The bars to be fixed across at the top and bottom on both sides may be made to end in a sharp edge on their inner sides, in order that any moisture might 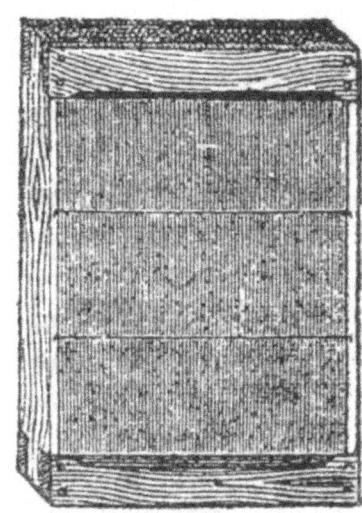 drain off more completely. It is also necessary that they should be let into the laths to the extent of their own thickness, so that there may be no projection and that the door may rest against the rabbet inside the hive all round. The straw will then become more compressed at the two ends by the bars on both sides, which may be pulled together still more by wire nails, driven through the middle of the bars. In order that any moisture collected may run down freely, it will be better to omit the bars in the middle of the door, and to use wire (brass wire in preference to any other) or split cane instead, which is drawn tight on both sides and then sewn up on opposite sides.

For doors of Twin-stocks, which are but short, it will be sufficient to sew the straw once only in the middle between the bars, but where the doors are of greater height, as in Ständer hives, this must of course be done more often, and at intervals of about six inches. But as the thin laths would easily become bent inwards, if the tension of the wires or cane were increased, it would be necessary to place several small, flat pieces of wood across from one lath to another, to keep the latter at their proper distance from each other.

If they were inserted where the wire or cane was intended to be placed, it would also be possible to draw the wire or cane towards these straw-covered pieces of wood by nails driven in from both sides.

If cane be used, it only requires to be nailed to the two laths like the bars, but it will be better to pass it through a hole bored through one of the laths, beginning the sewing there, and to wedge the two ends into the other lath.

The wire is, of course, passed through in a similar manner; it may also be doubled, and if a nail be put in between and twisted round several times, it will then bear a greater strain. If the laths and cross-bars have been properly fitted to the hive before the straw is put on, the door must necessarily fit, after the straw, sticking out at the two ends, has been cut away.

For the laths and bars a kind of wood is chosen which does not readily split, and in which the nails have a tolerably firm hold. The wood of the aspen, willow, and lime-tree, is particularly suitable.

The inner side of the door need not absolutely be covered with reed, still, if it has a covering of coarse reed, previously split and pressed flat, its appearance is very much improved. As such doors cannot shrink in the least, nor become extended in width, they will always fit exactly, and it would therefore hardly be necessary to make a rabbet for them inside the hive, as their getting into the hive too far might be prevented by the outer side of the cross-bar being made to project slightly. The bees propolise every corner in the hive, and a kind of rabbet will be found there in the course of time. But the joints of the doors might be stopped up with clay or putty.* It will be better, however, to have always a rabbet made for such doors.

Pasteboard may, of course, be used instead of reed for the exterior covering of the doors.

Every pile of Twin-stocks, whether consisting of two, four, six, or eight hives, should have a roof, which may be arranged according to taste.

A roof for Twin-stocks is easily constructed of two long and two short pieces of laths, and a few shingles, or thin pieces of board. The four pieces of lath are joined so as to form a kind

---

* The Author has evidently none of the English antipathy to propolis, and is not at all scared by the idea of its presence in a hive.—C. N. A.

of frame. The shingles, or pieces of board, are nailed to the two long pieces of lath (which should project about a foot beyond the hives on both sides) in such a way as to form a right angle, or, better still, an obtuse angle, at the top, so that a lath will not be

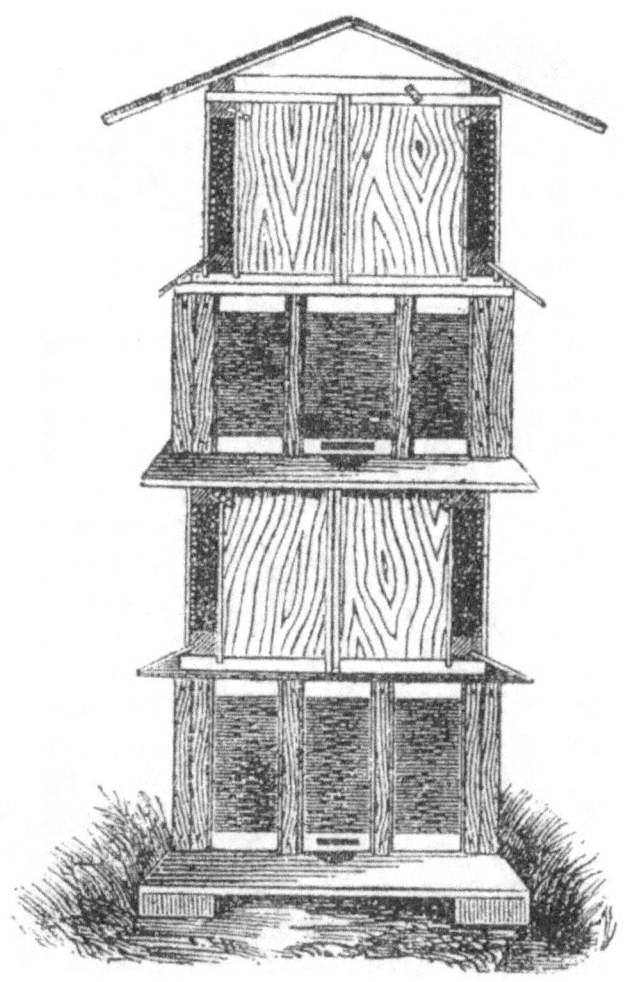

SIDE VIEW OF PILE OF TWIN-STOCKS.

needed there at all. One shingle is laid on after another, now on one side and now on the other, until the roof is complete. Beginning with a wide board on one side and a narrower one at the other the roof can be finished off much better at the top. The boards are joined by simply pressing one against the other, thin splits being placed under the line of junction. A coat of varnish, of course, renders the roof more durable, and costs but little. The cross-pieces must be placed at such a distance from each other that their outside parts are flush with the façade of the hives. The roof may be fixed to the hive by means of two wire nails, one of which is driven into the lath and the other into the crown-board, and some string; the latter being twisted

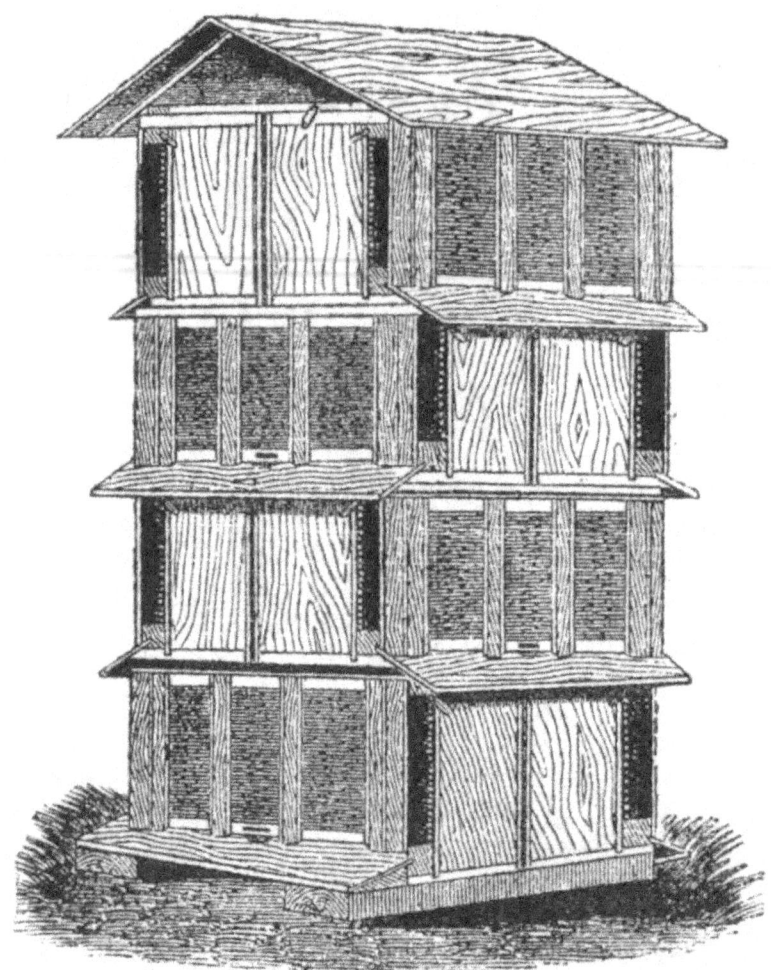

A PILE OF TWIN-STOCKS, LOOKED AT FROM ONE CORNER, SO THAT TWO SIDES ARE VISIBLE.

round the ends of the two nails, which are left to stand out a little; when fastened in this manner there is no possibility of the roof being blown off, even by the most violent storm.

There is not much fear of the pile being thrown down by a high wind. It might, perhaps, be possible, if the hives were empty, because those made in the manner just described are as light as straw-hives; but when they are stocked it would hardly be possible, and has never happened yet, as, on account of their broad basis, they have a very firm position.

But where Twin-stocks would be exposed to very violent storms they might be tied to each other in the same manner in which the roof is fastened to the upper pair of hives. As a precaution against thieves the hives might be united by wood screws, or in some other way, so as to make it impossible for the hives to be carried away separately.

But just because the hives are light, and can readily be removed, we are able to guard against robbery easily, by placing the hives into a suitable building late in autumn, allowing them to remain there during the winter.

Twin-stocks and double Lager-hives may be placed in the same pile together. By putting an indivisible double-hive at the bottom of the pile, instead of a pair of Twin-stocks, we obtain a firm foundation for the whole pile, and need not be very particular about the supports on which the hives are to rest; in case of need, for example, when the bees are removed to the heath for a short time, four stones, placed underneath the floor-board, will be quite sufficient; and if, instead of the fourth, or uppermost pair of Twin-stocks, a double-hive be placed at the top, this would give the pile of hives a firm position. In the case of removal into some building for the winter, the uppermost double-hive of the pile is placed at the bottom; but when the hives are returned to the apiary they are replaced in their former order.

The Double Lager-Hive.

It will hardly be necessary to enter into a full description of this hive, which, in every respect, is like two Twin-stocks placed against one another. The illustration on page 81 represents such a hive viewed from either of the two long sides; and looked at from either of the shorter sides it has the appearance as shown in the woodcut on page 87; the only difference being that instead of the two back walls the double-hive contains only one partition wall, of one inch in thickness. Floor-board and crown-board, of course, are continuous across the two compartments, and very much resemble a table-top three to four inches longer than it is wide. The floor-board and crown-board show the wooden or iron nails driven in to fasten them to the partition wall, unless it is inserted in grooves, which is not necessary. Grooves for the comb-bars having to be made on both sides of the partition wall, it should be, at least, one inch in thickness, and the fibres of the wood should run in a perpendicular direction. Between the two doors at either end the thickness of the partition becomes reduced to half an inch, because a quarter of an inch of wood is taken away on both sides to make rabbets for the doors. The two ends of the partition wall may at once be made shorter than the front and back walls of the hive by the thickness of the doors, which is about three inches (one inch and a half at each end), and a piece of wood, one inch and a half wide, but only half an inch thick, be joined to it afterwards; this would not only give us at once a rabbet for the doors on either side, but would also increase the strength of the hive, the piece added to the partition wall being let into the crown-board and floor-board, and fastened to them with wedges or nails.

Communication between the two compartments of a double-hive can be established in the same way as between two adjoining Twin-stocks. For this purpose a hole, three inches wide by one inch and a half high, is made in the partition wall near the bottom, at an equal distance from either end of the hive, and opposite the two entrances; this opening is closed by a small block of wood, which it should be possible to remove easily from either side. It could be made to project about half an inch on both sides, so that it might be extracted by means of a chisel. If very firmly cemented by the bees it may become necessary to use an iron bar passed through one of the flight-holes to force out the block of wood. But whoever is desirous of saving himself the trouble of removing the combs of one compartment as far as this passage—which would be unavoidable in such a case—might so arrange the wooden wedge as to be able to insert and withdraw it from below, through the floor-board. In order, however, that communication between the two adjoining compartments might not be established at a time when it should be carefully avoided, the bee-keeper would have to guard against the possibility of the wedge slipping out, and this might easily be done by means of a button. Twin-stocks are more convenient in this respect. When arranged in the pile they need only be moved a little apart at one side to allow of the removal or insertion of the wedge, according as communication between two hives is to be established or to be discontinued. But in order to prevent the bees from rushing out in large numbers on the removal of the wedges, and to save them from being crushed when the hives are pushed together, a sheet of paper, dipped in honey—which the bees would soon gnaw through—might be stuck before the opening temporarily. By doing so we should gain an additional advantage in uniting two colonies, as time would be given for the excitement caused by the disturbance to subside, while in the meantime it would become dark, and the two colonies would unite all the more peaceably, if only one of the queens had been previously removed.

The Advantages of the Twin-stocks

and of the last-mentioned double Lager-hives, which are almost like a pair of Twin-stocks joined to one another, are, indeed, so obvious that no intelligent bee-keeper can help perceiving them at once; and although these advantages have already been referred to, we will recapitulate them here once more, and when we come to speak of the different methods of bee-keeping we shall give some directions as to the management of bees in these hives, in order to realise the various advantages which they offer.

1. The Twin-stocks are very cheap. The material required is but trifling, and need not be of any special quality. The hives are also easily manufactured. They become still cheaper from our being able to dispense with a special bee-house to place the hives in for shelter, and they last a long time. The crown and floor-boards and the back wall are entirely withdrawn from the wasting influence of the atmosphere and the weather, and will last longer than a man's lifetime. The covering of the front wall and of the doors will certainly suffer in time, but can easily be renewed. The individual walls cannot come apart, because thin boards do not bend like planks, and in the pile there is no room for them to bend. Even if the glue were to give, the hives made in the manner described will not come to pieces if only they are well joined at the corners. It is well not to rely entirely upon the dovetails; a few French nails may be driven into them in addition, as well as into the floor-board and crown-board where two parts have been joined, to prevent the division caused by the

glue between these parts having accidentally become damp from getting wider.

2. Although these hives are very plain and cheap, nevertheless, they look quite handsome, and have been admired in the Carlsmarkt Apiary by many visitors, although the Author's hives are of very plain workmanship, and even without a coat of paint. Staining the external parts of the wood brown, or painting them green, but leaving the reed its natural colour, gives the hives a still more pleasing appearance; and if a nice roof be placed over a pile of such hives they will be an ornament in even the most beautifully arranged garden.

3. Both single and compound hives, or hives of more than one compartment have their advantages as well as their disadvantages. The latter class of hives, more particularly, have been reproached with being too complicated, making it impossible for a bee-keeper to dispose of a separate hive as he might wish, preventing his putting it into a different place, &c. The Twin-stocks, however, combine the advantages of single hives with those of compound ones, without possessing any of the drawbacks of the latter. They enable us, without having to lift off the hive above, to remove every hive of a pile and to put it back into the same place, or elsewhere, when its place can be taken by another hive, either empty or occupied by a colony. This is a great convenience and affords us a good many advantages. Towards the end of the winter, when, for a long time past, there has not been a favourable day to examine the bees, one hive after another may be taken into a room and cleansed conveniently, empty or dirty combs may be removed, and comb filled with honey, or clean comb, be inserted, a populous colony may be deprived of a quantity of bees, or brood, and a weak colony be strengthened, &c. Thus we are able, quite independently of the weather, to regulate the various operations as we please, as a considerable number of hives cannot all be attended to at the same time when a favourable day occurs.

4. But the advantages which Twin-stocks and double-hives offer are greatest when colonies are to be divided or reunited. While it is necessary to remove to a distant apiary artificial swarms taken from hives composed of several compartments, a second apiary may be dispensed with entirely when bees are kept

in Twin-stocks. Before the first flight of bees in spring it should be so arranged that there may always be an empty hive by the side of a well-stocked one, whose bees it is intended to divide afterwards. Such a change in the position of Twin-stocks may be made without causing any injury whatever, the more so, as the hives can be put into some suitable place for the winter, and kept there undisturbed for three or four months, after which, when the bees will pretty well have forgotten their former place, which is now occupied by an empty hive, we might let them enter their home through the latter. When the swarming season approaches the bees should become accustomed to fly out in two opposite directions, and, if necessary, the pair of Twin-stocks might be turned round, so that the empty compartment occupies the place of the parent hive, and the latter, that of the empty hive. If, now, the communication between the two compartments be interrupted, care being taken that some young brood, a royal cell, or a queen, be given to the colony which is without a queen, we shall have succeeded in making an artificial swarm with the greatest ease.

5. But if it be easy to divide a stock of bees the union of two colonies is effected still more readily. We have only to re-establish communication between the two compartments and to remove one of the queens. Of course, the two compartments will have to remain in communication until the winter, as the bees of the emptied compartment will always return from their excursions by the entrance they have become used to, and through the empty hive. In order to make a more convenient passage for the bees the moveable door (dummy) might be pushed inwards as far as the entrance and the opening by which communication is kept up between the two compartments, and a small board could be inserted on the other side, so as to form a kind of channel through which the bees are able to pass into the occupied compartment in the quickest way.

But not only can neighbouring communities be united easily and without the loss of even the smallest number of bees, it is also possible for any colony to unite with any compartment below it. The bees will, of course, return from their first excursion to the entrance they have been accustomed to, but, finding it closed, they will soon discover their new abode and enter it joyfully, if only an opening has been made in the projecting crown-board just large enough for one bee to slip through conveniently. The alighting-board, which covers this projection, and which rests on it, must, of course, temporarily be removed, or, at least, be withdrawn from the hive far enough to enable the bees to descend.* As it happens very frequently that two or more colonies are united in autumn, such openings, of the width of a finger, may at once be provided in the Twin-stocks and double-hives while in process of construction. These openings should be made in the corner of the crown-board near the door, and the back, or middle wall, at both ends, but they should be kept closed until colonies are to be united. But should the hives happen to have a perfectly square form, and to cover each other exactly, it would be necessary to make a temporary entrance for the bees in the door by not inserting it tightly, and taking away a little wood at the corner; this would answer the purpose sufficiently well until the bees had become familiar with their new entrance. But, although communities in Twin-stocks can be divided most conveniently and safely, yet it is by no means necessary to increase the number of one's stocks artificially. Swarms will issue from these hives just as well as from other hives, and even still more regularly, and earlier, because a certain degree of temperature will be reached in Twin-stocks sooner than even in straw-hives, if the space inside the hive is reduced by inserting a moveable door at one or both ends; the empty space might, perhaps, be filled up with straw.

6. On the other hand, it is possible to lower the temperature in the hive again, in order to prevent excessive swarming, and to stimulate(increase) the industry of the bees and raise the production of honey. This may be done by giving the bees more room, by establishing communication between the empty hive and a hive containing a strong colony, by separating two populous hives, to make a passage for the external air, &c. In a hot summer, when the temperature of atmospheric air reaches a high degree, the hives might be cooled still more by allowing air of the tempera-

* The crown-board of the lower hive is here supposed to form the alighting-board of the upper hive, and the direction is that a way should be made for the bees of the upper to pass through the crown-board of the lower, instead of congregating under their own alighting-board as they would usually do.—C. N. A.

ture of the ground to pass upwards between the hives. In that case the hives would have to be placed over a pit at least three feet deep. There would have to be a distance of about four inches between the hives of each pair, completely closed at both ends in order to exclude the hot external air. It is evident that it would not then be possible to have an indivisible double-hive in the pile, except, perhaps, in place of the uppermost pair of Twin-stocks. The cooling effect would, indeed, be felt through the thin boards of the back wall of the hives, but still more so, if the wedge in this wall were removed and the opening closed by a perforated piece of zinc.

In a severe winter, as in a hot summer, it might, in certain circumstances, be very beneficial to allow air of the temperature of the ground—which then would be warmer than the external air—to have access to the hives. Without removing the hives from their stand, they might, as it were, be (considered as) placed in a cellar, if the entrances were closed or very much narrowed and the air allowed to reach them from behind or from the enclosed cellar-like space. But bees do not only want air, they also require free exit at all times. If, therefore, we were to close entirely the entrances of the hives, facing south and west, because in those hives the danger is greatest that bees might be induced by the treacherous rays of the sun to leave their hive and die on the snow in large numbers, it would be necessary that the bees should have free exit at the back, in which case, however, the opening of the adjoining hive should be closed by a piece of perforated zinc, to prevent the bees from coming together. Some bees will occasionally leave the hive, and, in order to prevent them from falling down the kind of tunnel formed between the boxes and being lost, it would be advisable to place a piece of wire gauze or similar fabric over the hives horizontally. This arrangement, however, is somewhat troublesome, and the advantages to be derived from it would be but inconsiderable, except, perhaps, where extremes of heat and cold prevail, on which the temperature of the ground would exert a somewhat moderating influence. In our climate bees will always do well in Twin-stocks, even if the latter remain piled one against another in summer and winter, as neither heat nor cold is able to exert any influence, destructive to bees, upon a thick column formed by a set of Twin-stocks.

7. An empty hive may be joined to a hive stocked with bees and communication be established between the two, not only for the sake of making an artificial swarm, but also for many other purposes, for example, to supply the bees conveniently with honey, pollen, or water, or in case it became necessary to confine them to the hive on account of attacks upon them by robbers, or when making artificial swarms, to allow them to spend their rage in the empty hive, and even to cleanse themselves there. A hive without bees, but containing some comb, may be placed in communication with an inhabited hive to allow the bees to empty the cells of any honey contained in them, to consume any pollen left, to clean the combs, to preserve them from moths, or even to let them fill the combs with honey.

8. And, lastly, these bee-hives have the great advantage of being very convenient for transport; and on this account, and because they can be put up in any place, they deserve to be specially recommended to those who are in the habit of removing their bees from one place to another where pasture is more plentiful. Although an empty Twin-stock might perhaps weigh five pounds more than a straw-hive, the weight of a colony in the former can easily be reduced twenty pounds before it is removed by depriving the bees of some combs filled with honey; so that Twin-stocks have a considerable advantage over straw-hives as far as the weight to be conveyed is concerned. When straw-hives are placed in a new position many bees lose their way on their first excursion, they attack and kill each other, and many hives become quite depopulated, while bees in Twin-stocks will find their way home at once, provided the hives are placed in their former order, as the bees immediately recognise their hive in the pile.

No preparations are needed for a journey except a few slits in the crown-board, especially near the door, and these might be made with a tenon saw at the time the hives are being constructed, when it would only be necessary to remove the propolis from the slits. If instead of one of the ordinary doors a piece of board provided with slits, or even wire cloth, were used, any danger, even at the hottest time of the season, would have been guarded against. In order to be able to carry the hives conveniently they might have a handle at the top in the centre of the crown-board, and this handle could either be screwed into it or fixed by being inserted into a slit cut into the board, similarly as a key becomes fixed in a lock, by being slightly turned. The rather long but narrow cross-piece of such a handle might at the same time be used to remove the propolis from the slits. Two hives, provided with such handles, may be carried by a person as conveniently as two pails of water. It need hardly be mentioned that that part of the crown-board to which the handle is attached should be firmly united to the side walls, to prevent its breaking away when the hives are of considerable weight.

When the hives are removed to the heath for a short time it is not even requisite that the roof should accompany them, as the hive may be covered, if necessary, by a sheet of waterproof cardboard, a stone being placed on it to keep it in position, for it is only the hive at the top of the pile, or the uppermost pair of Twin-stocks, which has to be protected from rain, because the alighting-boards keep the wet off the lower hives. It is not exactly an essential of the Twin-stock that it should be a Lager-hive and extend more in length than in height. Whoever prefers the Ständer shape need certainly not hesitate to increase the height of the Twin-stock to about twenty inches, which would give the upper honey-room a height of eight inches. In that case a second pair of grooves would, of course, be wanted just under the crown-board. The entire height might also be divided into two equal parts, by which we should obtain the advantage of being able to insert any comb in any part of the hive without being obliged to shorten the combs, which cannot be done when closed frames are used. But of these Twin-stocks of considerably greater height we should not be able to place more than six at most in one pile, as otherwise it would become too high. The hives might also be arranged in such a manner as to have all the entrances towards south and north, and all the doors towards east and west, for the entrances would be at such a distance from each other that there would be no fear of any bees, and particularly young queens, losing their way.

The width of these higher Twin-stocks being nine inches, or more, they are certainly somewhat large; but this cannot exactly be called a fault, as it is possible to reduce the space by means of a moveable door (dummy). Such a Twin-stock, or each compartment of a double-hive, might also, temporarily, be **made** the abode of two separate colonies if the moveable door, or **else a** thin board, be pushed forward as far as the middle of the **hive**, where it must be fixed, and every passage, however small, especially in the grooves of the walls, be carefully closed. Two separate entrances would then, of course, be necessary, and we **might** either divide one entrance—which in this special case should be made about five inches long—into two equal parts by a wedge inserted in the middle, or the hive might at once be made to contain two distinct entrances, which would have to be separated externally by a projecting piece of board, as seen in the middle hive of a set of three Fourfold-hives on page 106.

This contrivance, in fact, converts a pair of Twin-stocks, or a double Lager-hive, considered as a whole, into a Fourfold-hive, in which four separate colonies, adjoining one another, and forming, as it were, one cluster, are able to winter.

It will be obvious to any intelligent bee-keeper that if two separate entrances are provided for such spacious compartments—because the colony is to be divided later on—it will also be advisable to have two separate openings in the back—or partition wall—one opposite each entrance.

In every Manifold hive, described hereafter, the advantages of the Twin-stocks may partly be secured, if an opening be made in the partition wall separating any two compartments, as near the flight-holes as possible; this passage, of course, remains carefully closed, but can easily be opened when in certain cases it is desired to establish communication between two colonies.

Having now, as I hope, given a very full account of Twin-stocks and double Lager-hives, which at the present time are considered the best bee-hives, and which—chiefly on account of their manufacture being easy and cheap—have been awarded the first prize at several annual meetings of the German Bee-keepers—of which I may mention the meetings at Dresden and Breslau—we might well bring the chapter on Bee-hives to a close, as every bee-keeper naturally will choose the most suitable hive. Suitability, however, is a relative term. It might, for example, be very desirable to one bee-keeper, who has his bees conveyed to different places for a change of pasture, that the removal of the hives should give as little trouble as possible, whereas another bee-master might be quite indifferent in this respect, and convenience for removal might even be undesirable in his case, because he never removes his hives from their place, but might, perhaps, have to take precaution against thieves.

We will, therefore, next describe other Lager-hives, of several compartments, and afterwards the different kinds of Ständers; as the latter also possess many good qualities, and will always have some advocates among bee-keepers.

The Lager Hive, with Four Compartments.

If we suppose the double Lager hive, already described, about as long again, and each of the two long compartments divided at half the length perpendicularly by a dividing-board, we have the idea of the Lager hive with four compartments. But because every compartment is only accessible from one side, and the taking out of combs would be inconvenient with a length of nearly thirty inches to the back, it is desirable to diminish the length of the compartments and to add to their height. The length of single compartments may therefore be about twenty-four inches and the height eighteen to twenty inches, of which, as in the Twin-stock, twelve can form the brood-nest and the remaining six or eight over it the honey-room. With a honey-room of this height it is advantageous to arrange the moveable comb-bars at one or one and a half inches from the cover, and to make the grooves at that distance. There is no need of a moveable door, as in the Twin-stock, because the space is sufficiently limited for a swarm by dividing off the honey-room, and the bees, even with superfluous space, are sufficiently warm, since four stocks here afford to each other mutual support. Whether the hives are made of suitable planks, or of thin boards wattled with straw outside, may depend on circumstances, and is a matter of indifference. The floor and cover need only be made of thin boards, because such hives can be set up as a pile of three one on the top of another, only the floor of the lowest and cover of the uppermost should be stronger or should be protected in the cold season. If we have planks and boards of the same breadth as the hives are high, eighteen to twenty inches, they could be used horizontally, while the walls can be hindered from shrinking and swelling by the perpendicular division-boards in the middle, and at the ends by strips or laths nailed on, which may also form the rabbets for the doors. This will answer excellently if the hive has an outer casing, the thin casing boards standing upright. Even if the walls have only been made of one-inch board they would be sufficiently retentive of heat if a thin layer of straw, rags, or paper, is inserted between the hive and casing.

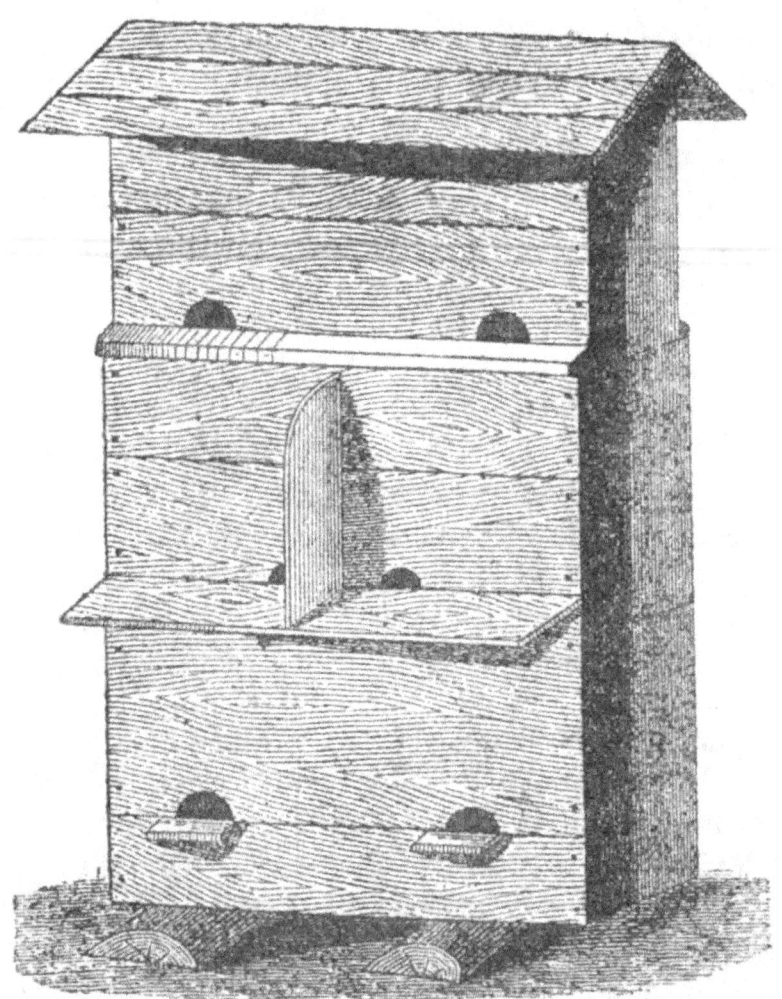

A PILE OF THREE LAGER HIVES, WITH FOUR COMPARTMENTS.

In the arrangement of entrances we have only to take care that the bees and queens cannot easily mistake their hives. This object may be attained by letting the position of the entrances in the middle hives of a pile be different from the upper and lower ones. These may have the entrances more towards the doors about six inches from them, while the middle ones may be near the dividing-wall, as shown in the accompanying figure.

The Lager Hive, with Four Compartments. 107

But because the two entrances of the middle hives always come quite close together, it is necessary that they should be parted by a board about ten inches wide, that the bees cannot run together outside. This division is supported below on the flight-board, which is put sloping beneath the entrances, and in order

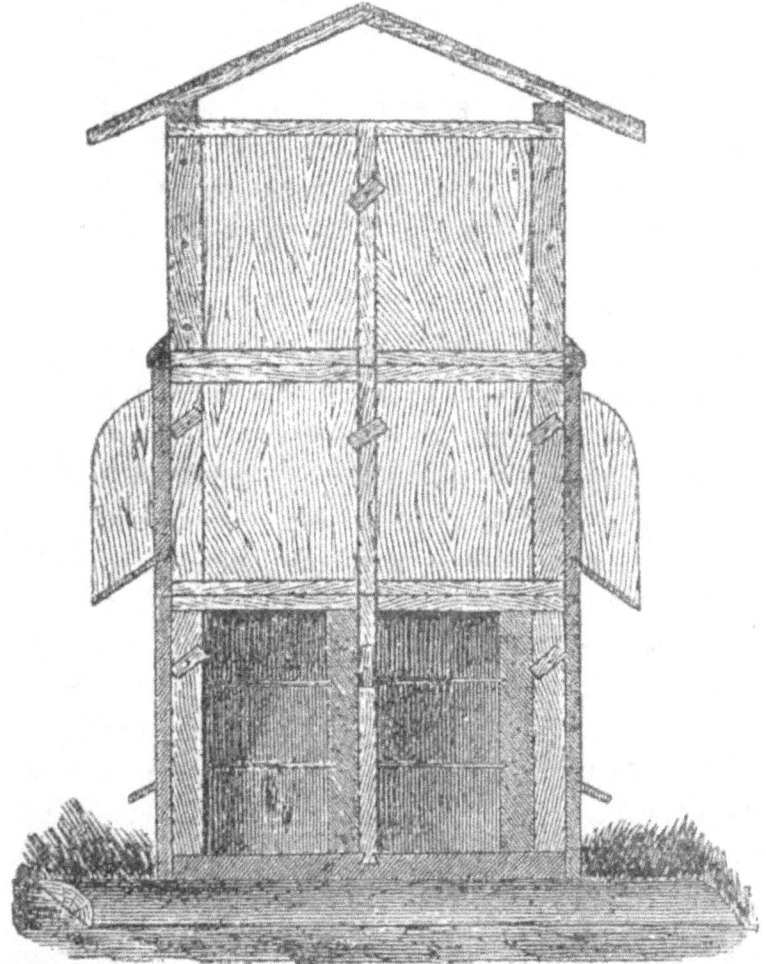

THE SAME PILE FROM THE SIDE, SHOWING THE DOORS, OF WHICH THE TWO LOWEST ARE OPEN.

that it may protect the lower hives from the rain should be continued the whole length. Below the other entrances it is sufficient to attach little boards or pieces of wood bevelled off.

In these fourfold Lager-hives the bees have an unusually warm winter dwelling, but this drawback may be mentioned, that sometimes a stock, when it has used up all the honey at the back, in severe weather will rather starve than make up its mind to leave the corner in which it enjoys the warmth of its neighbours, and will not even move to its stores. Even with an obvious superabundance the honey-room should never be emptied in autumn quite to the back, and honey-combs should in all cases be placed there without dispensing with the overlaying little boards, since in the warmth which the four stocks afford to the hinder part, a mere chink in the dividing wall suffices to entice the bees into the honey-room to consume the combs, or pieces of comb, inserted there. Water also can be poured into the cells of this, or another comb, so that the bees may suffer no lack of it, because these hives naturally keep unusually dry, for retentiveness of heat and dryness are always inseparably connected.

Ständer Hives.

Ständer-hives, or hives standing erect, are those which are higher than they are deep, or long; while with the Lager-hives, just described, the dimensions are reversed. In hives in which all the interior forms an uninterrupted whole, the difference of form gives rise to a difference of result. In Ständer-hives bees build but few combs, which, of course, soon become nearly all occupied with brood; most of the new cells which are formed when the combs are extended are immediately appropriated by the queen, and serve at first as brood-cells. Towards the lower part the bees change the construction to drone-comb, and do not then readily turn back again to worker-comb. Ständer-hives are, therefore, to be preferred as hives for raising brood, drones, and swarms. With Lager-hives it is otherwise. In these, when the combs are extended, the original ones are not lengthened, but new ones are continually begun. Every comb contains honey, at least, in the upper cells. The queen limits herself at last to a definite number of combs for brood, and gives up the others to the bees for the deposit of honey. As a rule, therefore, there is found in Lager-hives not only more honey but of finer quality, because a part of the combs has not previously served for brood. Lager-hives are, therefore, preferable as honey hives, although in good years for swarms they will furnish these as well. But with the hives with which we have to do here the difference between the erect and horizontal shape is not so considerable. In these, the Lager-hives' greater wealth in honey may be perfectly attained by the Ständer-hives, if the bees have a definite space allotted as brood-nest, and if access to the honey-room is so arranged that the bees cannot incorporate the two rooms into one large whole: it does not matter then whether the extended room set apart for honey is situated—as in the Lager-hive—at the side, or—as in the Ständer—above, or perhaps below. Whoever has chosen Ständers may remain content with them. They have even, in many respects, advantages over the Lagers, as will be shown in a fuller description of them.

The Ständer of One Compartment.

The simple Ständer is just as easy to set up as the simple Lager-hive. The necessary grooves for comb-carrying are cut in two planks on the sides intended for the interior, and on one side the rabbet is made for the door; crown-board and floor-board are nailed on; the one open side is boarded up, so that we have a box twenty-four to thirty inches high, fifteen to twenty inches long, and eight to ten inches wide, open only at the front,* and the simple Ständer is ready. The boarding-up of the side which lies opposite to the door must be done with some care. It is best to have a double wall, with its interior space of about an inch and a half filled with dry moss, fine shavings, or sawdust. The inner wall can be inserted in perpendicular grooves, and the outer one nailed at the corners of the planks. But since this obviously involves more work than putting a moveable door, it can be made with that, especially if the hive has considerable length. The one of the moveable doors, which will seldom be opened, near which the swarm would begin its comb, must, of course, be made retentive of heat, and it, or both, could be made of thin laths, strips, straw, and reed, in the way previously described. It may be well to describe how such a straw door and straw wall may be made easily, and of equal thickness throughout; for the long door required by Ständers, with the tightening up or sewing on both sides of five or six courses of split reed or wire, might appear a work that would take up much time. The mutual tightening up of the wires, stretched on each

* In England the bee entrance is always considered 'the front;' but in this case the end which can be opened, and which is twenty-four to thirty inches high and eight to ten inches wide, is called the front.—C. N. A.

side that they should not separate farther from one another than the proper thickness of the door, is best secured by wire clamps. In order to make these easily and of accurately equal length, say one inch and a half, drive in two nails one inch and a half apart, take wire, rather stiff, but not too thick, bend one end with wire-pliers, put the hook that is formed on one of the nails, tighten up the wire, bend it round the other nail, and break off the clamp you have now formed; and so go on till you have a sufficient number of clamps. On every pair of wires to be clamped, three or four clamps must be used, and if five wires are used on each side, fifteen or twenty clamps will be required. It will suffice if the surfaces of the straw doors are covered with reed to afford a wire to every six inches, otherwise one will be required every three or four inches, or else the straw would be liable to be displaced. In the first place, the cross-bars, which are on both sides at the top and bottom, have to be fixed on the two thin laths, which are only about one inch and a half wide, that is, they have to be let in so far that they do not project. On the one side these bars would at once be properly nailed down, and at the same time the wires from one lath to the other would be stretched at equal distances. The end of the wire is either bent under to a right angle and hammered in, or wound round a tack or nail and nailed fast. In the other lath the wires wound round the straw wall, which, of course, are long enough to reach on both sides, are drawn through small holes, bent down on the other side and plugged. The other end must not as yet be fastened, because this would hinder the straw being placed. Clean straw, as hard and stiff as possible, about two inches longer than the door, is now taken, and beginning at the lath through which the wires have been drawn, one wisp after another is placed, and at the same time about every two inches in the clamps are placed, though not to all the wires at the same time, because little furrows would be formed with ridges between. When an inch-wide layer has been placed, then let the clamps be applied to the first, third, and fifth wires, and so on. If at the beginning the straw wall is somewhat loose it will gain greater closeness the more straw is laid in from above, and the more so if it is at the same time pressed down with the clamps, and the upper wires that are still unfastened must be every now and then drawn up or tightened.

When it is pressed together it may be tied with pack-thread, and so held down for a while to prevent it again extending. If one or both sides are to be covered with reed this must either be put in at the same time with the straw or pushed under the wires while the straw wall is as yet loose. If the space from one lath to the other is now filled up so tightly with straw that no more can be pushed in, draw up the ends of the wires that are loose, bend them round and hammer them into the lath. The two cross-bars that have already been put in position may now be nailed fast at each end, fasten them also in the middle by long nails to the bar opposite, and lastly, cut off the projecting straw smoothly at both sides. If the length of the laths has been accurately fitted to the interior height of the hive, their breadth to the depth of the rabbet, and the length of the cross-bars to the breadth of the hive, then the doors that have been made must fit accurately.

The simplest and most durable outer casing for the straw doors would be afforded by wood splints, similar to what may be easily split off in quantity from shingle or thin wood for boxes. They may be pushed with the one end somewhat under the upper cross-bar, and with the other end under the lower one, and since, of course, they will bulge out with the straw in the middle they must be drawn together by wire nails of the requisite length, so that there is no further need of a reciprocal clamping, although clamps may be very well applied in the spaces between single splints. But in order that the nails may hold and pull properly, the little boards on the one side—perhaps the inside—must not be too thin, and the nails, however thin they may be, must be long enough for the points coming through on the opposite side to be clinched. Thin outside planks, with the smooth side turned outwards, could very well be used for this purpose. If one piece would not cover the entire side, two adjoining one another would serve. Smooth cardboard can also be used for the casing of the outer surface of the straw doors, and, when painted, it is very durable. But there is danger of the heads of the wire nails pulling through this. This may be obviated by putting, at the places required, either double wire, split Spanish cane, small strips of thin tin, or thin wooden bars, and let the nails be driven through these.

Since the making of such a door certainly does not require much more work than the boarding-up of the one open side of the simple Ständer with a heat-retaining double wall, it is by all means more convenient to provide it, like the Twin-stock, with such moveable doors at each side, although the one door might be but rarely opened, perhaps, to increase the space with a collateral box, to withdraw from the other side inaccessible honeycombs, to renew the brood-nest, to catch the queen, or the like. Ständers sufficiently deep could, like roomy Twin-stocks, be divided into two parts by a board interposed in the middle, and so be used as double-hives. In that case there must be a second entrance, which could be situated in one of the two doors. There must then be fitted into the straw a board, through which the entrance would be made.

Where there are a large number of straw doors or straw walls to be made it is well to make use of a machine, in which the straw is first pressed into the shape of the door or wall, and then secured by bars or wires. The hive-walls may be made two, two and a half, or even three inches thick, but for the doors a covering one inch and a half thick is enough. Two at once may be made on the same machine more quickly, since the mould for the straw may be narrowed as much as you like, by putting bars against one of the rows of pillars. On such a machine good-looking and suitable straw walls, with grooves for the comb-bars, may be made if a grooved lath is inlaid in the requisite position, and it need not be at all visible from the outside. Wooden bars may be also included in the straw wall, so as to be able to fasten the wires and bars in with wire nails if they bulge out.

The late lamented bee-master, Schmidt, of Ingolstadt, who made very convenient hives and apparatus, applied to such straw walls, on the outside, four wooden bars (as in the Twin-stock), but on the inside wire clamps from lath to lath. The bars must be pretty strong, so as not to bulge outwards. This arrangement made the inside wall-surface rather uneven. It might be better to enclose the straw wall with wooden bars only at the two ends (which on the inside must be quite thin), and between these thinner wire, which on that account must be taken double. On the inside, the wires from the lower to the upper lath could run over the grooves, since, if nailed fast into these, they could not easily be pulled away, and thus would not present any hindrance, especially if the grooved laths had saw notches made in them at the requisite places. Otherwise the wires must be fastened to the wooden bars hidden in the straw or be mutually clamped, though the clamps can be applied at the same time the straw is placed, as previously described. If two such straw walls are set up at such distance from one another as the hive is to be wide—say nine, nine and a half, or ten inches—the crown-board is nailed firmly across on the top, and the floor-board beneath, the doors, prepared in a similar way, are fitted to them; then the straw hive is ready.

Simple, and easy of construction, as the making of such a straw-hive is, yet it requires some time. Whoever has to be chary of this will prefer forming the side walls of his hive of thin boards; let floor-board and crown-board project over; fill up the space between with straw, keeping it in with bars in the way mentioned in the description of the Twin-stock, only there must, with the higher Ständer-stock, besides the laths below and above, be a third put in the middle in addition, in order to be able to nail to this the bars, which otherwise would bulge out. And where the entrance is to be there must be a lath or a piece of lath. Many would have the entrance of the Ständer-hive put on the floor, or immediately over it, and when the bees have built down to the floor that may be its most suitable position, but so long as the bees are clustered high up in the hive an entrance situated higher up, perhaps at half the height of the brood-room, will be by all means more convenient to them. As a rule, in the Log-hives the entrance is nearer to the crown than to the floor, and in the ordinary straw-hives the heath bee-keepers place it pretty high, and find it suitably situated there. In order to let the bees fly out at one time higher, at another lower, according to circumstances, two entrances may be made at different heights, one about an inch over the floor, and a second at about half the height of the brood-room. This would permit a reduction of temperature in the extreme heat of summer and ventilation in winter, the entrances being at the same time made considerably smaller. On which side the entrance is to be put depends on the situation of the hive. It is best in the wall opposite to the door, but if in one of the side walls it should be as much as possible at the back. For it is near to the entrance the bees cluster and have their brood-nest, and on the opposite side the store of honey, which can be conveniently taken out from the door.*

Just as the Lager-stock, which is longer than it is high, is divided by boards, doors, or glass windows inserted perpendicularly into two or three adjoining rooms, so the Ständer-hive, which is higher than it is long, is divided into at least two rooms lying one over another, like stories, forming a brood-nest below and a honey-room above.

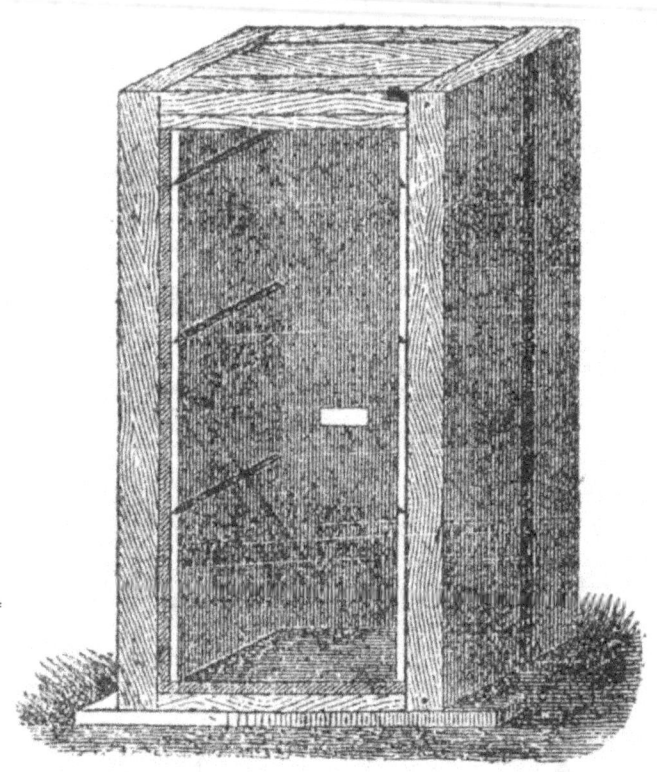

THE SIMPLE STÄNDER, WITH BACK DOOR REMOVED.

It has been thought to improve the Ständer-hive by inserting a fixed horizontal board, provided with a passage, as division between the brood-room and honey-room. But such an arrangement does not answer, because it makes many operations difficult, and, indeed, impossible. A cover made of several small boards adjoining one another, put on the bars or frames, perfectly

* This is exactly the principle that governs the construction of 'Abbott's Combination Hive,' now in such general use in England.—C. N. A.

answers the purpose of a fixed division-board, and it allows of being partially or entirely removed, in order to conveniently take out the combs below and put them in again, and to be able at all times to drive a swarm above, which could never be done through a comparatively small opening. As regards the height of the brood-nest and honey-room there is no occasion to be tied down to a definite measure. If the entire Ständer is not more than two feet high, a height of twelve inches will do perfectly well for the brood-nest, which in the case of a higher hive might be eighteen inches, which, if needed, could be easily diminished by a board inserted horizontally.

Whoever has in use frames of a definite height would have to make the brood-nest high enough for some space to be left on the floor, so that the hive may be conveniently cleansed, and a feeding-box, or comb with honey, or pollen, could be pushed under.

A higher brood-nest, especially with the use of simple bars, can be provided further at half the height with comb-bars, and the side-walls with corresponding grooves, because honey and brood-combs, especially when they are loosed from the sides and taken out, are liable to break off. There is no occasion to make actual use of the lower grooves immediately, or to insert comb-bars for a young swarm, by which only hindrances would be put in the way of taking out the combs quickly. It is better not to do this before the next spring after a corresponding shortening of the combs. But even then bars are better left out beneath the proper brood-nest, because bars present some hindrance to the extension of the brood. A comb-bar, therefore, is only inserted below and for the hindmost comb, because there is usually no brood there. Leave the space for four or five bars open, and then put them in regularly towards the door. For some support to the combs serving for brood, which by preference are continued without interruption, one or two bars rounded off are so laid that they become worked into the combs, lying at right angles to them, and giving to them some hold.*

---

* These are in the nature of the cross-sticks used in skeps in England; they are advised to be round and smooth, and would be better if tapering, so that when 'turned round a little,' as the Author suggests, they may be the more easily withdrawn.—C. N. A.

But when these round bars are grasped and turned round a little they can always be drawn out, and will not, therefore, hinder the taking out of all the combs.

There are many bee-keepers who use a double row of short frames in the brood-nest. This may afford much convenience to themselves, but the quantity of woodwork in the middle of the brood-nest is certainly very disadvantageous to the bees. Either whole frames ought to be exclusively used there, or, at least, every other one should be entire. There is nothing better in the proper brood-nest than simple bars or frames, open below, which offer no limit or hindrance to the bees in lengthening their combs nor to the queen in the appropriation of cells. Shorter frames may always be inserted in the neighbourhood of the door where in the brood-compartment of hives of some length there would never be anything but honey. There they do no harm, and are, at all events, very convenient.

As regards the height of the honey-room—if we are not bound by a definite height of frame there is no need to be so careful. A smaller honey-room, of course, becomes warm and fills with honey more quickly, but must be emptied oftener in good seasons. A somewhat high room could be divided, like the brood-room, into two equal parts, so that the bees should have access at one time to the whole, at another time to only half the room. When the bees had properly taken to the comb-foundation put in the lower division and begun to move further, the comb-foundation might then be put up higher, in order to have the whole room filled with comb, so that the bees would not have to take in hand new combs, which always occasion some loss of time. A fairly high Ständer may be divided into three equal divisions, of which the uppermost may be set apart for honey-room and the two lower for brood-nest. Since the grooves are so easy to make—a shallow channel cut with a saw is always sufficient—there might be made in a Ständer, thirty inches high, five pair of grooves (six inches apart), giving five divisions, of which two might be allotted for honey and three for brood. A larger number of grooves offer this convenience, that both high and low combs can be inserted in order, for example, to hinder the bees from the further construction of drone-comb, or to lengthen short combs, and so to complete an imperfect set, and render them warmer for the winter. When the grooves are not in use they do no harm, and can easily be stopped with clay. In lower Ständers the brood-nest and honey-room might be of equal height, so that any comb would fit in anywhere.

What has been said above on the doors, the position of the entrances, the interior divisions, and other arrangements of the simple Ständer, is fully applicable to every single division of the hive with several compartments about to be described, and, therefore, we will not dwell on these any longer.

If it is intended to place two simple Ständers close to one another, the walls adjoining each other may be thin, like those of two Twin-stocks placed side by side. But just as an inseparable double Lager-hive may be made instead of two Twin-stocks, so, instead of two single Ständers, there may be made

The Ständer, with Two Compartments.

This is done by dividing a hive more than twice as wide as the simple Ständer into two compartments, running side by side. The division-wall should be an inch thick, and provided with grooves on each side. It can be boarded up at the one end, or, if it has as great a length as twenty-four inches, it may be provided at both ends with moveable doors, and then, if each compartment is separated in the middle by the insertion of a dividing-board, it may, at times, be arranged for the accommodation of four different stocks.

Lower double Ständers of, at most, two feet in height, and, perhaps, equal outer breadth and length, could very well stand three stories high, but higher ones than that only in two one over the other.

In the double-hive the entrances are best put in the end walls, opposite to one another. But if it is preferred to have them both at one end, because the structure may have to stand in a hut, then care must be taken that the little block through which the entrances are made does not become loosened from the one or the other of the double walls, so as to enable the bees to run together through the chink. Each entrance might rather be made through a separate block, and the space between them carefully stopped up. It follows from what has been said in the description of the Lager-hive with four compartments, that the entrances must be divided exteriorly by a broad board resting upon the sloping flight-board. The division-board can be arranged to take off, so that in the union of two stocks, or removal, it can be easily put aside and fitted on again.

TWO DOUBLE STÄNDERS PILED ONE ON ANOTHER.

If two entrances are situated near to one another, and therefore close to each side of the division-board, in uniting two stocks it will be sufficient to move the division-board about a quarter of an inch, or to make an opening in it, that the bees may have access from one side to the other.

Such double Ständers may very well be placed one on the top of the other, as shown in the figures above, and with greater advantage if the entrances are thirty inches distant from one another, or are put on different sides. By this there is a saving of the roof to one hive. But there is a still further saving, both

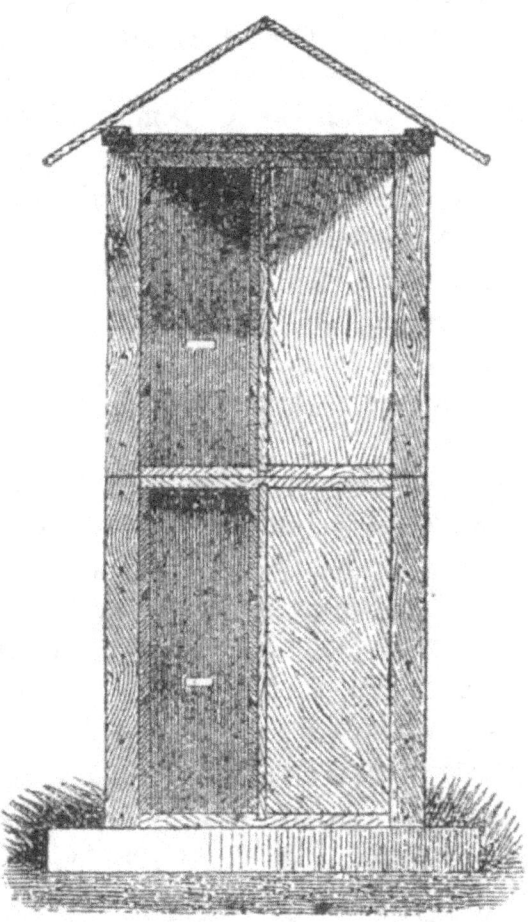

TWO DOUBLE STÄNDERS PILED ONE ON ANOTHER, SEEN FROM BEHIND. ONE PART HAS THE DOORS OPEN.

of work and material, if instead of two double Ständers placed one on the top of another we make

The Ständer, with Four Compartments.

Its construction is not essentially different from that of the double Ständer. A double Ständer is made of more than twice the ordinary height, so that when each of the two compartments is divided into two parts by a division-board inserted horizontally, we have four equal compartments, having a similar relation to one another as the four panes of a window, two above and two below. Such hives are easily made from planks fifteen or twenty inches wide. Two pieces of board, one inch to one and a half inches thick and about five feet long, and another, equally long and broad, will furnish the side walls and middle wall. The boards for the cover and floor must have the same depth, and their length depends upon the breadth of the two adjoining compartments and the thickness of the three upright walls. The floor and crown-board may be nailed on the three walls that are of exactly equal length. On the side on which the doors are situated the crown-board and floor may be joined to the side walls more securely by strips of tin being nailed on, or may be clamped together, by driving two nails in with wire wound round them several times and then hammered in. Joined in this way the hives often wear better than when secured with the tin strips.

In the middle, the side walls are prevented from warping inwardly by the horizontal dividing-boards. But in order that they may not be able to bend at the corners outwardly, and so give as to separate from one another, they may be bound round, at the height of the division-boards, with hoop-iron, which is not expensive, and contributes considerably to the firmness of the hive. The horizontal division-boards should rather be made too short than too long, because they would project if the side walls in the course of time should shrink. They are best laid across, so that they may, without disadvantage, be able to swell and shrink with these. Before fitting the whole together the rabbets for the doors must, of course, have been made on the side walls and middle wall on both sides.

The cutting out of the grooves may now follow, and these may be made with an ordinary broad wood saw. The upper angle of the grooves should be bevelled off a little with a chisel. Every thing must, of course, be done before the one open front side is boarded up, that is, if it is going to be boarded up, and not closed with moveable doors. If two such hives with four compartments, exactly alike, are to be set up on two long sills in common, with their walls against and adjoining one another, then the boarding up is easy, and can be done with thin boards nailed across on the three walls. The two hives put together will then appear on each of the two sides as shown in the following figure.

It is best to put the entrances at the back part, by the wall just mentioned. There will then always be two entrances close together, one belonging to one hive and the other to the other; and these must be divided by a board running from the roof to the floor. Sloping boards are fastened below the separate entrances. The division-boards must be accurately fitted quite up to the roof, so that the bees in front, that often resort to the roof and settle there, do not get together. The entrances might, however, be put at half the depth of the single compartments in the middle of each wall, and would then stand about one foot and a half from one another, in which case a careful separation between them would not be needful. In some circumstances the bees might, notwithstanding the distance, get fighting with one another. It will then be well to put in the place where the two hives join, some tow or strips of fur, over which the bees will not walk.

In addition, in front of the four single doors* of each of the four-fold hives there may be put a large door, which would serve to keep off the driving rain and the burning sunshine, as well as being a protection against the cold of winter. If it is made to lock up, or is fastened with strong wood screws, the hives could

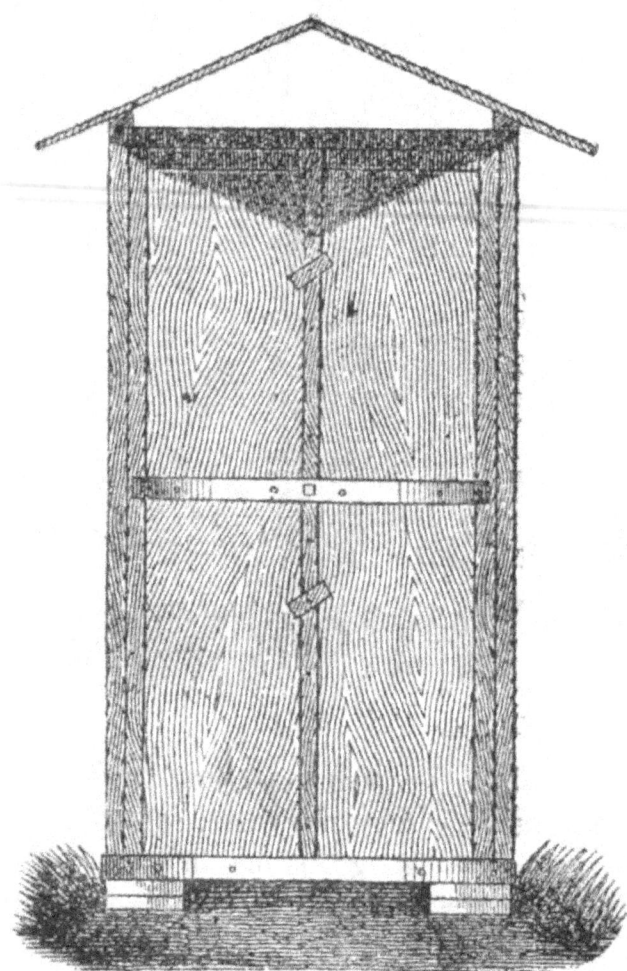

be rendered, in some measure, secure against robbery, where this need be guarded against. In summer, when the hives are often operated upon, and robbery not so likely to occur, these doors can be entirely removed.

The fourfold hive appears from the side as represented above.

It is a very suitable arrangement, economical of material, and

---

* That is, at the ends of the hives, not at the entrances.—C. N. A.

pleasant to the eye, if not only two but four similar Ständers with four compartments are united into one large whole, forming the so-called

Pavilion.

This is made in the shape of a cross. Suppose that two hives of the form just described are placed as far from one another as they are wide, and two others are moved up to them, so that the angles of the boarded-up front walls touching one another form a square empty space, we have then the form of the group.

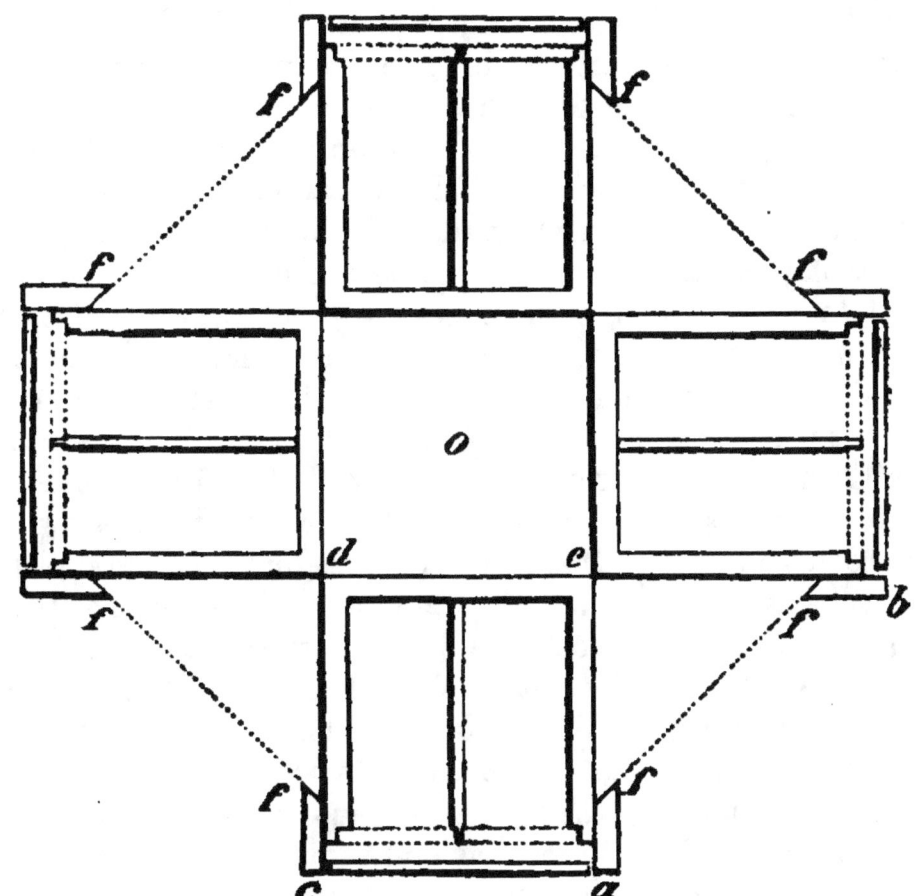

The accompanying figure, representing the ground-plan, makes this plain.

The two hives standing opposite rest upon the same sills. If the whole Pavilion is to stand on sills, and not on a foundation built of brick, the sills on both sides must be crossed and let into one another, so as to be level on the top. The four hives form, as it were, four wings for the whole, and their doors are directed towards the four points of the compass. The economy of material in this method consists in this, that the outer walls of the individual hives need only be made of one inch board, because they may be made unusually retentive of heat by fitting heat-retentive materials into the angles, building up the corners and filling up the enclosed space.

The greatest advantage consists in this, that the hives combined into a large whole are less influenced by the different changes of temperature than when they are set up in an isolated way, and the effect of extreme heat and cold can be alleviated by conducting air of the temperature of the earth between the hives, which would tell considerably upon them. It is only necessary to make a deep excavation beneath the open space, *o*, enclosed in the middle by the four hives. This excavation would be a kind of extension of the space downward, and would require protecting against collapse. If the entire space be closed above, it will form a small cellar, in which, both in summer and winter, about the same temperature will be found, which, of course, will exercise, on the hives forming the square, a cooling influence in summer and a warming one in winter. This influence will make itself felt by the bees, even through the one inch board wall, and the more so if clefts are cut into it a little over the floor, below the crown-board, and where the honey-room begins. In these places the propolis can be conveniently cleared away from them without taking out the comb from the brood-nest, or at most, if it reaches down to the floor, it would only have to be shortened. A further advantage of this grouping consists in this, that the bees, although sixteen stocks stand on a small space, do not lose their way and mistake their hives, because they have their entrances towards different directions and at different heights. Loss of queens, therefore, occurs just as rarely in the Pavilion-hives as in the Twin-stocks. In order that the neighbouring entrances, *f f*, may be placed at a suitable distance from one another, the hives must have a considerable depth—say about twenty inches. If this were not the case the hives would have to be moved farther apart, so as to give a greater circumference to the whole Pavilion, and, as the corners would then no longer touch one another, the interval would have to be joined by boards standing upright, so as to form in the middle an octagonal space instead of a square.

So far as the exterior is concerned, the Pavilion indisputably surpasses in this respect all other bee-dwellings.

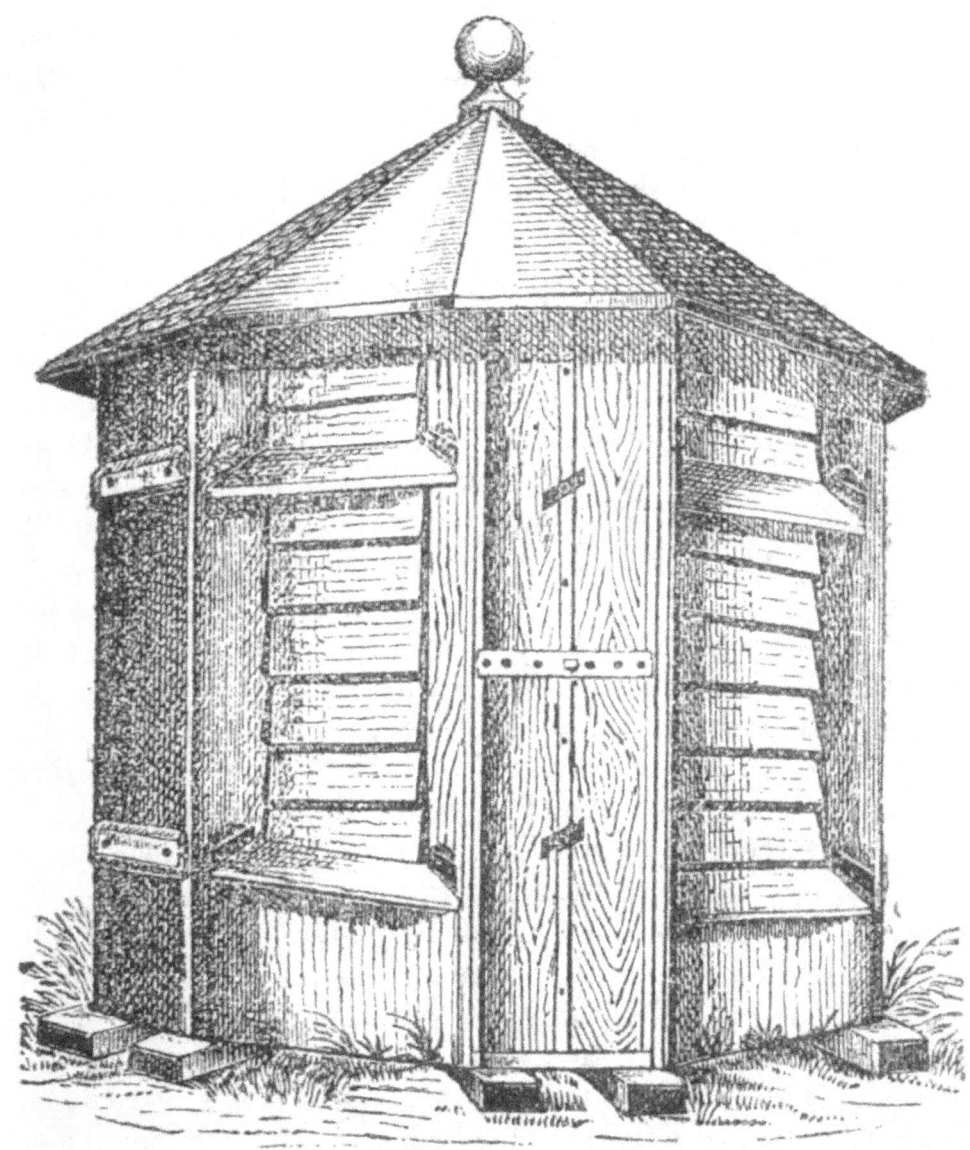

Even if the single hives are but roughly made, and simply nailed together, the whole presents a very pretty appearance if only the outside parts are planed smooth and painted, and it is covered above with a nice dome-shaped roof. The four thin board walls with which the corners, or angles, are built up, are especially visible. There only remains so much of the walls of the hives visible as is necessary for the placing of the entrances, viz., about four inches, since the entrances, of course, can begin only from the doors, and then go one inch and a half deep into the hives. In the upper compartments they are brought as high as possible, therefore, into the middle division directly under the upper groove, but in the lower compartment they are put close over the lower groove. If the single compartments have a height of twenty-seven to thirty inches the entrances over each other will stand wide enough apart.

In order to make those parts of the walls of the hive which yet project more retentive of heat, and so as to be able more conveniently to fix the walls already mentioned as coming in the four corners, strips of board are nailed right up to the top on both sides of the hives. The corners of these boards are bevelled off in the direction of the walls mentioned, so that the walls only need to be pushed behind the casing-boards that have been nailed on, to become at once firmly fixed in position. Compare *a f* and *b f*, page 123. The entrances necessarily go through the casing. The casing-boards project on the outside, as shown in the ground-plan. The purpose of this is, that a large door, which it is very convenient to have fitted to each of the four wings, can be put in and pushed between it in the same way that an inside window shutter is pushed between the window casing. The amount of projection, therefore, is determined by the thickness of the doors that have to be placed there. These must not be fastened there by a fixed connexion, because, when opened, the bees would be disturbed in their flight. It is better simply to put them in and bore holes, that they may be fastened by pegs or buttons. In operating, the doors are lifted out and put aside, or, at certain seasons, quite put away, when frequent examination is necessary. When in, they greatly help to retain the heat, especially if the small doors are rather thin, and they give a better appearance to the Pavilion, of which no more can now be seen than these doors, as well as the small casing-boards going down both sides and the four walls by which the corners are built up. The figure on page 125 represents the wing in front with the door taken away, and the one on the left with it in position.

As regards the four walls that build up the corners, they may either be formed by shutters previously made entire and then pushed behind the casing-boards, or of little boards, or shingles, which must be adjusted with the joints downwards, so that the walls will have the appearance of Venetian blinds. The filling up of the space behind could at the same time be conveniently done, so that, as single boards were fitted in above, there might at the same time be put in a fresh layer of moss, fine shavings, flax waste, sawdust, or the like. When ants are abundant and troublesome to the bees the space may be made distasteful to them by putting layers of ashes at intervals with the other material. Should the wall made in this way in the course of time settle or shrink, new boards, or shingles, can, by way of addition, be pushed in above.

Boards about four inches broad are fastened on these walls from wing to wing close under the upper and lower entrances, sloping, so as to be more convenient for alighting, and these boards, with the cross-bars of the doors, will form two string-courses round the entire pavilion. The cross-bars of the doors, which of course come on the outside, are put at a corresponding height, and they must also project a little on both sides, so that the doors may be lifted by them, and be more conveniently taken out and put in again.

Every structure looks better, and more natural, if it is somewhat broader at the base, as, for example, a stove. In the Pavilion this is accompanied by another advantage, that the larger the area of its foundation is, the more does it gain from the temperature of the earth. The sloping cornice at the foot of the Pavilion where the circumference is diminished, can be suitably formed by the sloping flight-boards put beneath the lower entrances. These can be put on first. They must be rather broad, and be well secured; for otherwise they would be liable gradually to be pressed down by the thin board walls standing above them. The sloping flight, or cornice-boards, must, on that account, extend a little under the walls, so that all the rain dripping down from these may be conducted outwards. The space beneath the lower entrances is now completely boarded up, for which a board about a foot wide will suffice. That this must be some inches longer, because it is considerably more prominent than the upper wall, is obvious, though the sloping board must always project a little over it. It is evident that the four spaces so built up must be shut off below either by bricks or boards of the shape of right-angled isosceles triangles, so that mice may not be able to get in, that is, if the entire hive does not rest on a built foundation, but only stands on sills.

The Pavilion would then be ready as far as the roof, which every one may make to his own taste, either of zinc, tin, cardboard, boards, shingles, or straw. It might be best to make first a large octagonal disk of boards, to cover the open space in the middle, as well as the four triangular spaces that have been filled up. This should project a little on all sides, and then the roof may be constructed upon it. It may either run up to a point or be truncated. If there is a large circular piece of zinc or tin, at any rate, on the top, there will be a considerable saving of boards, or other material. The space under the roof can also be filled up with moss, or the like, in order to keep off from above the influence of cold and heat, not merely from the entire hive, but specially from the cellar-like space in the middle.

Individual bee-keepers have erected still larger Pavilions, pursuing the idea of making a room that shall always be frost-proof, by the grouping of several Manifold hives in the interior. These Pavilions have been so arranged that the individual compartments are opened out from the inner room, that is accessible by a door, and the bees are manipulated there. Such an arrangement, however, is not practical, because the inner room is both too limited and too dark for any one to do every thing conveniently and surely. And when some operation has to be undertaken in cool weather many bees that come out are lost, because they are chilled before they get into the open, find their bearings, and reach their hive again. It is further unpractical, because the individual compartments are only closed by a glass door without this being darkened by a separate wooden door. When an operation is going on in one compartment hundreds and thousands of bees flutter at the glass doors of the other compartments and try to get out there. The bees often at such times break open the honey-cells already sealed and seek to conceal the honey and place it in security, because the light breaking in leads them to believe that the hive has been opened.

The Manifold hives hitherto spoken of are not to be considered as other than a further grouping of the double hive, because only two compartments lie beside one another in them. In a broader whole three compartments can be put alongside of one another, as hives for three separate stocks. This is especially suitable, because then the three entrances of the three compartments can be directed to three different points of the compass, while the three doors are situated on the fourth side. The hive has also then a greater breadth and a firmer position, since, of course, it can only be set up in the open.

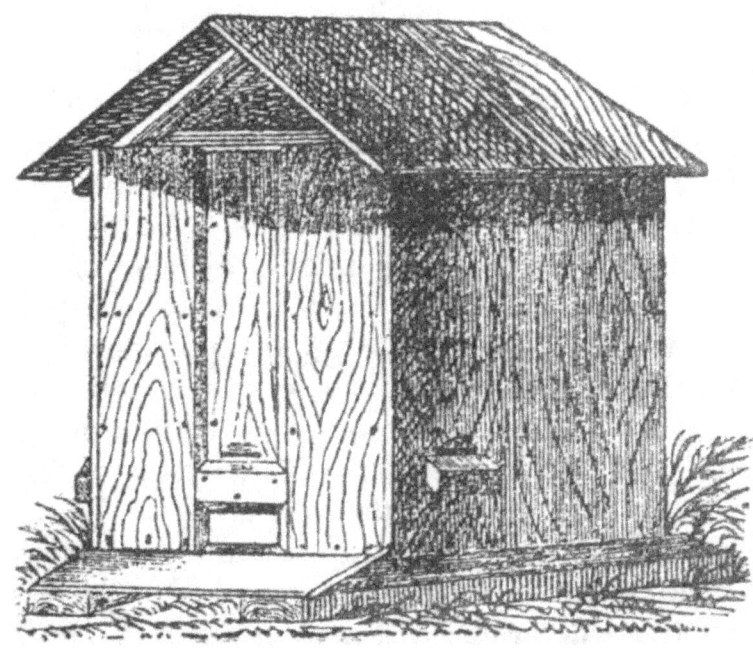

The Hive of Three Compartments.

The triple hive is made just like the double hive, only it must be correspondingly broader, and instead of one perpendicular division-board, one inch or one inch and a half in thickness, two of these are inserted, forming compartments of equal breadth. Its exterior breadth, or the length of the boards forming the floor and crown-board, will, therefore, consist of so many inches as the breadth of all the three compartments and all the four perpendicular walls amounts to. If the three compartments are to be made accessible from both sides, the entrance of the middle compartment must be situated in one of its doors. But, as a rule, the one front side is boarded up, which must be done with care, as has already been said respecting the double hive. It looks better if two such hives are put one on the top of another, and there is a saving of roofing. The pile then has very much the appearance of an ordinary clothes-press. But instead of two separate triple hives, if there is no probability of the hive having to be removed, it may be preferred to make directly a sixfold hive, when it is wished to insure as much as possible against robbery.

The Sixfold Press Hive.

What has been said in the description of the fourfold Ständer holds good of course here. The two are only distinguished by this, that one contains two compartments more—the two middle ones. The entrances for these are put in the front wall, that is boarded up last, while the entrances of the side compartments are put in the side walls, as near as possible to the front wall, and as far away from the doors as may be. As regards their height, they may be placed in the middle division of the three equal parts into which the hive is divided, but in the upper compartments a little more towards the top, in the lower ones a little more towards the floor, so that they may be, at least, three feet apart. To put the ones in the lower stage farther down would only be allowable if the hive is placed rather high, so that they would not come to stand too near the ground.

The front wall is best boarded up in the following way. A thin board wall is first nailed on crosswise. If this casing is nailed on to the side walls, and not inserted in grooves, the hive is less liable to give way later on. All four walls must then have the same width; the middle ones might even project a little, as the casing would then fit better, which is as it should be. This is yet better attained by putting on laths on the outside and attaching them strongly to the walls by long nails that will hold well. Instead of the two middle laths an entire board, somewhat thick, is nailed up in the middle to the whole length or height, and in this the two entrances of the middle compartment are cut. It must, of course, be so broad that it reaches somewhat over the two middle walls, to which it is firmly nailed through the casing. Two boards, planed and bevelled off on the inside, are now nailed on both sides, on the one side to this middle board, on the other to the side laths, as well as to the corners of crown-board and floor. These must be chosen of such breadth that they cover the entire side, only leaving visible in the middle three or four inches of the middle board, or the breadth of the entrances. These will then come to stand in a kind of channel, as shown in the following figure. The rain beating upon

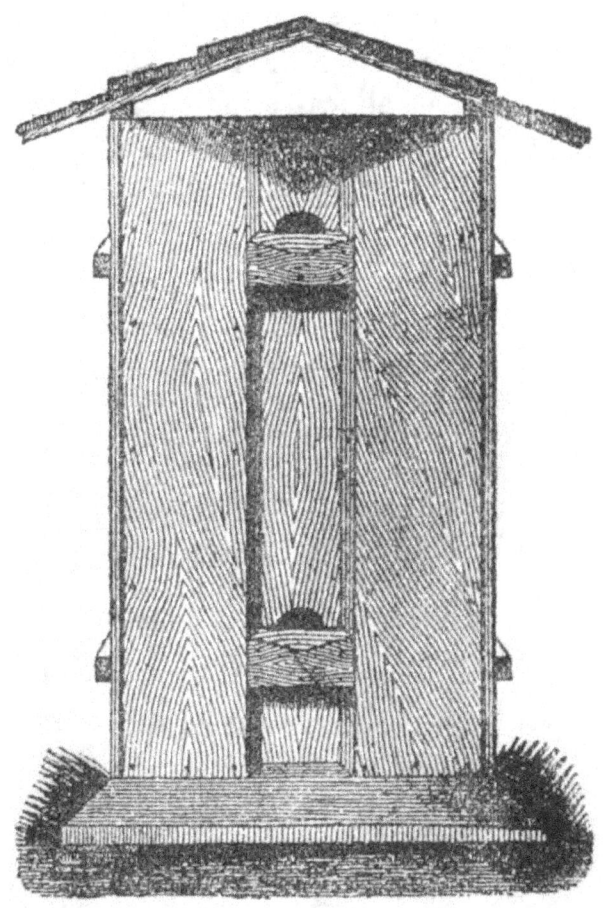

the hive glides better off the boards going downwards, and in shrinking there will be no clefts, because the two boards last nailed only have a little play over the ones beneath. The hollow space under them is filled up before they are nailed fast, the shavings and the like being mixed with ashes, especially at the bottom, so that ants shall not make their nests there. If the two side walls should not be sufficiently retentive of heat they may be cased over with thin boards. The casing-boards can then partly be laid over one another, that they may contract and swell without forming clefts, and it might then be so arranged that the entrances should be placed in a similar channel or depression, as on the front side. A piece of board, bevelled off above, is fixed under every entrance, for the sake of more convenient alighting.

Because the floor is liable to rot, especially in those places which come in contact with the sills on which the hive stands, it is better that some thin knotty board should be fixed in the requisite places, or contact should be prevented with strips of tin. These should, in all cases, be coated with varnish, that they may not attract the damp, and so be the longer preserved from rotting. And for the floor itself quite knotty boards should be selected.

The sixfold hives do very well for putting up in pairs on

TWO SIXFOLD HIVES REPRESENTED AT AN ANGLE.

sills in common, and under a common roof; but in a way quite different from what has been recommended and described for the fourfold hive.

They cannot be placed with their front sides one against another, because the entrances of the middle compartments are there. On the contrary, they are placed with these sides outwards, and with the doors opposite one another, but in summer so far apart that there is room to stand between and conveniently attend to everything, as may be seen in the following figure:—

In autumn they may be moved close together for the sake of mutual warmth, and effort must be made so to arrange them that exteriorly they are shut in closely, but further in, from the floor upwards, there remains a small open space between them. If a small excavation is made in the required position before they are moved together, warm air streams up continuously between the hives, which in severe cold is of much advantage to the bees. This advantage will be increased by surrounding the hives at their base with moss, litter from the woods, leaves, &c., so the frost is prevented from penetrating to the floor.

In order to be able to conduct the temperature of the ground in winter to the sixfold hives set up in pairs, in the same way as to the twin-stocks set up in pile, the doors are provided with some clefts, as may be seen in the figure on page 132. These incisions in the doors have in summer this advantage, that in great heat cool air streams in between the hives, while the hot air is driven out by the bees from the entrance by ventilation.

The greatest advantage of the press hives, perhaps consists in this, that they more easily afford security against robbery by men, since they can be moved close together and combined securely in ways not generally known. This can be done, for instance, by driving into the corners of the crown-board that touch one another, on the one side a hook and on the other side an eye. If, then, in moving the hives together while the one hive is inclined toward the other, the hook is allowed to fall into the eye, and the hives are then pushed together at the bottom, they will be so combined that they will not easily be separated. There may be some earth, too, heaped up around the base, so that when frost has set in they will be the less liable to be forced apart.

When the hives have to be guarded from thieves in summer, the space between can be built up by two doors, one of which is fixed and the other moveable and made to lock up. If the space at the top between the hives were built up as well, and there were formed on all sides a mound of moss or litter, the bees would pass the winter in perfect rest and fairly equable temperature, as if they were placed in a cellar. But, because the ordinary entrances would then be stopped up, openings for entrances in addition to the clefts would have to be made in the doors, and these must be simply sufficient for the bees, and not large enough for a mouse to go in. For when a strong stock accidentally gets disturbed, and finds all the entrances closed, it tries to make a way out by force, and in a little while the bees worry themselves to death. But if they find a way out, and the space before it perfectly dark, as would always be the case here, they return again soon to perfect rest, thinking it is night.

Four and Eightfold Hives.

It scarcely needs to be remarked that there may be made, instead of threefold and sixfold hives, fourfold and eightfold hives, that is, such Ständer hives in which four compartments of equal height lie alongside, and that the eightfold can be set up by pairs, with the doors opposite one another, as with the sixfold hive. The difference is only this, that two entrances open out in the front wall from the upper stage and two from the lower one. If the two adjoining entrances were put as wide as possible apart, a division-board between them might very well be spared. It might, however, be more suitable to put them close together, and separate them in the same way as in the two fourfold hives (see figure, page 121), for the division-board, projecting so far in front, has this advantage, that the bees, even in strong winds, can fly in quietly.

Hives for Queen Raising

are often mentioned and recommended in works on bee-culture. These are small boxes, hives in duodecimo we might call them, which are meant to serve as temporary dwellings for small stocks, so that young queens may be raised in them. The bee-keeper who works in the old-fashioned way, with the ordinary straw or log hives, or boxes similar to them with immoveable combs, has most need of such hives for the purpose mentioned. But whoever has introduced the box hives with moveable combs, described last, whether they be Lagers or Ständers, simple or compound, has no need of them. Every compartment yet unoccupied is capable of serving for such a small hive when the superfluous space above, below, and at the sides, is shut off and filled with warm material, and in this a small stock will be the more warmly settled, because it is protected on all sides. Out of one of the hives accessible from both sides, like the Twin-stock, two hives of half the depth can be immediately made by inserting a board fixed firmly in the middle, in the way previously described. Higher Ständers could be divided horizontally into two parts at half the height or where the honey-room begins, and each part could be used as a separate hive. The one entrance could also be put in the door, which, of course, must be constructed in two divisions of corresponding height. If there is a choice of several empty hives at command, such compartments will be used for queen-raising whose entrances are as distant as may be from that of the others occupied with bees, so that young queens on their wedding flight may not easily lose their way, and so perish.

In the warmer season of the year, from about May onward, the superfluous space does no harm, and small stocks can be lodged in every hive of normal size, and later on, when the stocks have raised a queen and she has become fertile, they may easily be formed into complete stocks by the addition of brood or bees, and this can be done without removing them into another hive, as must be the case if the queens are raised in small hives for that special purpose.

Now that the problem has been solved of emptying full honey-combs and giving them back to the bees again to fill, even the normal hives do not any longer require to be made so large as formerly: for if the emptying machine, or slinger, which will be spoken of further on, is used, a great quantity of honey can, little by little, be gained from a small hive. Collaterals or supers can be given to smaller hives in good seasons, when the bees begin to be short of room for storing honey, and these will, of course, be taken away again in autumn. An internal capacity of about 2000 cubic inches may very well suffice for moderately favourable districts, meanwhile a size of from 3000 to 4000 cubic inches would certainly be no fault in hives which can be so easily diminished and parted into two divisions as dwellings for two separate stocks. But in carrying on the travelling system of bee-culture the latter size would be an evident drawback, because the transport of hives would be thereby rendered more difficult and costly.

The Observation Hive.

Many will expect in a bee book some guidance as to how an Observation-hive may be best arranged. The most varied observations may be made in every hive with moveable comb without its being specially arranged for the purpose. Since it is possible from time to time to look into every cell, the beginning of the brood deposit, its gradual extension, the foundation and progress of queen-cells, can all be followed if the requisite combs are every now and then taken out and investigated. But as the Twin-stock still remains the best of all the hives previously described, though they are all good, so it is, and continues to be, the most convenient Observation-hive. The circumstance that it is accessible from two sides* makes it suited for observation. If it is not wished to take out combs, but only quietly to look on at the activity of the bees, the moveable glass doors previously described should be applied at these opposite sides. The outer door must fit easily, that it may be taken out without disturbance. The bees undisturbed by any draught will then continue their work, and the queen her egg-laying, and our own eyes will with carefully continued observation convince us, if it is so arranged that a drone-comb is brought up to the glass door, that she lays not only the worker eggs but the drone eggs. And the comb on which it is wished to observe something, as, for in-  stance, one containing drone-cells for which the time of occupation has just arrived, or queen-cells just begun, can be put directly by one of the glass doors, filling its place by the one previously there, or putting the first one altogether away for a time. A single comb can be left to a weak stock, and on this can be observed through the two glass doors every thing that is going on in the hive. Not only can our theoretical knowledge be extended by the possibility of extended observation, but many an end of practical importance can be attained. We may have, for instance, a brood-comb with several queen-cells nearly ready for the queens to come out, which cannot well be detached without injuring them or destroying many brood-cells. If this is put to a driven swarm, or stock otherwise queenless, so that an oversight of all the queen-cells may be maintained through the glass doors, the coming out of the first queen may be quietly awaited. This queen, or the remaining cells, are then quickly removed and otherwise used. Whoever would like to have an insight into the Twin-stock from a third side might put a larger or smaller pane into the back wall. In a hive standing singly it would, of course, have to be darkened, but in one standing in the pile this would not be necessary: but in order to be able to look, for instance, to see whether sufficient stores were still there, the hives would have to be moved apart. If the panes were taken out of

* That is, from both ends.—C. N. A.

both Twin-stocks a second communicating passage would be gained, by which the bees could be united in severe weather, or could fetch honey from the neighbouring compartment. Whoever wishes them can put in the back wall of the Twin-stocks either glass panes about five inches broad by two inches and a half in height, or moveable doors. They can do no harm, and in many cases would be very serviceable.

ON THE DIFFERENT METHODS OF BEE-CULTURE.

There are generally two different ways of obtaining honey and wax, the valuable products of the bee. Either the whole contents are taken from a part of the hives at the time when they have the largest store and no brood, which is generally the case in the middle or toward the end of September, and the stock is cashiered either by sulphuring the bees or dividing them among the stocks that are to be kept on; or the hives have taken from them the honey and wax they can do without, partly in summer, partly in autumn, and partly in the following spring.

In order to be able to harvest much in autumn by the first method of treatment, matters must be so arranged that the number of the stocks may be as much increased as possible by swarms in the course of the spring and summer. This method of treatment is therefore called the 'Swarm Method.' The other is called the 'Zeidel' method, because the taking away of the overplus, the cutting of the combs of the bees, in the language of bee-keepers is called 'Zeideln.'*

The Swarm Method.

The swarm method is applicable when the bees have a pasture continuing some time, even if not of extraordinary richness; but it must especially not be deficient in autumn. Only in such a country can the mother stocks recover themselves again, become populous and stored with honey, and the swarms derived from them fill their hives with comb and collect sufficient winter stores. The increased number of stocks will then in autumn have

* A word equivalent in sense and meaning to the English term 'depriving.'

carried in more than if a division of the stocks of bees had not occurred, no matter whether it had been by a natural or artificial method. For in the increased number of hives, where formerly there was but one queen, there are now two, three, or more, so that the number of workers is largely increased, and these will at the same time show a greater industry than if they were crowded together in one hive in too large number. In such case they will, as is well known, often, even at the time of most abundant pasture, inactively cluster in front of the hive, on account of the excessive heat and lack of space. Moreover, a stock diminishes considerably in industry when it has completed a considerable amount of comb and stored up much provision. The increase of weight will no longer remain in proportion to its strength, on the contrary, new swarms show marvellous industry both in building and storing, as is well known to every bee-keeper. There can, therefore, by this method be not only more wax and honey harvested if in autumn the increase of hives is reduced to the original number, but this advantage is further obtained, that the stock hives for the next year can be chosen with respect to the fertility of the queen, age of the comb, and quantity of provision, while in the Zeidel method the stocks must be wintered as they are.

But with these brighter features the swarm method has its dark side.

In all produce, and especially in the products of bee-culture, it is not merely a question of quantity but of quality. In the swarm method it is just the worst honey that is harvested, that is especially honey from the well-known rich autumn pasture, the heather, which is neither pleasant to our taste nor yet wholesome winter food for the bees, whereas the most valuable honey from the spring and summer flowers will have been mostly used up for comb-building and brood. For in the small swarming-hives the bees cannot, even in the richest pastures, accumulate any considerable store for want of room, because the brood takes up nearly the entire hive. And the swarms thrown off take with them part of the stores collected and expend it immediately in comb-building.* The largest part of the honey which the new, as well

---

* This appears one of the strongest arguments in favour of the use comb-foundation. Instead of consuming the early fruit-blossom honey

as the old, stocks possess in autumn can only be heather honey. It will be difficult, then, by this method to get the handsome sealed combs containing the honey so much in demand and commanding the highest price.

Even in quantity there is sometimes less harvested by the swarm method; indeed, in certain circumstances, both parent stocks and swarms will be entirely empty of honey in autumn, and the whole apiary will come to ruin if the owner does not have recourse to costly and troublesome feeding, while Zeidel stocks will have carried it beyond their own requirements if the summer has been pretty favourable and the autumn deficient.

THE ZEIDEL METHOD.

The Zeidel, or depriving, method is the more certain of the two, as, in a district not entirely unfavourable, it would rarely occur that the stocks would not gather sufficient for their own requirements. In any year there would scarcely be wanting a few favourable days, which would enable stores to be collected by stocks maintained in such strength as these would be and housed in hives of such capacity.

The Zeidel method alone can afford any profit in districts in which the bee-pasture is limited perhaps to the flowers of the rape, clover, white clover, or buckwheat, which, plentiful as they may be at their blooming time, last but a little while. For both old and young stocks would towards winter be poor in honey if the short honey period were used in unsuitable preparations for swarming and the honey acquired was expended again for comb-building and brood. But since a decline of stocks cannot be prevented there will have to be in this method provision made for a regular increase to replace deficiencies. The bee-pasture may be considerably prolonged in certain circumstances in a district, which otherwise affords but a short pasturage, if an early and specially favourable spring sets in, if honey-yielding plants are by chance more abundantly cultivated, or if weeds producing

in making combs, the bees store it forthwith in the new cells given to them, and virgin honey can thus be had from swarms so treated long before those without foundations would have any comb to put honey into.—C. N. A.

honey grow in greater abundance than usual. On the other hand, the productiveness of a district, which usually continues longer, may be limited to a short time by unfavourable weather. The judicious bee-keeper will not, therefore, tie himself down to either of the methods mentioned, but according to the difference of weather will use now the one plan and at another time the other.

Rational Bee-Culture.

The way to carry on bee-culture rationally is to be armed by a thorough knowledge of the entire nature of bees, and to treat the stocks so that there is derived from them surely and continuously the greatest advantage which can be obtained under the prevailing conditions of yield and weather. The bee-keeper who has not learned better just lets the bees do as they like, and if he has any thing to do about his hives does it after the manner of his father and grandfather before him It is otherwise with the rational bee-keeper. He conducts his business systematically, interposing in the economy of the bees, at one time stimulating, at another time checking, and having for everything that he does a reason why. He will proceed in one way so long as he is increasing his apiary, and in another when the desired number of stocks is completed. If he is looking to derive his chief profits from the sale of stocks the treatment will have to be changed if these find no buyer and the harvest is to consist of honey and wax. He will not complain, as we often hear complaints made, that the bees at one time are swarming too much and at another time are clustering in inactivity in front of the hive and will not swarm. For he knows how to check immoderate swarming and how to make artificially the natural swarms, that hang outside, when he thinks that the division of strong stocks into swarms will be of advantage under existing circumstances. He will, at most, only complain of that which cannot be remedied by any power of man, as of continuous unfavourable weather, which may make all exertions vain in every department, especially in that of agriculture, and which may bring to nought the fruits of all human activity.

From what has been just said of the different methods of treatment it may be seen that a distinction has to be made between bee-hives and methods of culture. Different methods may be carried on with the same hive, and essentially one and the same method may be pursued with different hives. So far there lies a measure of truth in the conclusion that is often advanced that rational culture may be carried on in every hive, even in the ordinary straw one. But with what an expenditure of time and trouble would the end be attained in the one case, and with what certainty of success in the other? That is another question. For instance, the queen of a hive has to be caught because she is defective, or because some other important end has to be gained by it. In the ordinary log or straw hive this is extremely difficult, takes up time, and is often quite impossible without disturbing all the comb, whereas in hives with moveable comb the work is done with certainty and takes but a few minutes. It may therefore be replied to the conclusion mentioned above that bee-culture can only be carried on rationally in hives with moveable combs, because it is not rational to try to do in a long, difficult, and uncertain way that which may be attained by a short, easy, and unfailing method. If, therefore, other hives widely used like the ordinary log and straw hives should not be left quite unconsidered in this book we shall still speak by preference of the treatment of bees in the hives with moveable comb already described. For it may be expected that these hives, even if made of straw in countries where wood is scarce, and whether round, angular, or door-shaped, will more and more replace the inconvenient hives having fixed combs.

Passing over to the treatment of bees proper we find ourselves in some embarrassment as to the order in which every thing should be spoken of and as to where we should begin and leave off. Perhaps some instruction as to the way in which stocks are put into new hives might most appropriately follow the chapter on the manufacture and arrangement of the most suitable hives.

On the Occupation of the New Hives.

It would be superfluous to speak of the season when the occupation should take place if many bee-keepers in their eagerness to introduce these hives into their apiaries did not come to grief by transferring stocks from log or straw-hives in autumn—a time quite unsuitable. They not only thereby do themselves harm, but create prejudices against the new hives, because the unfavourable result is ascribed to the hives rather than to the unsuitable time when the changes are made.

New hives should be occupied when the bees are naturally ready for moving. There is no need to await the full arrival of swarming time, as in the case of other empty hives, because we can provide comb for the bees beforehand so that eggs may be laid immediately, and thus hives may, at any rate, be stocked in April. But the weather must be such that the bees can prepare wax for filling up gaps in combs and for building new ones, and that chinks in the hive may be filled and the bees be able to settle down comfortably and arrange a warm brood-nest. There is no need to be over-hasty in the introduction of new hives. It would be foolish to do away with log or straw hives while they are still good and the combs in them have not become too old. They should be made use of as long as possible, and the swarms from them, whether natural or artificial, be put into the new hives. It has been already mentioned in speaking of the straw hives (page 69) how brood-combs for the new hives may be taken from straw-hives. Log-hives can be dealt with in a similar way. If at the top the combs do not hang from moveable bars, the latter can be fitted up below, so that later on brood-combs can be taken away for artificial swarms, or for the strengthening of swarms put into new hives. If the breadth of the log-hives should differ too much from that of the boxes, super-hives provided with moveable bars can be applied to the log-hives in order that the combs, when filled with brood or honey, may be put into the new hives.

If, further on, it is thought desirable to do away with the old hives altogether it should either be done in autumn when there is no brood present, everything being cut out and the bees divided among other stocks, or at the beginning of the swarming time if it is wished to preserve the stocks. This is done by taking away the old queen, with or without a swarm, and three weeks later, when all the brood has come out, the stock meanwhile having, perhaps, yielded a second swarm, the whole of the comb is cut out and the bees with the young queen, perhaps already fertile, are put into a new hive.* If the comb is required to be entirely free from brood there must be a delay of some twenty-five days after the

* This is excellent teaching, and 'rational' indeed.—C. N. A.

removal of the queen, as many eggs may be a little delayed in development, and the drones require some four days more for their development. But since every day let go by is one lost to the new stock, and since the individual sealed brood-cells will not be lost if the combs are attached to bars, so far as they are suitable, and are again given to the bees and the other irregular pieces of comb put in for a little while, then the cutting out of the combs may be proceeded with as soon as the greater part of the brood has come out.

In log-hives the cutting out of combs is made easier by reversing them, and the bees are driven, as the combs are cut out one by one, into the empty space that is now found above, where they will cluster like an ordinary swarm. From here they may be scooped out with a box, tin-pan, or ladle, and put into the new hive, or into the transport-hive if they are to be put on a distant stand, which latter must be done if the new hive is not to have the old one's place, since the bees of course would only fly to the old familiar stand.

Whoever possesses a supply of full and empty combs, and can put together from them a complete set of combs, may put a stock into a new hive, even in early spring before much brood has been deposited. Since such an operation ordinarily takes up some time, and in cool weather bees are chilled and in hot weather robbers are attracted, the transferring, especially out of straw hives, might be done indoors. The bees, that would possibly fly off, fly immediately to the window (which might, however, be darkened), and collect, after flying about some time, on pieces of comb placed for them, especially if they contain some brood, and can then be conveniently added to the others. If it is cool in the room, or it is allowed to become cool, the single bees scattered at or near to the windows chill, and may be swept with a feather on to a piece of paper and be put to the others, and those left behind in the hive that is to be emptied may be allowed to chill, so that they may be more conveniently collected. If it is wished to finish the matter off quickly, quite cold water can be poured into the vessel, tin-pan, or box, with which the chief mass of the bees is scooped out, and the individual bees yet remaining behind may be brushed into it, when they immediately chill, and can then be conveniently put to the others, so that not a bee gets lost.

The Furnishing of the Hive that is to be occupied.

This hardly needs to be specially mentioned, as its arrangement may be gathered from what has gone before. Some observations upon it, however, may be of use to the beginner.

Beginning at the back part of the hive the comb-bars are put directly where the bees should make their nest for brood, and for the winter, therefore, in Ständer-hives, not in the uppermost, but in the next pair of grooves, which divides the brood-nest from the honey-room. The inserted bars are of course to be provided with longer or shorter beginnings of combs as much as possible along their entire length. The combs given must be proportionally larger according as the occupation is longer before or after the proper swarming-time. Even at this time longer beginnings are very desirable, though not directly necessary, because the bees themselves are then in the position to extend them rapidly.

The bees do not usually extend all the combs equally, but their comb has mostly, so long as it does not reach to the floor, the form of a grape cluster—a form which the bee cluster itself usually takes, whether it is hanging from a tree or in the hive. The hindmost combs are, as a rule, the longest, the front ones, on the contrary, become shorter the nearer they stand toward the door, as is made plain by the following figure, which represents a Twin-stock nearly, but not altogether, filled with comb. The cover-boards are here shown removed, so that the comb-bars may be seen. We may imitate the bees in this arrangement, giving them the longest combs behind and the shorter ones in front.* This has the further advantage that it may at once be seen from below whether the comb beginnings have been inserted at proper intervals, standing barely a half inch, or the width of two worker cells, apart from one another. It is especially desirable quite at the back to put a comb that is not merely long but broad, filling up the corners of the hive as much as possible. For a swarm that is weak, or thrown off too late, often does not thoroughly complete its comb in the same year, and has then in winter but a cold nest, because in the vacancies left the heat escapes from it.

---

* That is, nearest the end at which the door has been opened.— C. N. A.

The long comb put in behind stimulates the bees to make the others equal to it as quickly as possible.

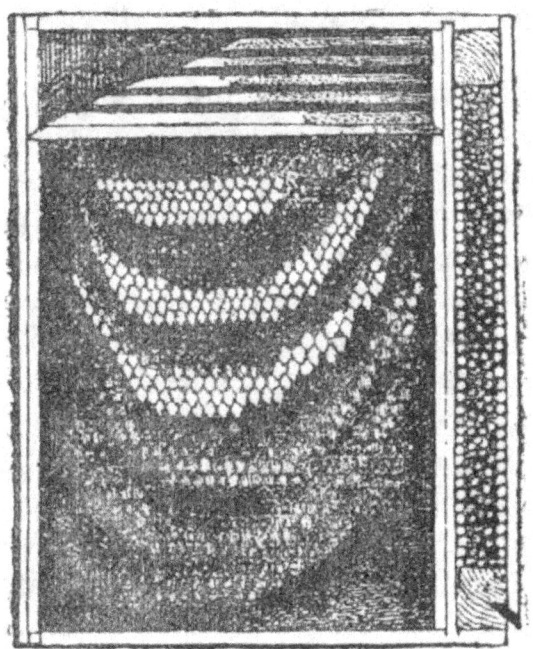

It is evident that there is no need to put in immediately so many bars and beginnings of combs as there is room for in the hive, but only so many as the bees can cover and soon extend. Especially there must not be too many comb beginnings put in to a late swarm. It is better that five combs should be fairly finished than that the extension of eight should be begun and all be left but short at last.

A recently occupied hive is, as is well known, sometimes forsaken by the stock that has been put in, no matter whether it is a natural or artificial swarm, or an old stock transferred. This may be prevented by the insertion of a comb, or a piece of comb, containing brood, having among it some unsealed larvæ. The brood given to them impels the bees at once to greater industry, and further insures them if they have a young queen still unfertilised against queenlessness in case she should get lost on her wedding flight.

By the introduction of a comb, or piece of comb, with brood, the place where the brood-nest is desired is at once indicated to the bees, and the queen immediately betakes herself to it, settles there, and begins to extend the brood from thence outwards on all sides.

The lower or deeper the bees arrange their brood-nest, the greater the advantage, because they will have the more space and opportunity to accumulate honey above it. A comb should therefore be chosen which at the top is empty, or only contains honey, and has brood in it lower down, or a short brood-comb is put in Ständers in the lower grooves, and over it, as extension upwards, a comb that is empty or only contains honey; or, lastly, a brood-comb may be set up with the help of supports on the floor of the hive, or a bar is put in at a suitable height, so that it may by building be united with the empty one put above it. It is evident that the bees must be able, if cool weather sets in, properly to cover the brood-comb that has been inserted. For if the bees in clustering are obliged to forsake it and it should perish, the stock, instead of being attached to its hive by it, would be the rather induced to forsake it.

The comb-bars inserted are of course carefully covered on the top with the small covering-boards, and in addition, in cold weather, with dry moss, hay, straw, or an ordinary straw-mat laid upon them. At the back the cover-boards can be laid on parallel with the length of the hive, therefore in a direction opposite to that of the bars. In Manifold hives the board adjoining the outer wall is laid on first, and then the one touching the mutual dividing wall.

If it is desired to let the bees into the honey-room this or the next year, it is only necessary to remove the last board, or to move it a little away from the wall and let it overlap the first cover-board more. There might further be left immediately by this wall a little chink, which could be temporarily closed by a smooth bar, which later on could easily be removed. As is well known, the bees in winter like to cluster by the party wall, because they enjoy the mutual heat, so that even in extreme cold they could ascend through this passage and bring down honey if it were exhausted below.

In most cases, in Ständer hives with three or more divisions, it would be quite useless, or indeed injurious, to put up a second series of bars lower down if the stock is not very strong, and pasture would not be holding out much longer. The bees would be hindered in building, and if the height of single divisions were only about eight inches they would have for the winter a nest that would be too short, and therefore too cold. With a 10-inch division it would be less injurious, since such a division, closely built up, would always afford the bees a warm nest for winter. The insertion of the combs into the lower division can generally be put off till the next year. Whatever parts of the combs project below the lower grooves are cut off, and make capital beginnings for the comb-bars. It has been already mentioned that the bars are not to be the whole depth of the hive here, but only need be put from the door to about the middle, but at the back they are to be quite omitted, and the combs then can have more hold given to them in another way.

It is somewhat different if the stock is strong, the pasture excellent, and promising to continue for some time. Then the comb-bars can be put in not only below, but in the honey-room at the top after the bees have arranged their brood-nest in the middle. The large stock, which otherwise might be burdensome to itself, can then divide itself better, and at the same time build at the top, in the middle, and below.

General rules, therefore, cannot be given on this point. Judgment and experience must suggest to every one how many combs he will let the bees build, and when he shall give them access to the honey-room. In Lagers, especially in Twinstocks, the bees may at times be allowed to build above, although the brood-nest has not long been completely built. For, if the bees have only partly built the upper space and filled it with honey, they will winter all the better for having over them abundant provision, which will be accessible even in severe cold.

In Ständers, however, it does not do to be over hasty in opening the honey-room, so that the bees in any circumstances, even if the pasturage should unexpectedly cease, should not only make an adequate winter nest, but should accumulate in it sufficient stores for the winter. The warmth does not depend so much on the length of the combs as on how closely the space is built up, and twelve inches of closely-built-up comb afford the bees a warmer nest than fifteen to eighteen inches of loose and imperfect construction.

Our purpose in getting the bees to build closer, and at

the same time to have more honey left, is better attained by hindering their lengthening the combs by inserting a horizontal board, which will oblige them to fill up the whole breadth, and to build up all the corners completely. In the description of the Ständer hives there is certainly no mention made of such a board by which the space below can be limited, as it may in the Twin-stock, laterally, by the dummy; it is not absolutely necessary, but still many advantages may be attained by it.

Besides compelling the bees to build up the limited space more compactly, as already mentioned, in this way a warmer winter nest is made for them, as the board keeps the heat together above, and keeps cold air from rushing in below. This will be more completely attained if the board is placed even higher than the entrance. The cold air outside would not then so directly penetrate into the bees' winter-nest, and they would not be so easily enticed by the sun on to the snow. In that case, of course, the board will have to stand a little away from the wall in which the entrance is situated, or at least must have a hole cut in it as a passage for the bees.

By means of the board inserted horizontally, which can rest on four nails or on two bars stretched across, there is assigned to the combs a definite length, just as it may be desired, independent of the height of the hive. It will then be possible to remove brood from Ständers into Lagers like the Twin-stocks, which are much lower, without being obliged to shorten the combs.

With its help the cleaning of the hive is made much easier, because all the refuse can be taken out with the board. As the dead bees and all the rubbish will fall on it, it might in autumn be put a little lower.

The food to be given to the bees, whether it consists of barley-sugar, pieces of honeycomb, or candied honey in pieces, could be put close under the bees' nest on the board above mentioned, so that even in a moderately cold temperature they could come down and consume it.

At the recommencement of building the board can be let down to within an inch of the floor, where it can rest on four nails driven into it like a little bench on four feet. In this way an empty space is reserved below, which is very serviceable in different operations, especially in the driving or smoking out, to be described later on.

If the board is left standing a little higher, a honey-room below may be formed by it. With sufficient strength and good pasturage, the bees will so much the sooner begin to fill the lower space with comb, if comb beginnings are attached to the lower surface of the board, which should not be put across, but parallel with the depth of the hive or length of the board, so that it may be more conveniently drawn out and given to a stock poor in honey, which in this way would be supplied in the easiest way with its necessary winter food.

In a strong stock there will be continually large masses of bees hanging under the board, of which a large number can easily be taken at once and used for nuclei or for the strengthening of weak stocks. The board is carefully drawn out, and the bees are either allowed to run into the new hive, or shaken provisionally into some vessel.

If the board furnished with comb beginnings is put into the upper honey-room, an artificial swarm driven into it may be at once taken out, and the trouble of scooping out the bees prevented. Even in a honey-room that is quite low, as in the Twin-stock, this way of catching the queen is generally successful, doing away with the necessity of taking out the separate combs when they are of unusual weight, which requires care, and takes up time. There is no need to have a special board for every hive for these temporary uses, but one and the same is applicable for all that have the same width.

The Swarms.

The question has been suggested whether the hen or the egg came into being first, and the same question may occur with regard to the parent stock and the swarm, although every parent stock was first a swarm, and as such, moved into its present dwelling, and young swarms become complete, populous stocks, which, in their turn, throw off swarms again. The different phenomena in the life of the bee run through a continuous cycle, but, as is well known, the circular line has neither beginning nor end, or the beginning is just where the first stroke is made.

Since we have just spoken of the preparations that have to be made in new hives that are to be occupied, or that are already stocked, it may be more convenient and orderly if we conceive of the swarm as the earlier in the field.

The Different Kinds of Swarms, their Origin, and Treatment.

A swarm is a stock of bees temporarily homeless, which, with its queen, has forsaken its former hive in order to move into a new one and establish a fresh independent colony. If the entire stock forsakes a hive that has been long occupied and completely filled with combs, it is usually called 'a pauper swarm,' it being occasioned by hunger, or other causes.

Pauper Swarms.

Necessity or hunger is not the only incitement to this kind of swarming, since dirtiness or coldness of hive, or of comb, or the prevalence of wax-moths, will induce it. A weak stock will readily forsake its hive if it has little or no brood, and especially if there be no unsealed brood, and therefore, as a rule, the pauper swarms occur most frequently in spring, and again towards autumn.

Many call such swarms pauper swarms which appear at an unusual time, either long before or after the ordinary swarming time. These are occasioned by the queen of a hive dying and of the young queens raised, one, as a rule the first that hatches, swarms with a part of the stock. But the term is by no means rightly applied to such swarms. They are always proper swarms, more accurately second or 'tüt' swarms.

Regular or Proper Swarms.

As such we describe those stocks of bees which migrate from populous stocks, so that only a part of the inhabitants leaves the hive with the queen, the other part remaining behind. The old stock which has thrown off a swarm is with propriety called its parent stock, because the swarm is indebted for its existence to it.

The First Swarm.

This comes out with the old fertile queen when she is still living, and it is really the queen-cells already partly sealed which make the old queen uncomfortable in her present dwelling and incite her to move, just in the same way as poverty, uncongenial temperature, dirt, or moth, occasions pauper or hunger swarms. In consequence of the querulous tone of the queen the bees are seized with a certain excitement, ardour, or craving to swarm, and at about noon some fine day she with her following leave the hive as a swarm. Swarming has its foundation in the queen's jealousy and intolerance of another queen, or even of sealed queen-cells. The impulse to it proceeds from the queen, but the moment of departure is determined by the populace, which alone can know when the proper time for departure has arrived, and cases have occurred in which a swarm has come out, although the queen has been taken from the hive some time previously. It very rarely occurs that a first swarm is thrown off simply on account of the heat having suddenly risen to an intolerable degree without queen-cells having been previously commenced. A certain innate instinct impels the bees to form queen-cells, and this finds expression when the brood has become considerably extended—drone-brood being present, and the hive having attained a high degree of temperature. That which is favourable to the brood and increases the heat of the hive, as warm, moist weather, the opportunity of continually being able to carry water from a neighbouring pool, or a warm southern aspect for the apiary, will all promote swarming. They may be partly incited to it artificially by cutting little notches at the corners of the brood-combs below or at the sides, in which the bees readily lay the foundations of queen-cells.

It has often happened to me that a swarm has come out of the one compartment of a double-hive without any previous preparation, simply because a cleft had accidentally opened in the party-wall, not, indeed, such that the bees could pass through, but sufficient to make the queens aware of each other's presence, so that one of them moved off in fear for her life. Perhaps it might be possible to cause natural swarms artificially, especially in Twin-stocks, if the two neighbouring stocks were separated only by a piece of wire gauze, or a perforated board, that could be pushed in or drawn out. Bee-keepers who have the leisure could experiment in this direction and send the results to the editor of the *Bienen Zeitung* at Eichstädt.

Swarms have before now been gained by letting a queen, casually obtained, run into a strong stock. Perhaps these experiments only succeeded where a stock had accidentally lost its old queen and only possessed queen-cells. In this case the bees would readily accept an unfertilised queen. But if the bees are bent upon swarming they will not let the queen-cells already commenced be destroyed, but continue to care for them and compel the queen to leave the hive with the following she may have gained. The same thing often results if a queen-cell nearly ripe is introduced into a strong stock robbed of its queen. The young queen emerging from this cell comes away in a few days with a swarm if the bees do not let the queen-cells be destroyed which they themselves have begun, but continue to care for them. In weak stocks, *i.e.* those that temporarily have been much weakened, by swarming, the addition of a queen or of a sealed queen-cell results in the queen-cells already begun being destroyed, and no further swarm comes off.

Stocks which have thrown off the first swarm are like stocks robbed of their queen, and are exactly similar in every respect to those from which she has been taken, or in which she has accidentally died. If, then, to-morrow there is added to a parent stock, which to-day has thrown off its first swarm, a young queen, either in or out of the cell, in most cases the second swarms, which are more disadvantageous than useful, will be prevented. The designation, 'First Swarm,' already indicates that as a rule a stock throws off not simply one but several swarms.

After Swarms.

The succeeding swarms are called, as is well known, after swarms, casts, or 'tüt' swarms, because they are preceded by the piping of the young queens, and they announce themselves by this means generally some days previously. These are occasioned by the bees forming not simply one queen-cell but several, often as many as twenty. The queen first coming out has nothing more urgent on hand than the seeking out of the queen-cells still present to bite them open and to give a fatal bite or sting to her rivals contained in them, whereupon the worker bees complete the work of destruction. But if the weather is favourable and the bees are strong and eager to swarm they keep the queen off, who then runs about the royal cells like a cat about hot soup, and gives voice from time to time to the well-known distinct notes expressive of jealousy and anxiety, whereupon the other queens already ripe but still concealed in the cells answer her with 'quack, quack.' If the first of the young queens has quitted the hive with the first of the after-swarms, which follows the first swarm in about nine days, then her successor undertakes her part. So it is possible, with the quantity of young bees daily coming out of the cells that there may be about every two days a second, third, even a fourth, or fifth, after-swarm come off. At last one of the young queens attaches herself to the hive, the others are killed off or expelled, and the ones not yet quite ripe are thrown out of the cells.

Whether the old queen has died, has been taken out, or has gone away with the first swarm, only makes this difference with regard to the after-swarms, that these come off after the departure of the old queen with a first swarm perhaps five days earlier, say in about nine days; otherwise they do not come off till fourteen days have elapsed, because in the one case queen-cells were already in progress, but in the other are commenced as an after addition.

First swarms can never be foreseen with entire certainty unless the necessary investigation within shows that queen-cells and the old queen are both present. After swarms or tüt swarms are announced not only by the piping of the queen, but may be made with tolerable certainty. If any one would like to have the pleasure of seeing a swarm come away on Midsummer Day, the 24th of June, he must on the 10th of June take away the queen from a stock without weakening it much, or in case he had at the same time taken away from it an entire swarm with the queen, it must be made strong again by letting it change places with a populous stock, in the way to be described later on.

The swarms appear, as is well known, at the best time of the year, and at the warmest part of the day, though they may, especially if their flight has been hindered by rainy weather on previous days and the queen is very urgent, come off as early as eight o'clock in the morning, or as late as five in the afternoon. But the time when the bees generally play in front of the hive is the ordinary time for swarming.

The cases are rare in which the swarm thrown off immediately betakes itself to a distance. As a rule, it settles on a tree, bush, hedge, or other object, and the sooner the bees do this the better. For the longer the bees swarm about, the sooner will the queen, especially if old, become tired, fall to the ground, and get lost, the more bees go back again to the parent stock or, in a large apiary, alight on other stocks, where they may be stung to death, or even, in their excitement, they may sting the queen of the hive to death, or, if that does not happen, they will at any rate not be so useful there as in the swarms.

There are, however, means of inducing the gathering and settlement of a swarm, without which a weak swarm would be overpowered by the hum of the bees playing before other hives, and might perhaps be entirely destroyed. The bees settle readily on an object that is dark or brown, having the idea that a cluster of bees is already settled there. They go, therefore, readily into a basket of brown willow twigs, set up on a pole and turned with the opening to the side if in it is put an old blackish brown wax comb, or a piece of brown oak bark. The smell of balm is very enticing to the bees, and it is well to keep up a stock of this sweet-smelling perennial plant in the neighbourhood of the apiary. If the place where the swarm is desired is rubbed with it, or a twig of it put there, the swarming bees generally settle there immediately, and remain together some time, although the queen might not be among them, so that time is gained to add them to old stocks that need strengthening, or to young swarms. A new hive, ready for occupation can be rubbed with it, for it may be seen how pleasant the fragrance is to the bees from this, that a hive which has only had its entrance rubbed with it is taken possession of by the bees sent out as scouts, and of course is occupied by the swarm connected with them if it has not been already hived.

The swarming bees, however, are mostly guided by the sense of hearing, and collect immediately when they hear a loud, enticing hum. If, therefore, there is danger of a swarming stock being destroyed and causing disaster in other stocks, which is quite possible in swarms caused by hunger, it will be well to excite such a hum in the neighbourhood. A weak hive, for instance, is opened, the bees are shaken from one of the combs taken out partly on to the ground, partly before the entrance, or they are thrown into the air. While they are now betaking themselves homeward humming and enticing, they attract the swarming bees, which will settle, following them, and humming as they go. It is then at least known where the queen is to be looked for if it is wished to save her. The queen of the hive, because she is in great danger of being stung to death by the intruders, must, of course, be protected against the danger threatening by caging her. It is, therefore, best in such a case to entice the bees to an artificial swarm, which possesses only queen-cells, and where no further damage can be done if mutual stinging to death is prevented by vigorous smoking.

But a swarm may be induced by the hum of other bees to betake itself where it is not wanted, as to a swarm that is just hived and is strong enough without it, or to one still hanging on a tree. This must then be quickly covered with a cloth, or the bees that fly to it must be driven back by smoke.

Smoke can further be maintained about the parent stock, so that too many swarm bees may not go back, to which they are much inclined.

The queen is the principal thing to keep one's eye on. It might happen that she was not able to come away, because she tried to find an exit through a narrow cleft in the shrunken door of a box-hive. The door may, therefore, be opened so wide that everything can come away that will. Young queens often turn back again, occasioning the return of the whole swarm, which is always very unpleasant, because not only is a day lost to the swarm, but many bees, in the confusion, alight on neighbouring hives and get stung to death. The queen can be hindered from this return by sticking a reed in the entrance, and letting it project outside, or a tube made out of a piece of paper would serve the same purpose.

It is possible for the queen to fall to the ground on account of defective or injured wings. A young queen would then be of no use, because she cannot become fertile, and might as well be immediately put aside. In the case of the fertile queen of the first swarm, that is of no consequence. Indeed a wing may be cut off from every queen as soon as she has begun to lay, so as to prevent the loss of the swarms; only when the swarm comes out, whether it is in the same or the next year, she must be carefully looked after and caught if she falls to the ground or creeps about on the hive, enclosed in a cage and brought to the swarm, which in the meantime has generally begun to settle somewhere. If the queen is seen when the swarm comes off, especially if it is a first swarm, it is always safest to catch her. If, in consequence of the recent disturbance and excitement, she has nearly given over laying, and has so become lighter and more capable of flight, yet she is always a little clumsy, and can easily, in windy weather, be beaten down and get lost. When the queen is caught, the whole swarm is in our power, and if it will not settle at all, or only in an inconvenient position, an empty hive can be quickly put in the place of the parent stock, the imprisoned queen be put in, and the returning swarm be left to go in of itself.

As we have in the queen-cage a prison for the queen, so, when the swarm method is followed, we have in the so-called swarm-net or sack, an enclosure to catch the whole swarm when it comes out of the hive, and to prevent its flying away. It is made of a transparent material as airy as possible, and is of about the size and shape of an ordinary corn sack, kept extended by means of hoops, and must allow of being quickly fitted at one end to the swarming stock. After the exit of the swarm it is taken away and hung in a cool place till the bees have quieted down and clustered.

The Securing of the Swarms.

The catching or putting the swarm into the hive intended for it is then, of course, very convenient. With an ordinary straw hive the swarm may at once be shaken into it. But since in hot weather the swarm might again fly apart, ascend, and perhaps fly right away without settling again, it is better to put it gradually with a little box into its hive. If one part is put in first the others stream in of themselves, enticed by the joyful hum, and thereby collect the stragglers together. The remainder may be left to go in through the entrance, since the bees will thus be taught something of their new way home. They can then be shaken on to a cloth spread out and fastened below the entrance by two nails. The longer this moving in lasts, the more completely will all the bees that belong to the swarm be attracted from all sides. It is only when another swarm is expected, or is at the point of coming off that we must cut it short and seek to shake the swarm into its hive as quickly and completely as possible.

If the swarm has settled on an object that can be brought near to the hive, for instance, on a basket prepared for the purpose, on a piece of bark, or on a thin branch that can be cut off without much disturbance, the securing of the swarm is managed in the same way. But if the swarm is hanging from a thick branch, or from the trunk of a tree, or on a hedge, the special circumstances must teach every one how it may best be dealt with. At the outset it is well, especially in great heat and hot sunshine, to sprinkle the swarm a little with water, either with a syringe or a sprinkling-brush, in order to reduce the swarming heat, and to hinder the bees from flying up too much. They must not, however, be made too wet, as novices sometimes will make them. The work is then only made more difficult, and many bees, and even the queen, may get lost. It is only when the swarm is obliged to remain hanging out longer, and especially exposed to the hot sunbeams, that it can be more freely and repeatedly sprinkled, so that it may not fly up and away. Since in the ordinary way the hive which the swarm is to occupy does not usually permit its being brought to the swarm, even in simple box-hives, this is not advisable, because the combs in it may become disarranged, so that the swarm is immediately put into the basket used for catching, or into a transport-hive, or a sieve, in which the place of the wire is occupied by linen. Whether this is best done by shaking, or a violent jerk, by brushing them off, or smoking, must depend on circumstances, as whether the object is moveable or immoveable, smooth or rough, accessible or inaccessible. Often the bees have first to be driven by strong smoke to a more convenient place, as on to branches put over them, in order to shake into the skep, or to put them in with the branches. If one part has been got in the other soon follows for company, especially if the skep has been rubbed with balm and if smoke is kept up at the place where the swarm is. If by chance we have been able to see the queen, and have caged her and put her in the skep, the bees will gradually leave that place and betake themselves to the queen. On the following day, towards evening, the queen can be released. It is only in places where the swarms are very liable to be lost, because of the number of hollow trees in the woods in the neighbourhood, that she should be kept caged a longer time—say three or four days. Though even then it is better to attach the swarm to the hive by putting in a brood-comb and setting the queen at liberty early, so that she may begin to lay eggs, or, if she is a young one, that she may take an early wedding flight. For in its young days the swarm plays before the hive more than usual, so as to learn the way to the new home, and the queen will then be less liable to miss her hive. To make more sure, it is well to give to swarms with young queens a position that is recognisable, and as much isolated as possible, and not to bring them too near to other hives.* The entrance can be shaded for a couple of days by boards leaning against it or by overhanging branches, so that the queen of the swarm may not easily lose her way. Nothing must be changed on the outside of the hive, still less must it be moved or its position changed till the queen has become fertile. For in her later excursions she does not take note of her hive by any means so particularly as at the first time, and thus easily gets lost if in the meantime any thing has been changed. The stopping up of a crack through which the queen may have flown out the first time may result in the loss of the queen. To put the swarm on a different stand, if the queen has already made one excursion, is dangerous. Even with a young queen already fertilised, if she is put on a fresh stand, precaution must be used that she does not fly when the stock is put in or when she is liberated from the cage, or she will rise immediately to look for the earlier position well known to her, from which she had taken her wedding flight, and if she does not find this she will in most cases be lost.

* This direction very poorly accords with the Author's commendation of the piling system under the heading 'Advantages of the Twinstocks,' in which the hives, being in a heap, cannot be considered isolated by any freak of imagination.—C. N. A.

In large apiaries it is often not possible, even with the greatest care, to prevent two or more swarms flying together. If they are of the same kind, first swarms only or after-swarms only, the union has generally no ultimate disadvantage. In this case the search for and catching of at least a part of the queens cannot be avoided. This is facilitated by throwing down the mass of bees on the grass or on a cloth spread out, bringing a skep or box, into which the bees will soon begin to move, when the queen can be easily noted and caged. If only two swarms had flown together the bees can be pretty equally divided in two hives and the caged queen put to the division which begins to be restless, indicating that it has no queen. The bees generally take to an old fertile queen immediately, even if it is not the one they have had before. But after-swarms with young queens only attach themselves to the queen that has led them out of the parent stock, whose voice is well known to them, and take little or no notice of other young queens, although they may be from the same parent stock, and perhaps have been reared at the same time with her. But a fertile queen, though a stranger, they take to joyfully. If it is desired then to introduce the gentle and industrious Italian bees, the simplest and surest way is to get a fertile Italian queen at the time of the after-swarms. The bees of a first swarm, and indeed of every stock with an old queen, must first, by longer* queenlessness, be disposed to receive such an one, whereas after-swarms, as well as parent stocks, are glad to receive them.†

Because the bees of the first swarm not only do not take to a young unfertilised queen, but generally immediately attack and seek to sting her to death, it is therefore bad if a first swarm and an after-swarm fly together. The bees which have flown together are certainly then more harmonious than usual, on account of the

* This is decidedly against our experience, as we have invariably found bees that have been long queenless most *indisposed* to receive a queen.—C. N. A.

† Introducing valuable queens in the way here suggested requires a very considerable amount of care, because it so often happens that after-swarms have two or more young queens with them when swarming, and are likely to have others join them when hived; and in either of these cases there is danger that one or more may be left with the bees when the valued queen is introduced, when she would be almost certainly sacrificed.—C. N. A.

confusion in which they are found, and do not perhaps attack one another, but a desperate battle generally takes place between the queens, which may easily result in the death of them both. The old queen easily succumbs to a sting from the active young queen, superior to her in battle, and this one again to the stings of the workers, who avenge on her the death of their sovereign. The securing of both queens as quickly as possible is necessary to avert this misfortune. If it is only thought desirable to occupy one hive with the double swarm, because it perhaps is not too strong and has come off somewhat late, the fertile queen is given to it, and peace is soon restored. But if the swarm is to be divided it must not be done in the uncertain way in which two united first swarms are dealt with. The part with the old queen would perhaps behave peacefully, but the first swarm bees, because they have been used to an old fertile queen, would hardly take to the young queen. They would with tolerable certainty leave her and return to where they were reared. At any rate the bees could only be kept together by the speedy introduction of a brood-comb, and the queen be protected from death or injury by a long confinement. It is preferable then to leave it to the bees themselves to divide by turning them into a large trough, or vessel, and putting both queens into it, as far as possible one from another, and letting the whole stay overnight in a cool and dark place. In the morning the two swarms will be found separated, each one collected around its own queen.

In such a case the convenience of the Twin-stock is conspicuous. Here the double swarm can divide as it likes. For if communication is opened between the two compartments till the next morning the bees divided according to their own wish will have ranged themselves each part about its own queen. But because the party of the young queen, as may be anticipated, will turn out the smaller, a larger following may be obtained for her by the speedy insertion of a brood-comb in her compartment, since brood has in all circumstances a great attraction for the bees, although it may be only a piece of otherwise useless drone-brood that is temporarily put on the floor of the hive. It scarcely needs to be said that two united first swarms or after-swarms can be treated in the same way. An after-swarm, however, will easily accustom itself to a strange young queen if she is caged, say over night, with it in a cool place; with a brood-comb it would generally settle even without a queen. Such a brood-comb must therefore be quickly given to an after-swarm whose queen has gone back to the parent stock, or has otherwise got lost, or the surplus queen of another after-swarm is put in, so as to prevent the repeated vacation of the hive, by which the bees waste valuable time, and cause the bee-keeper much disagreeable work.

In order to prevent repeated after-swarming, the first after-swarm, especially if it is too weak, can be put in the place of the parent stock, which can have a new place assigned to it. The latter will then think no more of swarming, but will proceed to do away with the superfluous young queens. To put the first swarm in the place of the parent stock is not always judicious, though it is advised by many, since a good stock may very well yield two swarms.

Unpleasant as many of the accidents that have been partially touched upon above may be, which are always liable to occur at swarming-time, yet the most disagreeable of all is when a swarm has been allowed to escape, or when it immediately rises without settling, flies away, and is never found again, or is found to have taken refuge in the hollow of a tree, rock, wall, or building. If its extraction is not combined with manifest risk of life or danger from fire, it may well be attempted, but a favourable issue may only be expected when the swarm has but just moved in, and has as yet built no comb and deposited no brood. If these have been begun, the bees, and especially the queen, will hardly be moved by the strongest smoke, and the whole stock can only be got out by laying their nest and comb quite bare. Otherwise, strong smoke alone will be sufficient to drive out the swarm. This succeeds best if the smoke can be applied from below, allowing the bees to escape by an opening either found or made above their cluster. Especially do the bees run before the smoke of touchwood soaked in strong solution of saltpetre, and dried again, or if wax, or especially assafoetida, is caused to evaporate by being put on burning charcoal. If we could succeed in getting them to come out voluntarily there would be no further trouble and no bees lost even if the occupied cavity is situated near to the apiary. This could, perhaps, be attained by the application of an offensive volatile oil or old strong-smelling cheese, while the bees are shut in for some time and allowed to heat without going so far as to suffocate them. Swarms that have been shut up some time, as when travelling, like to come out when they regain their liberty. Perhaps the swarm that it is desired to extract might be moved to come out in the same way.

When a swarm is recovered again that has previously flown away, it is very liable to go off again. Brood is therefore given to it, the queen is kept confined for a longer time, or one of her wings is cut off if she is old and fertile.

If it is wished to fell the tree that has been occupied by an escaped swarm, it is perhaps better to wait till autumn, and then to take the honey and wax and add the bees to others needing strengthening. For the immediate extraction of the swarm does not always succeed, and even in successful instances is very often not worth the trouble it takes.

Many a one gives himself a great deal of trouble to get out a swarm that has escaped from him, or that he has discovered, and does not take the swarms which he might obtain so much more easily, by parting the stocks that are too strong, and that are clustering idly in front of the hives. But there is too much anxiety and inexperience in this respect, though he only can be called a genuine bee-keeper who conducts the increase of his stocks systematically, takes them in hand himself, and does not leave it to chance, and the whim of the bees, who only too often will not swarm when it would be advantageous, and swarm immoderately when it is disadvantageous, to themselves and their owner. If the different, unfortunate, and disagreeable accidents mentioned above are considered, which may occur when a swarm has actually come off, such as uniting with another swarm, alighting on another hive, settling inconveniently, rising and flying right away, or losing its queen, there will be no waiting for uncertain swarms, but we shall make them ourselves as soon as the proper time has arrived. In order that there might be a proper beginning, and that damage may not be brought about instead of advantage, we propose to give in the following chapter on artificial swarms the necessary teaching and guidance for making them.

Artificial Swarms.

The objection that voluntary swarming, as being the natural method of increase, is to be preferred to artificial swarming, is not sound, since we do not keep bees for themselves, that they may follow their own instincts, but for the sake of the profit, that we may gain from them as much honey and wax as is possible. The farmer does not leave his horses and cattle to their instincts. With one part he delays the satisfaction of the reproductive instinct that it may express itself more perfectly; another he robs of reproductive capacity altogether, that it may be more docile and willing for draught, or more fitted for feeding; and only a fool would conclude that this is not appropriate because not in accordance with nature. Just as foolish is the objection which has often been raised against a driven swarm as being an artificial division. The most short-sighted must see that if at the right time the rearing of several young queens is brought about, then more eggs can be laid, and more workers reared, and with continuous pasture more must be brought in. It is evident that even in districts with only a short season of pasture, where the swarm method cannot be followed with advantage, there must every year be some young, vigorous queens raised and new comb built, not for the increase of the apiary, but that it may be maintained complete and efficient. But if it is wished to increase the number of the stocks, this purpose can only be quickly and certainly attained by artificial swarms or division. As there are different ways of multiplying plants and improving fruit trees, as by inserting a bud of a better strain, by grafting, or by bringing a graft into union with a piece of root, so there are different ways of founding new bee-colonies and making fresh stocks. What belongs generally to the founding of a separate colony follows from what has been said in the theoretical part of the economy and reproduction of bees. There is required a fertile queen, or the possibility of one being raised and attaining fertility. Young brood, eggs, or small larvæ yet unsealed, or an occupied queen-cell from another hive, are sufficient or will be required for the purpose.

The fertilising takes place, as is well known, through the drones high in the air. Since the queen and the drones do not make inquiry whether they belong to the same hive or to different stocks,* there is no need to trouble at all about the drones. Care, however, has to be taken that the young queen, on her excursions, does meet drones in the air, although at the time then present there may be none to be found. The queen begins these excursions some days after hatching out, and repeats them for three weeks, till fertilisation has taken place. The time of the year must be such that fine warm days can be reckoned on with 70 to 80 degrees Fahr. in the shade.

The queen-bee alone, however, is not capable of founding a colony as a queen-wasp or humble bee, but she must continually be surrounded by a certain number of workers. These are therefore essentially requisite for an artificial swarm. Their number need not be very large, especially for the early swarms to be made in April, and more especially if a queen-bee has to be reared and fertilised. The stock which is small at the beginning can be strengthened with brood and bees later on, when larger quantities of bees are present in the other hives, and when the queen of the swarm has become fertile, and brood deposit and building can go on properly.

But how are the necessary bees gained for the artificial swarms?

It is very easy, especially in hives with moveable combs, to bring as many bees as we like either directly into the new hive or to collect them in a transport-hive and out of this to put them into the new hive, in the same way as a swarm is made to move out of the vessel in which it was caught into its new dwelling. In summer there will, by this time, be many bees to be found on the door. These are shaken or brushed off, or let to run in of themselves. The brushing off with a feather, not with an entire goose-wing, is used only when the article occupied by the bees cannot well be shaken, and cannot be thrown open so

* This is an assertion which experience tends to contradict, for there is little doubt in our mind but that the queen is exceedingly particular in her choice of a mate, and is apparently very averse to mating with one of her own kin. With twenty Ligurian stocks each containing plenty of drones, and one stock of other bees also having drones, the probability is that of twenty young Ligurian queens nineteen would mate with drones of the non-Ligurian stock—at least, it has always been so in our experience, tending to the belief that 'natural selection' is more than a myth even with bees.—C. N. A.

that the bees fall down from it, because the brushing off irritates them and causes many to fly. The brushing off of bees from the door may be quite avoided, if it is put in and one or two combs cleared of bees leaned against it. The bees run immediately on to them, and they can be shaken off from them again quite conveniently. In order to facilitate the taking away of bees from the strong stocks, which can afford in May and June moderate swarms, every week, for nuclei, old combs—empty, and as firm as possible, not liable to be broken by the movement required—are put by the door, after removing any honey-combs that may have been previously there. The attachments, or ties, by which the bees fasten them to the walls are always to be cut away. They can then, next time, be quickly taken out without using the knife, and will have on them the more bees, as they will not have had time to move off. If the combs in the brood-room are already too heavy for shaking, perhaps those in the honey-room will be still light and convenient for the purpose. If sufficient bees are not already to be found on them, a proper quantity can be previously driven there by smoke or drumming. If there are no combs at the top the bees can be scooped up with a box, a ladle, or a tin-pan, just as those clustering in front can be taken off, always applying the vessel below and working it upwards. The bees, however, must be continually quieted by smoke, as the ones clustering in front are especially irritable and ready to sting. The bees already shaken in must, from time to time, be smoked, especially before fresh ones are shaken in, so that the former may be excited to a hum, inducing the others the more willingly to run in. If this is done from time to time, and fresh ones continually shaken in, the bees will remain quiet without a queen, although the operation may last for an hour. They will not think of biting one another because of the confusion into which they get with the shaking and sudden falling down, although they may have been taken from the most different stocks. If fighting should occur in particular instances let them be jolted, with the vessel containing them, forcibly on the ground, so that the bees fall down and get into fresh confusion, or let them be sprinkled with thinned honey, so that they may lick one another and so become reconciled. In order that they may be the better able to cluster, one or more firm combs are inclined obliquely in the transport hive, so that the bees may betake themselves behind them, and with which they can most conveniently be put again into the new hive. From other kinds of hives bees may always be obtained for artificial swarms by stupifying them either with puff-ball, smoke, damp gunpowder, or touchwood soaked in saltpetre, and when they have fallen down on a sheet of paper laid underneath, they may be put wherever we like.

In other operations, as, for instance, when taking honey away, bees for artificial swarms, or for the strengthening of weak stocks, may be obtained by brushing off the bees into a Transport hive standing by. This is advisable even if the bees are finally to be returned to their own hive, because with the brushing off the bees into their hive the others are much irritated, and the ones falling on the ground may be chilled, whereas in the Transport hive they keep one another warm. With the Magazine-hive, which is here and there still in use, a super or eke may be given after this itself, as well as its comb, has been smeared with honey, and when a good many bees have gone in it may, by preference, in the evening be placed on or under the hive, or be only put close to the hive to which it is wished to assign them. During the night the bees will certainly have moved in if brood is there, and no other hive is in the neighbourhood into which they could move.

The question occurs, will the bees remain with artificial swarms formed in this way? They will remain only till the next morning, at most till their next outward flight, and will then as certainly alight on their previous hive as the pigeon which one neighbour buys from another will, without doubt, fly back at its first outward flight to its old cote. The artificial swarm so formed must be put on a stand that is at a distance of two thousand paces or more, so that the bees do not get to see their previous position. They return then to the place from which they have flown, and so form a new colony. Failing an apiary at a distance, two bee-keepers, whose apiaries are a couple of miles or more apart, can either exchange artificial swarms, or only the bees, required for them. But whoever will carry on apiculture in a larger way must try to establish two apiaries at a distance from one another. This affords advantage not only in making artificial swarms but in other ways which will be spoken of further on.

A further question arises, is it not possible to make the artificial swarm in such a way that it may remain in the same apiary as well as a natural swarm? This is possible; and may best be done with the Twin-stock, if an empty hive is situated alongside of the full one that is to be divided. It is only necessary, then, towards the swarming time to put the two neighbouring hives — or if it is an inseparable double-hive the two compartments—into communication, so that the bees pass over and become accustomed to fly in and out of the entrances situated at opposite ends. In order to attain to this more quickly the Twin-pair, or Double-hive, is turned round, and so the two compartments and entrances exchange positions. The old bees will now fly in through the new entrance that is occupying the old position, but the young ones will keep more to the old entrance for their first outward flight and play in front of the hive, and in a short time the flight will be pretty equal on both sides. If, now, the communication between the two compartments is closed, and care is taken that the queenless part receives a queen, or has the possibility of rearing one, then there has been made, with the least trouble, an artificial swarm which cannot turn out a failure. There is no necessity immediately to shut off, hermetically, the opening between the compartments. There may for a time be put there a perforated sliding-board or a piece of wire gauze, so that the bees hearing the peaceful hum in the neighbouring hive will remain quiet, and will miss the queen but little, or not at all. But they will, notwithstanding that, soon proceed to the forming of queen-cells, like a stock does whose queen is caged and cannot move freely among the bees. The only accident that can happen to the artificial swarm is, that the young queen may be lost on her wedding flight, which cannot, however, easily occur with the great distance of the different entrances, and can be easily made good again by the introduction of another queen, a queen-cell, or a fresh brood-comb.

If both neighbouring compartments are already occupied by old strong stocks, and one or both of them are capable of division, artificial swarms can be made from them in a similar way in the compartments situated over them, if they do not already occupy the topmost position in the pile. But here a second entrance is placed by the side-door, while the door is somewhat raised, or the corner of the door nearest to the entrance situated over it is cut away. But not to spoil a good door another piece of board can be temporarily fitted to it. If, besides this, the old entrance is now narrowed, the young bees, attracted by the strong light breaking in, will play before the new entrance more than before the old one, and when the opening is shut up again will go in at the entrance situated over it, and associate with the artificial swarm that has been put in there. There is very much gained if the artificial swarm has or receives only several hundred bees, which, either because brought from another apiary, or because already used to their flight, are to be depended on, and at the beginning can fetch water, which, for the preparation of brood-food, cannot be done without. From the brood which is soon given to it, or put in by way of addition, the stock becomes stronger every day, and by the continued introduction of brood it can, as if by steam, be made as strong as you like. It is plain enough that when it simply requires strengthening, brood as old as possible should be chosen, but no more combs should be introduced at once than the bees can really cover.

Another way of making artificial swarms in the same apiary, but applicable only to single hives that can be easily moved, is this: the former position is divided between the old stock and the artificial swarm, putting the one to the right of it and the other to the left, so that the bees are divided, half flying to the old stock and half to the young one. Which of the two keeps the old queen does not matter in stocks with moveable comb, because a complete set of combs with brood, honey and empty cells can be put together for the queenless one, so that it is situated like an old stock robbed of its queen. The matter stands differently in hives in which a large series of combs cannot be put together for the bees, as in ordinary straw or log-hives, where at most a brood-comb is with difficulty secured. Here the queen must be given to the part which ought first to form a set of combs, because without the queen little is built, or what is built is mostly drone-comb, which is generally very disadvantageous. On the contrary, in the old stock, which is already mostly or entirely furnished with comb, and is busied with filling the combs already made, the temporary loss of the queen has not only no disadvantageous consequences, but has even the double advantage of affording more cells for the storing of honey, and of affording in fourteen days perhaps one or more after-swarms, from which we can get capital young stocks, whether they come off of themselves or are driven again in the same way the old queen with the first artificial swarm was driven.

Driving.

The purpose of driving is to get the queen, with a part of the bees, out of a hive from which they cannot otherwise be taken because the comb is immoveable, or so full of honey that a separation of the combs would be difficult or dangerous. But how is the driving, or drumming, as it is called, to be done? It deserves to be accurately described because it is an operation which is often used for other purposes, as, for instance, when it is wished to clear the bees from a set of combs in order to preserve them for future use, or to break them up, or to change a defective queen for a good one, or an ordinary black queen for a yellow Italian, or to get bees for artificial swarms, or for strengthening. In smoking, rapping, and drumming, it is especially the heat which incites the bees to move from their comb, as we see this shown in the clustering in the front of the hive. The afternoon of a sultry day, as free as possible from wind, should be chosen for driving. In a cool temperature the bees stick too firmly and immoveably to their nest and are by no means disposed to come out, at least it is so with the queen, on whose account the driving is principally or alone undertaken.

If the bees with their queen are to be driven out of the comb there must either be an empty space in the hive, or such must be made or applied, into which the bees can go and cluster. The empty space can only be above or below, but during the operation it must be at the top, because the bees are always driven upwards and go there most willingly, clustering there like a swarm. If empty space not yet built in is at the bottom, as in the ordinary Ständer log-hive, or if it can only be applied at the bottom, as in the straw hive or other similar hives that are only open there; the hive must be reversed or set upon its head. In this way the bees go out most easily and quickly, because in this direction the passages between the combs are wide, but at the top are more contracted through the honey-combs being thick. The bees may be got out in this direction by simply smoking them strongly. In different hives, therefore, different treatment will have to be adapted.

In Ständer log-hives, without bar arrangement, driving is only possible when a space about ten inches high is yet empty, or can be made below by cutting off the ends of the combs, or by taking out a board that has been previously inserted horizontally. The order of proceeding is to open the hive, smoke the bees, clean the floor, sprinkle everything with water so that the comb does not get covered with dust, and reverse the hive, placing it on its head. The reversal must be so managed that the combs never come to lie flat but stand on their edges. If a log-hive (or any empty hive) is previously rolled to the place, and the hive is tilted over it, the reversal and setting up again of heavy log-hives is much facilitated. In calm, warm weather the hive can remain open during the driving, so that the queen can be easily seen as the bees go up. In a cooler temperature with wind it is better, at least for a time, to close the hive, because puffs of wind, or even a breath, directed against the bees, cause them to turn back. At first the bees are strongly smoked and driven away into their combs ; this causes them to become heated and soon to move up the more tumultuously, while knocks are made on the outside with a hammer or a little axe,* blow upon blow, in continuous succession. With a strong, firm log-hive the blows must be stronger, but with a weak one, weaker. At the beginning the bees attack the honey and gorge themselves, and delay their march upwards. To save time the hive should be opened and the bees smoked a little beforehand, then giving a few blows to the hive while yet standing will make the bees get ready for travel and provide themselves with the necessary food.

The stronger the stock is, the larger will be the stream flowing upwards and the quicker will it be driven, although the catching of the queen is often somewhat more difficult. In that case, the principal mass of the bees must generally be scooped

---

* The hives, it should be remembered, are log-hives, formed of hollow trees or limbs of trees. A hammer or an axe applied to skeps would smash the combs.—C. N. A.

out into a Transport hive, or on to a sieve; the bees must be afresh strongly smoked and drummed until success attends the effort to scoop out the queen with them, or to see her running up the wall of the hive, when she can be caught and caged. As she so easily conceals herself among the bees it is best to brush her down with a feather into a box or tin pan held underneath, or to scoop her in with the bees surrounding her. For this a sharp and practised eye is certainly necessary, and whoever has not this must decide more from the behaviour of the bees scooped out whether the queen is among them or not. He will, perhaps, be obliged to try to get out the queen more certainly by putting empty combs at the top to lean over the set of combs, and when they are covered by bees taking them out and shaking them. This is a successful way of catching a queen that has become weak or lame, and that can no longer climb up the wall of the hive. Many queens are shy of the light and only come up when the hive is closed, or when they are attacked vigorously from below with smoke. A smoker machine, with a long tube at the nozzle, by which the smoke can be brought deeply into the hive, in such case affords good service. The bees and the queen may be moved out too by simply blowing upon them with the mouth, or by the draught caused by a small pair of bellows.

There are individual queens which will not forsake the comb at all, and are obstinate as sheep which will rather be burnt to death than run out of the stall. It does not do to repeat the operation if it has once failed, because it might be quite possible that the queen had accidentally gone right away, although young brood and even eggs are to be seen. If it is not possible to quickly replace her by a fertile queen from another hive, and there is no desire to make an artificial swarm from brood,* which in log-hives has been already noted as a disadvantageous kind of artificial swarm, there is nothing to be done but to let the driven bees go back into their hive after it has been put in its former position. The most that could be done after the chief bulk of the bees have flown off is to let the young bees, which as yet are

* That is to say, from a queenless hive having brood, part of which would be given to the swarm, and from which the bees would have to rear a queen.—C. N. A.

partially incapable of flight, be used for strengthening a weak stock. But if we had, along with log-hives, box-hives with moveable comb, all that would be needed would be to put together quickly a set of brood-combs and honey-combs and let the bees run in. The young bees will at any rate remain as a driven swarm, and of the old ones a part could be kept back by confinement, or putting in the dark, till the stock had made increase from the brood put in. It would be the best, certainly, to put the artificial swarm on a stand at a distance. There would then be a protection against the danger of bees being stung to death through going back to the parent stock.

If the attempt to get the queen out of the log-hive has been successful — which it will be, generally, with the treatment described,— or if the bees scooped out announce the presence of the queen among them by their continued quiet behaviour, then the driven swarm is hived like an ordinary swarm already on the sieve. The driven swarm is similar to an ordinary swarm with this exception, that the latter has come away from its previous hive naturally and can be placed wherever you like, but the bees of the artificial swarm, not knowing what has occurred to them, keep only to their old place and fly back to it. If the driven swarm could be left to swarm out from the sieve, transport hive, or skep, it would then be changed into a natural swarm, and could be put wherever wished. And it is not very difficult to accomplish this. If many a driven swarm forsakes the proper hive allotted to it as a swarm, how much more will it be inclined to forsake the vessel in which it finds itself as its queen, clumsy at the beginning, becomes lighter. If the bees are put out on the following day in the hot sunbeams, and sprinkled with liquid honey, there would not be any occasion to wait long for their swarming. If a second queen, or a second driven swarm. were brought so near that both queens, separated only by a wire lattice, became aware of each other's presence, or if a bad smell was caused by the means previously mentioned, the swarming would follow so much the more promptly and certainly.

But all this is not necessary if the locality allows of such displacement of the hives as is required. The driven swarm is put into a new hive, the former position is divided between the old stock and the young one, taking care that the bees are divided as equally as possible. The weaker half may be helped by moving it more toward the old place.*

If it is wished to obtain a specially strong swarm because of the new hive being very large, the young stock should have exactly the old position given to it. If the old stock has very many young bees just come out, or at the point of coming out, which must all, if possible, be left with it, or if they had been driven must be returned, then a quite new position can be allotted to it. Only care must be taken that it is not weakened too much, otherwise there is a danger not only of the unsealed brood starving and dying, but that no queen-cells may be formed, thus leaving the stock without queen. In order to avoid this it will be well to put to it again at night a couple of boxes full of bees† out of the driven swarm, if they are not to be obtained from another apiary, which would certainly be better because they would all remain with it. Let the stock be supplied with thinned honey, so that the weakened bees are in a position to prepare the necessary food for the brood till the stock gradually recovers its strength.

There is, though, no necessity to strengthen the old stock from which the driven swarm has been taken if it is put entirely, or partly, in the place of another stock; there fly to it of themselves, such a quantity of bees that it not only gathers much honey because more cells are continually being emptied, but a large number of queen-cells will be formed, and in fourteen days it will with tolerable certainty throw off after-swarms. There is no need to fear that the driven or swarmed stock will be much weakened if an entirely new position is allotted to it, for it does not lose a single young bee, and of the other bees many remain with it, because they, like the quite young bees, play in front of the hive when they fly out, and by so doing note the fresh position. The displacement of hives otherwise is not advisable, because it endangers the life of the queens. But here no queen can die,

---

* In practice we find it safest (on a fine day when bees are working vigorously, and 'fit' for swarming) to put the queen, with a 'handful' of bees into a new hive properly prepared, and to set the latter in the place of the old stock, when the bees returning from the fields will form the swarm.—C. N. A.

† This is rather vague as to quantity; but in a general way, half to three-quarters of a pound weight of bees will be ample.—C. N. A.

because there is none present. The additional bees that have flown in assist very much in replacing the loss of the queen, and in forming queen-cells. Such a treatment is, therefore, strongly advised for log-hives as well as for straw-skeps, with which it is still more easily accomplished.

As regards the driving of straw skeps, the treatment does not need to be more particularly described, since it remains essentially the same, the only changes being dependent on the construction of the hive. The skep to be driven is, of course, reversed, strongly smoked, an empty skep put on the top, the swarm is driven up by knocks with a little stick or hammer-handle, and when it has clustered at the top is taken off with the skep. If smoke is applied to the bees for a time through the bunghole below, or through a trocar-like tube thrust through the straw wall at the side, they will then go up more quickly and certainly. There is no need to put a black cloth or board beneath them in order to learn from the eggs deposited on it whether the queen is amongst them or not. That is soon enough known by the behaviour of the bees. If they remain quiet and continue to cluster together they have the queen with them, and the driving has succeeded. Their further treatment has been spoken of at some length above.

In Lager-hives, both log as well as straw, the driving of a swarm is only possible when the combs do not run across, but run by accident parallel with the length of the hive. In log-hives, the space as yet not built up, without which no driving would be possible for them, is put towards the top; in straw-hives, after they have been set up, a skep is put on the top—for the rest they are treated like Ständers.

With combs running across it would be difficult to get the queen, although bees might be got for an artificial swarm if, without setting the hive on end, which would be a dangerous thing to do, the bees were set moving toward the empty space by smoking and knocking, and then scooped out or taken out with combs that had previously been put in there. Out of Ständer log-hives it is easy to get bees alone in quantity without tilting the hive if the bees are driven from below to the top by smoke, drumming being kept up at the same time. This obliges them to go outside the entrance, where they cluster in front, and can

Driving. 177

then be taken off. Another way is to drive them from the top to the bottom, from which they are scooped out, or they may be taken out on combs that have been put in for them to cluster on.

If driven swarms are made from box-hives with moveable comb the treatment will be different. The most certain way of attaining the end is to take out comb by comb till the queen is found, when as many bees are added to her as is thought well. When the combs are very heavy the end may be more easily gained by smoking the bees up into the honey-room after the cover-boards have been taken off. The floor-board must only be gently tapped if a manifold hive is being operated on so as not to disturb the other stocks.* The bees and the queen must go upwards if there is a small space below still not built up, or if such space has been reserved by a board inserted horizontally, permitting smoke to ascend into all the passages. The whole swarm can be taken out at once by inserting previously, upside down, a box made of thin board, something like a cigar-box, in which the bees would cluster. The bottom of the box, which now will form the cover, must be furnished with comb beginnings, so that the bees may not break off and fall down. Then the bees have a surer hold, and the queen is more willing to remain there.

In Lager-hives provided with two moveable doors the capture of the queen is much more convenient than in compartments accessible only from one side. For the queen may have withdrawn herself quite to the background when the combs are taken out one by one, so that the combs have to be taken out right to the end. But if she is gradually driven backwards, in hives with two doors, by smoke and knocking, and the hive is suddenly opened from the far end, she will with tolerable certainty be surprised on one of the first combs. This should be loosened previously, so that there may be no loss of time.† It is well

---

* This is practically strong testimony in favour of using moveable frames, particularly in 'Twin-stocks,' so that drumming may not be at all necessary, and the combs and bees removeable without disturbance.— C. N. A.

† We have not yet discovered any advantage in either of the Author's hives over our ordinary English frame-hives save that obtained by using bars only for the brood-nest, which causes the bees to build against the front and back of the hive and prevents loss of heat by circulation round the ends of the combs. Nevertheless, we do not approve of the simple bar for this purpose. But when it comes to the question of catching a

known that the queen, as a rule, is to be found on a brood-comb, and, therefore, these must be particularly looked at. We may, therefore, fitly remark here that by inserting a comb with brood, especially young brood, we may easily get to see a young queen in a stock, otherwise without brood; or easily catch a queen that is unfruitful, or otherwise defective. Where the brood is all life and activity is concentrated, here the greatest warmth is produced and maintained, and at the next investigation the queen will be certainly met with here, and can easily be taken if she has not been frightened away by the disturbance. If it is a fertile worker that is sought for, not to be distinguished from the other bees, then all the bees which are found on the brood-comb must be done away with. Of course there must not be any other brood, nor even any drone-brood, to be found elsewhere in the hive.

In box-hives there is often no need to make any special search for the queen. She is seen often enough in casually looking over the bees, or in taking bees away. She can then be caught, caged, and an artificial swarm be made either immediately or as opportunity may occur.

But how are the artificial swarms to be treated further on? Simple box-hives, like the Twin-stocks, can be treated as straw skeps, since they are easily displaced. The artificial swarm can be put into a new Twin-stock, and inserted in the place of the parent stock, and to this can be assigned the place of another strong stock. From a manifold hive the swarm is taken to a stand at a distance, and the parent stock is left undisturbed.

What can be done by the help of a Fertile Queen.

With the help of a fertile queen, which may be often got opportunely from a pauper-swarm that has flown to the apiary, or from a half-starved stock, made-up artificial swarms may be formed very early, at any rate in April. The method is to add bees as they may be got from the different stocks by taking away, brushing, or shaking, and the swarm made in this way is removed to a distant apiary into another sphere of flight.

queen, and the Author confesses to the necessity for so much trouble and precaution, while with a modern frame-hive one can catch a queen without difficulty in a few minutes, the advantages of the British system are strikingly manifest.—C. N. A.

It does not matter whether the swarm is put immediately into its proper hive and sent away in it, or whether it is put into a transport-hive and then moved into a hive already there. The latter would probably be the simpler. Two or more swarms of bees can be sent together in one large transport-hive to the other stand, though it is advisable only to have one queen in the same box; the other can be reserved apart, or be sent beforehand to the other apiary if there are already other stocks there.

Although the bees in the mutual confusion about the queen generally do nothing if she, though a stranger, is mature and fertile, yet it is well to keep her caged for one, two, or three days and longer, if the bees do not quietly cluster about her cage but surround it, making a hissing noise, trying to get in to the queen. If it is not possible to go one's self on the second or third day to set the queen at liberty, the bees will do it themselves if a thin sheet of wax has been stuck on the opening of the queen-cage. The queen-cage must be put so that the queen remains encircled by the bees, although these cluster together closely in a cool night. In a cold* season it can be laid at the top, over the comb, and under a covering-board. To put in a brood-comb soon to the artificial swarm is very good in all circumstances, even if it should have to be taken from another artificial swarm made earlier the same year. The swarm is thereby attached to its hive and excited to greater activity, and the queen is then less liable to danger.

In every old stock that is purchased, or otherwise displaced, a good deal depends upon the condition of the weather when the bees make their first outward flight. At one time a great many may be lost, and at another, perhaps, not a single bee. If it is cold, windy, and changeable, keep the bees as much as possible at rest, and restrain them from flight by blowing smoke in at the entrance, sprinkling the entrance, and the bees that would come out, with cold water; or, lastly, by entirely confining them. Their confinement may be arranged by fixing to the entrance a perforated sliding-board, a little sieve, or a wire-gauze bee-hat. But if a favourable time has set in, or appears to be just at hand,

* In a cold season we should infinitely prefer to place the queen-cage amongst the brood and clustering bees—the previous sentence giving the reason. Bees will not always cluster round, or attend to an alien queen, if the weather necessitates their clustering for self-preservation, or for the protection of their brood.—C. N. A.

let the bees be stimulated to play in front of the hive by thinned food, either sugar-water put in to them or syringed through the entrance, so that they may learn the new flight and begin to carry in. If the bees have been first in this way brought into activity there will be no need to be in any anxiety about them later on, even if continuous unfavourable weather sets in, although in that case they must be again assisted with food.

If the queen that is to be used for an artificial swarm is one produced the same year, though already fertilised, and bees are added to her which have had an older queen, great care will have to be exercised, and she must be caged longer, so that she may not be stung to death. To a queen so young, and on that account so much more valuable, it is best to add bees which have either a young queen, or no queen at all, but only queen-cells; or if bees can only be taken from an old queen it is best to keep them without queen, perhaps, for a day before, or at least over night, providing them with the requisite food, making them by that means disposed to receive a younger queen. To stupify the bees by one of the means already mentioned can do no harm, but is only advisable as a last resource if the bees are taken from a hive with an old queen.

How to get Fertile Queens.

But the question may be asked, How are fertile queens to be obtained for use in artificial swarms, or for the improvement of queenless stocks? As already mentioned, a good queen may sometimes be obtained by chance when a stock has come off from one's own, or somebody else's, apiary and dispersed itself, or when a stock, half-starved, has been pillaged by robbers. In the late summer and autumn such would be of little value, and could only be used at most for a stock which either has no queen, or one over three years old. No prudent man will form a separate colony at this time, even if he could put together for it a complete set of combs.* In the spring, on the contrary, a fertile queen is

---

* This seems to imply that late swarms—'condemned bees'—are not worth keeping, or that it is imprudent to attempt to form them into stocks, but the experience of hundreds of English bee-keepers will disprove this, now that comb-foundation is cheap and so readily obtainable.—C. N. A.

worth half a swarm, especially to any one who possesses some quite strong stocks. Although a couple of thousand bees be taken from these for an artificial swarm it would do them no injury, and might, indeed, have the advantage of hindering them from building so much drone-comb and rearing so many drones. But the earlier an artificial swarm is made, the more complete must be the set of full and empty combs put together for it.

But in order to have surplus fertile queens in spring with certainty, they must be kept from the autumn previous, through the winter. But this is rather troublesome, uncertain, and somewhat costly, because a small stock of about two thousand bees, by which a queen must be continually surrounded, always consume a couple of pounds of honey through the winter, and may swarm out and go off on one of the first fine spring days. It is therefore only worth while to attempt to winter quite superior queens. This is done in the honey-room of a box or cylinder hive, into which some heat from the brood-nest continually streams up. The heat in the background of a fourfold hive, occupied by four strong stocks, is especially high. Here the wintering will, at any rate, succeed if the room is shut off from the one below, so that the bees cannot get together. A small set of full and empty combs is put together, and the little stock is allowed to take possession. But it is not to be supposed that the bees can be completely confined here. This might be done for a couple of days with a queen surrounded by about a hundred bees, although it would be even then more convenient to put the queen into a little box and to insert this with her. A small stock more numerous would soon worry itself to death, or get out of health with dysentery. It must by all means be packed closely to keep the cold out, but an opening must be left for free exit and flight, that the bees on a mild day may play and enjoy a cleansing flight. It is evident that the bees have either to be from an apiary at a distance, if not they would fly back to their previous position, or have had their previous entrance in or about the same place where the new one is situated. This would be so if they had lived in the Twin-stock situated above the new hive, and now fly in and out through an opening made in the crown-board close under the former entrance. The slanting crown-board must then be for a time removed.

But success would be uncertain, and the little swarm might in spring easily be out and away before an opportunity might be found of using the queen. It is better, then, to produce young fertile queens in the spring as soon as possible. Undoubtedly they cannot be obtained so soon then, perhaps not before the beginning of May, but the trouble and expense of wintering will be avoided. Preparation can be made when the bees begin to deposit drone-brood, which is usually the case when they can carry pollen in abundantly, mostly at the time of the willow bloom. Whoever keeps bees only in log-hives or skeps must provide for himself at least some boxes or small hives with moveable comb. Whoever has such boxes already can use every compartment for the purpose, especially every Twin-stock, since he can diminish the space nearly one-half by pushing forward the moveable door up to the entrance, and the taking out and putting in of the combs can be made very convenient.

First of all, an artificial swarm is made so that it may form queen-cells, and, if possible, several of them, since it only requires one queen for itself, and thus will prepare the way for the swarms to be made later on. Since it has to do this at a time of the year that is cool, it should not be made too weak. A moderately strong stock may be used for the purpose, if the queen be taken away from it and applied further to a swarm to be made in the way just described, or for the improvement of a queenless stock. To manage in this way is better than to make the swarm first that is to form the queen-cells. For such a swarm is naturally in the early season unsettled, the bees very easily get into the adjoining hives, or are lost in other ways, and at a cold time of the year may be seen lying about here and there chilled. In order to avoid that, the bees should be confined for from twenty-four to forty-eight hours, or should be put into a dark place till they are pacified and have made preparation for the formation of queen-cells. It hardly needs to be remarked that the bees of a swarm newly made should be put upon a distant stand if they have not been derived from a distant apiary, or must be put into half of, if not the entire position of the stock from which they have been taken. Bee-writers certainly come to the conclusion that the bees of an artificial swarm when they have formed queen-cells keep in any new place that may be allotted to them. But experience teaches otherwise. The bees mostly go back to their previous hive, and the artificial swarm would be nearly empty of bees, if it had not received so much brood just ready to come out that the loss was made up again. By delaying the spring comb-cutting a little, those who have log-hives can then get sufficient suitable brood-combs for artificial swarms, fastening them immediately on bars, so as to put together a set of combs part empty and part containing honey. The bees can be got by shaking the combs or taking them away with a box, since the bees generally hang down in abundance beneath the shortened combs.*

QUEEN-CELLS.

In order conveniently to get queen-cells for cutting out, such a brood-comb is selected for the artificial swarm, which, at least, with regard to one of the edges at the side, where the bees are most disposed to form queen-cells, has cells not crowded behind one another, but here and there containing small larvæ or eggs. Queen-cells will be found formed there after some days with tolerable certainty. On the second or third day it is well to look, and if several royal cells are found begun close together some should be destroyed, as, for instance, the middle one of three, so as to make their cutting out later on more convenient. Later in the summer, when the bees come lower down, they are specially disposed to form queen-cells when a brood-comb is so far shortened as to be cut up to cells with small larvæ. At this cut the bees often form one cell after another, so that they stand there like organ-pipes. In this position their formation is convenient to the bees, and the greatest activity always prevails on a shortened comb. The piece of brood-comb cut away will, of course, contain eggs and small larvæ on its cut surface. This can be quickly fastened to a bar, of course in a reversed position, and inserted in the same or another artificial swarm; and later on several, perhaps the most, queen-cells will be found formed on it. Repeated feeding at first, even if stores are abundant, keeps the bees more lively, and occasions the formation of more queen-cells and a more

* We cannot imagine a cruder method of treatment throughout.—C. N. A.

abundant nourishment of the royal larvæ. Most of the queen-cells will, about the twelfth day after the making of the artificial swarm, or the removal of the old queen, be so far ripe that the young queens come out by biting with their sharp mandibles through the cover, which then, as a token of the successful coming-out, falls down on the floor of the hive as a little cap, if it does not perchance remain hanging on the forsaken cell, or does not spring back into its previous position, so that the cell may deceive us and still look as if it were full. From the size of the lid, or of the opening in the queen-cell, the size of the queen herself can be inferred, for they generally entirely correspond. Meanwhile, it is possible that a queen may come out in ten days if a larva already six days old had been selected for the purpose. The queen then immediately begins to destroy the other cells.

A second and third artificial swarm must now in the same way be begun from the beginning, and wait about twelve days for the coming out of the queen. Trouble and time may be saved to the later artificial swarms, as well as to stocks otherwise queenless, by giving them one of the surplus queen-cells. By this they gain a queen some ten days earlier, which is a period of some significance in the development of a stock of bees. This introduction of queen-cells, sealed and just ripe for the queen to come out, is of very great advantage. In this convenience lies the chief advantage of artificial division over natural swarming. The matter, therefore, deserves to be more particularly spoken of.

First of all, the day should be particularly noted on which the artificial swarm is made, or the queen is taken away from a hive, and should in all cases be written on the door of the hive with chalk. If it should have been forgotten, or never have been known, because of the queen having by chance gone off of herself, the time for the cutting out of the cells will have about arrived, when all the brood-cells on the comb or in the hive are sealed. Because many of them will have only contained an egg at the departure of the queen, a week at least must meanwhile have elapsed, and it is better the cells should be cut out one or two days earlier than one or two late, when they have been already bitten open. There is no need to cut all out at the same time but only the oldest, and the younger can be left as much longer as they have been later sealed. If the artificial swarm is looked into on the third or fourth day it will be easy to conclude from the circumstances in what order they attain maturity and must be cut out, which can be duly noted. Some cells will be already sealed, some unsealed; and of these some will have a large larva, others a middle-sized one, and another a quite small one. If the comb is held against the sun, and the queen is seen through the cell already dark, or even to move her limbs, she will come out in a few hours, and it is time to cut out the cell so as not to risk the destruction of the others.

The cutting out must be done with care, so that the cells may not be at all injured. Meanwhile, a trivial injury does no harm, and the cell itself may be opened with the point of a penknife to see whether it is good and not rotten, and how far the queen is developed; but the opening must be closed again, and thick wax be applied to the place with the heated point of the knife. The cell can be so inserted that the injured place is not accessible to the bees, by putting it either by the wall of the hive or sticking fresh white wax in considerable quantity in front. The cells must be inserted where the greatest heat prevails, where brood is either present or would be; either on the side edge of a thickly covered comb, or in the middle of it in an opening already there, or that may be made. A depression may be made with the little finger, and the cell put in it as it might be formed there by the bees themselves, and it is made so fast with softened wax that it cannot fall down.

In the heat of summer queen-cells, especially when just ripe for the queen to come out, in which she is perhaps already moving, can be put at the top under a cover-board, either lying on a comb-bar or between two combs over the brood-nest, so that the point where the queen must come out remains free. In skeps the cell could always be inserted through the open bunghole.

The introduction is easiest to accomplish when a surplus queen-cell hangs from a separate comb. The cell is put in with the comb and the bees that are on it. In other cases it is best to wait till the bees have realised their queenless condition and have themselves begun to build queen-cells. If this is not done the cell may be easily destroyed, and even the young queen in coming out may be stung to death. If the entire comb with a party of bees are given together the introduction may be ventured upon earlier, because the bees introduced continue to care for the cell, and the others soon follow their example. When all the queen-cells are on one comb they can be introduced elsewhere, and a cell can be cut out and put to the artificial swarm, since there is less reason to fear that this will do an injury to its own cell. If one has time to look often, the coming out of the young queen can be awaited and she can be left with the artificial swarm, and the comb with the other cells be introduced elsewhere. In the same way the young queen herself can be caught and otherwise used. If the queen-cells, as often happens, are built so close that they cannot be separated without considerable injury, there is nothing else to be done but to await the coming out of one of them, and then quickly to remove either the other cells or the queen.

When a stock bites open a good, uninjured queen-cell that has been introduced, it does not do to give it a second, because that usually has the same fate; but a young queen just come out is given to it from another artificial swarm to which a queen-cell is given, or the former is left to itself to rear a queen from brood. Care must be taken not to be deceived. Sometimes the cell appears to have been destroyed by the bees, as it has a hole in it at the side, and yet the young queen has come out either at the side opening which the bees had made, or at the opening made by herself at the top which had closed again, and been built up still more securely by the bees. If the royal nymph, therefore, is not found in or in front of the hive, there must be more particular investigation, so that queens may not be sacrificed unnecessarily. It is well to examine the young queen to see that she has uninjured limbs—especially sound wings—and that she is not too small and weak, because such queens generally attain to fertility later than the strong ones. Many queens come out deformed and crippled from short cells made close on the floor of the hive. These should be removed immediately, but should not be thrown in front of the apiary. They might creep into one of the hives and cause damage there, for instance, by destroying queen-cells, or causing a commotion in which the good queen of the stock might be sacrificed.

If all the surplus queen-cells that are cut out cannot be

immediately used, they can be kept for one day, or for several days, by putting them into a little box and keeping it in a warm place, for instance, under the cover-boards of a strong stock, in a hatching machine, or in the nest of a sitting hen. The queen-cells can further be secured in little boxes with a small honey-comb, and a party of bees shaken to them from the artificial swarm, and be put into the honey-room of a hive to await the coming out of the queen. A royal cell must not be let fall or be violently shaken, for the tender nymph might be killed or maimed. In warm weather one might be gently carried for two or three miles, but to send queen-cells to a distance by post, as is often done with perfect queens, without danger, is not to be thought of.

When bees have been prepared by longer queenlessness to receive a queen, and one is introduced that is still young and tender, she is not caged, but allowed to run at large among the bees where they are most crowded, perhaps on the brood-comb, her natural abode; because in other places where she did not belong she would be more liable to be seized and stung to death. She can be smeared a little with honey beforehand, or as soon as she is put on the comb, so that the bees in licking her may become more friendly and not be enraged by the odour attaching to her, perhaps derived from contact with one's hands. At the same time the bees might be stupified a little. A young queen can be much sooner added after another unfertilised one that has perhaps got lost on her wedding flight, than after an old fertilised one, whose loss the bees only get over and forget with difficulty. The introduction may be made much sooner, at any rate on the following day, if the bees by being driven into an empty skep or box are brought into greater confusion. If the young queen has been allowed to run among the somewhat stupified bees, they can be left till the next day with the requisite food in the vessel, so that they may become better acquainted with the queen. They are then taken to another apiary, if they have not already been brought from it and hived similarly to other swarms. Artificial swarms can thus be made with young surplus queens and bees that have been driven or otherwise collected.

An artificial swarm is not to be considered a success, nor is a parent stock that has given off a swarm voluntarily or by compulsion to be looked on as in proper order, until the young queen has been fertilised and has begun to lay worker-eggs.

There is much that may be done by compulsion and scheming, but fertilisation, which is dependent on weather and other accidents, must be awaited in patience. In the cool time of the year, when there are fewer drones, fertilisation may be delayed for two, three, or several weeks; but in the heat of summer it usually is effected within eight days. With increased heat and activity, therefore, in artificial swarms to which brood has been given, or in after-swarms that build vigorously, the queen attains to maturity and fertility sooner than in other cases. In spring, and again in autumn, when drones are only present in some hives, the most that can be done by art to promote fertilisation is to stimulate the stocks with queens ready for fertilisation and others with the most drones. This is done by giving them thinned food, or syringing it through the entrance at the warm time of the day, inciting the bees to play, and making it the more sure that queens and drones shall meet in the air. After two days, evidence of successful fertilisation may be seen in the queen, as in consequence of the development of the ovaries filled with egg-germs, her abdomen enlarges considerably, but the freshly laid eggs which may be found in the cells on the third day afford palpable proof. Now the queen has first attained to her full value and become a complete fertile mother, and since she does not fly out any more, excepting with the entire swarm, can be caught and be otherwise applied for the formation of an artificial swarm, if it is wished to use further the artificial swarm that reared her for the raising of new queens. If the swarm robbed of its queen is quickly assisted by a queen-cell cut out elsewhere, a fertile queen may again be found in it in about ten to fourteen days.

In this way a number of fertile queens can be obtained in the course of the summer from weak artificial swarms, and stocks otherwise weak, even with foul brood, if good queen-cells are introduced to them. Since in that case there would be no need to rob the other strong stocks of their queens, and the rearing of brood would suffer no disturbance or interruption, the more bees could gradually be withdrawn from them, perhaps as many as from four to six swarms, and the number of stocks could be very largely increased.

If there cannot be so many queens reared as swarms can be taken from the stocks, a queen must be given to the swarms made, at least at the beginning, and left till the bees have learned their way home and are properly settled. The queen can be caged, and in Lager hives put on the floor; in Ständers under the nest of the bees. Her presence has this effect that the bees, which otherwise would not remain quiet with simple brood, a queen-cell, or an unfertilised queen, but perhaps would become scattered and alight on other hives, keep together and accommodate themselves to their position. An old weak queen, otherwise valueless, can be used as such a decoy queen. After about two days she is no longer necessary to the artificial swarm and can undertake for another the same service of keeping together the dissatisfied. Drone brood contributes a good deal to the pacifying of a stock still queenless, because the bees feel instinctively the necessity of drones for the fertilisation of the young queen that has to be reared. They, therefore, take as much care of the cells of the drone-brood as of the royal cell. But they then form fewer queen-cells, because their attention is directed more to the drone-brood, and the queen-cells are often by mistake formed on drone-comb. It is not, therefore, well to introduce drone-brood if the forming of many queen-cells is desired.

It might be the proper place here, when speaking of the increase of stocks, to say something of the propagation of Italian bees. This race is worth propagating and extending, as it possesses distinct advantages over the ordinary grey or black bees. It is not only more handsome and gentle, but more industrious and capable of defending itself. Their culture, therefore, not only affords greater pleasure but produces larger profits. Even the hybrids or mongrels of this race are distinguished by extraordinary industry, so their introduction is not without value and profit, even if there may be no success in preserving them pure and genuine.

The Introduction of Italian Bees.

This is much facilitated since there is no need to send for an entire stock from a long distance—a thing that is always risky, as a single violent jerk may ruin all the comb, and a strong stock if kept confined for a long time suffers a good deal, and may even be suffocated by too great heat and want of air. To make an Italian stock it is quite enough to get a fertile queen of this race and introduce her to an ordinary stock or swarm from which its own queen has been removed. A queen, accompanied by only a small company of bees, can be sent by post to the most distant place in our quarter of the globe without risk; and the sending only costs a trifle. The Italian queen can be introduced at any time when a mild temperature allows of the opening of the hives, a period ranging from about March to October.

As regards the introduction, the conditions of success may be gathered from what has been previously said. The introduction succeeds best in stocks that have a young queen yet unfertilised, as in after-swarms and parent stocks that have swarmed. These are well pleased with the exchange, and in them the introduction can take place. Immediately the bees have noted the loss of their previous queen and begun to become restless, the fertile queen may be allowed to run among the bees after she has been smeared with a little honey. The painting of the wings of the queen with honey is advisable, because that hinders her from flying, to which a young and flighty though fertile queen is disposed; and because it conceals any strong odour that might attach to her, which might for the moment enrage the bees. If, however, any of the bees should seize the queen and try to sting her, which in this case would seldom occur, she must be caged for about twenty-four hours.

The Italian queen is just as readily received by a stock that has not had a queen for a week, but only possesses queen-cells and is therefore welcomed by a swarm made by means of brood-combs, or a parent stock that has thrown off a swarm or had one driven from it. The queen-cells that are begun are in most cases immediately destroyed by the bees themselves; but the bee-keeper does well to do it himself. For there might be a twofold undesirable accident happen. If the bees were tolerably strong, eager to swarm, and continued to care for the queen-cells that had been begun, the newly introduced queen might be moved to leave the hive and swarm out with what following she might have gained; and if this accident did not occur, one or another queen-cell might easily be left undestroyed, and the fertile queen might after several days be stung to death by the young queen that had come out.

For the jealousy that impels the queen to seek out and destroy all queen-cells that may be found in a hive is only specially strong in her early days. The queen that is already fertilised has her attention more taken up with filling the brood-cells, and may easily leave unnoticed a cell that is situated near the hive side, and which conceals a rival dangerous to her. In hives with moveable comb the destruction or cutting out of all the queen-cells is easy enough, but in log-hives it is an utter impossibility, and even in skeps only accessible from below it is difficult and uncertain, since a cell situated in the crown may very easily be passed over. In these hives, therefore, the time must be awaited for a young queen to come out and for the bees to destroy the remaining queen-cells. If, then, the young queen were drummed out or taken away when starting on her wedding flight there would be a certainty of no more danger being present for the Italian queen to be introduced. If it were wished to assign a new hive furnished with comb to the stock to be Italianised, it would only require to be driven when all the queen-cells are sealed, and the reception of the queen by the driven bees would be insured. If a swarm made in this way were put into combs without bees, but full of brood and without queen-cells, the Italian stock would be made very simply and certainly. If a small swarm, whose transport is not so risky, is obtained with the Italian queen the latter course can be safely adopted. It may be allowed to enter a set of combs full of brood, or brood-combs may be given to it gradually.* In the heat of summer the sealed brood comes out of itself without further care, and although some brood in unsealed cells perishes, the damage is not great. If it is wished to avoid even this, combs are selected from a stock that has been deprived of its queen for a week. All the brood-cells will then be sealed. At first, until the stock has increased in strength, it must be well looked after, and placed on exceptionally cold nights in a warm place, or the bees should be stimulated to the production of greater heat by food being given to them every evening.

* Exactly the principle we 'invented' and adopted nearly a dozen years since, and have always recommended as the safest and best means of Ligurianizing, and which American bee-keepers have since learned to follow.—C. N. A.

The introduction of the queen is more difficult and dangerous when the stock to which she is to be added had previously possessed an old queen, and the older she was the greater difficulty would it have in forgetting and getting over her loss. The bees must, then, be left queenless for at least three days, and the queen that is introduced be kept caged for about two days. Puff-ball (*Lycoperdon bovista*) is an infallible means, that has been communicated by Mr. Hübler, Court apothecary at Altenburg, to the meeting of bee-keepers in Hanover. In the evening, when all the bees are at home, the stock—not the queen—is stupified by fumigation with burning puff-ball till there is no more movement to be perceived. The hive is then opened, so that the smoke can be replaced by fresh air, and the queen to be added is put into the heap of bees, into which she immediately creeps and behaves quietly. The caging of the queen would be, according to this method, unnecessary. Although generally I do not think much of puff-balling, it may in this special case be of good service, inasmuch as the queen will have acquired the odour of the hive, and made herself better acquainted with the stock, before the bees have recovered from their stupefaction.*

If, now, an ordinary stock has had given to it a fertile Italian queen by one of the methods named, it is to be considered as Italianised, just as a wild fruit stock is looked upon as improved as soon as a better graft or bud has been united to it. Just as the tree now produces different foliage and other fruit, so does the stock produce a generation of bees of different colour, and the former one by degrees passes away. If the introduction of the Italian queen takes place in autumn, there will be black bees found in the hive until late in the following spring, because in autumn the deposit of brood is limited, or ceases altogether. But if the queen is added in spring, after about two months it will be difficult to find a black bee in the hive.

But the Italian stock often remains purely Italian only so long as the introduced queen lives. If she goes away with a first swarm, the pure Italian breed will be propagated in it. Whether the parent stock, and the after-swarms that perhaps will follow, become genuine Italians, depends on whether their young queens are fertilised by Italian or other drones.

* This method of introduction is by no means safe or certain.—C.N.A.

The Propagation of Italian Bees.

There has been much discussion whether the natural method of increase by swarms, or the artificial methods of division, deserve the preference. I have been from long ago greatly in favour of the latter method. When, in 1853, I received the first Italian stock, I made the following statement in the *Bienenzeitung*:— 'Herr von Baldenstein, who had first called attention to the Italian bees, has left his Italian bees to increase naturally, but I, on the contrary, shall make use of the artificial method for this race as well. The result may furnish some answer to the question, which method of propagation attains its end most surely and quickly.' Herr von Baldenstein, although he wished very much to propagate this race of bees, on account of their decided advantages, which he set forth at some length in the *Bienenzeitung*, No. 11, 1851, possessed during eight years but one pure Italian stock, and after nine years he had not one at all. On the other hand, I had in the autumn of the first year twenty-seven Italian stocks, and have since that time sent away thousands of Italian stocks, swarms, and fertile queens, to all the countries of Europe, and even to America. By artificial division, extraordinary results may be attained. The possession of one genuine Italian stock enables an entire apiary of one hundred stocks to be Italianised in the course of a summer, because a queen can be raised from every female egg laid by the Italian queen, and the introduction of a queen-cell suffices to transform an ordinary stock into an Italian. Whether it will be pure Italian depends upon whether the young queen coming out of the cell meets with an Italian drone on her marriage flight.

The chief endeavour, first of all, must be directed towards getting Italian drones as early and in as great numbers as possible. But since only strong stocks make preparation for drone-brood, the stocks must be quite strong early in the season, which may be attained by the introduction from other stocks of brood ready to come out, as the addition of bees is not advisable, because of the risk to the valuable queen. Young bees are better than old ones, because they take a more lively interest in brood-rearing. The stock should then already possess drone-combs in the brood-nest, or should be provided with them at the beginning of pollen-collecting. The best combs to use are those that have worker-cells by the side of the drone-cells; for as the queen immediately occupies the former, and is thus on the comb, she will the sooner be induced to occupy the drone-cells. The introduction of combs with much pollen and feeding with honey stimulates the queen to deposit brood more freely, and therefore furthers the commencement of drone-brood. For the preparation of brood-food the bees require much water. If they are hindered by the weather from carrying in what is needed, there must be no delay in providing them with it, perhaps in a comb put in below, or at the side. When the bees of a district have no opportunity of carrying pollen early from the hazel and alder, wheat and rye-flour, or, best of all, oatmeal, should be put for them in combs in a sunny place in the garden, free from wind, and they should be enticed to it by a little honey water. They will immediately form pellets of it upon their legs, and be excited to greater activity in breeding.

But a few genial spring days may be followed by still more severe cold and otherwise ungenial weather which will not allow the bees to make even short excursions, and renders them again out of heart. Breeding is in such case checked, and very soon all the drone-brood is sacrificed. Not only the eggs laid in drone-cells are ejected, but the small larvæ are sucked out, and even the sealed nymphs are thrown out. It should not be allowed to come to this, but the combs in which a number of drone-cells are found occupied with eggs, should be taken out and put into a queenless stock, or may be put to a strong (queenless) artificial swarm, to be hatched out, and to the Italian hive is given in its place another drone-comb, so that when it is for the most part occupied it may be treated in the same way as the first. In this way thousands of drones may be gained from an Italian hive if the queen shows her willingness to fill again the inserted drone-cells, which is not always the case. And there will be hope of a genuine succession if the breeding of drones is hindered as much as possible in the other hives. As soon as drone larvæ are observed in tolerable numbers, preparation may be made for breeding young queens, so as to reap the advantage of the early drone-breeding; for the young Italian queens will be the more certainly genuine if they go on their marriage trips before the other stocks have drones, and a young fertile queen has the greater value the earlier it is obtained. If the young queens should hatch out a little earlier than the drones, it will not matter, for at this time of year, yet cool, there might be eight or ten days before the young queens take their flight, and they repeat it through two, three, or more weeks. Meanwhile, the drones will be so far developed as to be able to take their flight when suitable days set in. Unfortunately, the first queen-breeding too often fails for want of such days, and in Silesia one cannot hope for the fertilisation of young queens with any certainty before May.

Since many queens get lost on their wedding excursions, and many get mated with black drones when their hives are in the same place or neighbourhood, there must therefore be as many small artificial swarms as possible made, in the way before described; let queen-cells be continually formed, of which as many as possible should be lodged in the stocks that have been deprived of their queens or are otherwise without them, so that the largest number of fertile Italian queens may be gained. Among the many there will always be some genuine ones to be found, and even the ones not genuinely mated, if they are of pure Italian descent, make the propagation easier for the next year, as they only produce Italian drones, fertilisation exercising no influence on the drone eggs or progeny.*

Since it does not matter whether the young Italian queens are reared by yellow or black bees, brood must be taken for the artificial swarms out of the Italian hives, but the bees can be taken from the ordinary stocks. Not to weaken the Italian stock too much by continuous withdrawal of brood-combs, just as much brood from other hives can be put back as compensation. So that the stock does not lose anything, but rather gains, if a comb with sealed brood is inserted when one with eggs and small larvæ is withdrawn. If the brood-comb put back into the Italian hive contains eggs, it should not be used for an artificial swarm before the lapse of some six days. It is better, therefore, to put in combs

* This is a point on which there is a great diversity of opinion. It is known that a purely white hen, having once mated with a black rooster, never afterwards breeds purely white chickens, the black influence having pervaded her being; and it is no stretch of imagination to believe that a similar effect is produced in queen-bees, and results tend to convince one that it is so.—C. N. A.

which contain only sealed brood which have either been taken from an earlier artificial swarm or from a stock otherwise deprived of its queen. The introduction of such a comb is, further, the means of exciting the fertility of the queen to the highest degree, because she is in haste to fill all the emptying cells, which cause vacancies between the full ones, so as to complete the brood. In a few days all the sealed brood that had been present is hatched out, and the comb is then only occupied by small larvæ and eggs. By cutting through the comb horizontally, and fastening the lower piece reversed on a separate bar, two brood-combs are obtained, as good as can be desired, for gaining a quantity of royal cells conveniently situated for division. It is well, although not absolutely necessary, to put in a brood-comb to every artificial swarm or stock in which a young queen has hatched out. With the increased activity and warmth then prevailing in the hive, the queen will sooner attain maturity and undertake her marriage flight, and at the same time more certainly find her hive again, because of the bees playing before it in greater numbers. Whether the young Italian queens will be mated with Italian drones is a matter of chance, especially in the height of summer, when drones and queens go a long distance from the hive. Meanwhile, the bee-keeper may have some influence in effecting true mating, by exciting with liquid food given or syringed into them, some of the hives that have most Italian drones and queens ready for flight, thus inciting them to play in front of the hives earlier, while the other drones are still resting quietly in their hives. It is evident, then, that it is better to have the Italian drones in few hives but in large numbers. By making the stocks queenless, the drones can be preserved far on into the autumn, so as to have some ground for hoping for the genuine mating of the queens that take their marriage flight after the extinction of the other drones. To set up Italian stocks in an entirely isolated apiary is really not necessary, nor even advantageous. The more hives there are at the same apiary, the greater is the opportunity of assisting the Italians with brood-combs and bees. The more young Italian queens can be lodged out, so many the more genuine ones will there be among them. But drones must not be allowed to be reared in the other hives. This is attained by removing drone-comb and brood as much as possible, and driving the stocks at times, which will stop drone-rearing for the year. If a number of drones come out, they may be caught by the so-called drone-traps. It will suffice for this purpose to have a tube of wire, tin, or partially of wood, with openings at the side that will allow of the passage of workers but not of drones. If this is fitted to the entrance at the time the bees begin to play, so that it projects outside, the drones will come out of the hive but will not return, and can then easily be killed. To hinder the drones from flying off, and to save the trouble of killing them individually, the tube must open into a wire vessel and project a little into the interior. The workers must be able to escape on all sides from the vessel, which may have the form of a ball or a cube.* To facilitate the removal of the drones that are killed in hot water, it should be provided with a lid or a little door. The figure of such a drone-trap will be found later on among the implements belonging to bee-culture.

When there are brought from a distant stand swarms or bees for artificial swarms, with a large number of black drones, which it would be tedious and waste time to kill individually at the time of hiving, the method of treatment may be as follows:—The queen that is in the transport hive (of course caged) is taken out and put with a part of the bees into the hive that is to be occupied, the transport hive is brought close to the entrance, and the lid lifted high enough for the workers to come out and go into the hive, but not high enough for the drones. Another way is to make beforehand, toward the upper margin of the transport hive, one or two slots that afford passage only for workers, and not for drones, to come out and go into the hive. The slots are now opened and put near to the entrance. The bees will slowly move in, and the drones will be left behind. If only bees have been brought from a distant stand to make artificial swarms, or to strengthen those already made, let them be left for the night (for the return home is generally in the evening) to move into the first best artificial swarm, but it must be one without queen, having only royal cells, so that no damage may be occasioned by the bees.

* This is exactly the principle of the drone-traps commonly used in England, but now-a-days it is found much easier and cheaper to prevent the production of drones by filling (nearly) the frames of hives with worker-comb foundation only, thus preventing the production of drone-cells, except in small degree.—C. N. A.

On the following morning, before flight has begun, the bees are distributed at pleasure where required. The added ones at least stay in the place from which they take their first flight, and will receive a young queen, as they are already well disposed toward queen-cells. Though it is necessary, in order to make sure of this, to leave the bees a little longer with the artificial swarm, and either to confine them so as to hinder the artificial swarm from flight, or to put it into a dark place.

To make a new artificial swarm, it would only be necessary to put a part of the combs, with the bees on them, into a new hive. The comb with the queen-cells must be given with them, even if only one were found in the swarm, so that the bees may be kept quiet and attached to the new position. This part must have the majority of the bees, since the ones already familiar with the stand fly back again to the other part, and would strengthen it. Even if this has to give up all the queen-cells it does not do it any injury, for most of the bees are familiar with their way home, and do not become scattered, even if a temporary disturbance breaks out in the hive, but soon make preparation for rearing queens again from the fresh brood given to them.

If there is not time and opportunity for parting and using the bees immediately, they can be reserved for future occasions, as, for instance, for making a swarm immediately, when required, by adding to a fertile queen that has been gained a corresponding number of bees. But the swarm could neither be put on the present nor yet on the previous stand from which the bees have been brought—it must have a third provided for it. There might be some expectation of a voluntary swarm being thrown off in fourteen days after its formation; indeed, in the swarming time, with queen-cells present and many bees, it might be expected with considerable certainty.

Bee-keepers that still keep their bees in straw-skeps might get a large number of swarms with young Italian queens, by putting the Italian stock with the genuine queen at one time into one and at another time into another skep that is furnished with comb, and, where possible, already containing brood. When the queen has deposited brood in it, this is put into the place of a skep with a strong stock, whose bees repopulate the combs made empty of bees, form queen-cells, and probably would divide themselves in fourteen days into several swarms, led by Italian queens. Yet more of the swarms can be gained by effort through strengthening the stock afresh after it has weakened itself, by moving it nearly into the place of its strong neighbour. If queen-cells are still in the hive, fresh swarms may be expected in a favourable season. In the separate swarms several queens may probably be found, and the surplus ones can be caught and otherwise used. When the parent stock has ceased swarming, its surplus queens can be drummed out and applied as required. They may be preserved with a small company of bees for some time in a little box, which can be conveniently inserted in the honey-room of a hive till an opportunity is found of giving them a home.

It is possible to manage in skeps and log-hives in cases of need, but not with that convenience and certainty we have in hives with moveable combs, whose advantages in respect to propagation alone are so exceedingly great that they must be evident to every one free from prejudice. Increase progresses marvellously in these hives where there are a number of strong stocks and there is at command a store of combs and the season is moderately favourable. Indeed, this is so much the case, that there is difficulty in lodging in the apiary, the colonies almost increasing against one's will, so that new hives have to be built, whereas other apiaries are desolate. If there should be the necessity of making a considerable reduction of stock in a bad year, the apiary can again quickly be made complete in a favourable year following, if the combs of the cashiered stocks have been preserved. What to do with them has, I hope, been sufficiently described in detail in what has gone before. If this description should appear to any one too detailed and diffuse, he must remember gifts are different, and one finds even the most detailed description unintelligible, where to another a mere hint is enough.

But because artificial division is so convenient in hives with moveable comb, the beginner especially may be easily led astray into pushing it too far. Though it is so useful an operation in itself, it may, followed immoderately, result in great injury, and ultimately ruin the entire apiary. The conditions of weather must be well considered, and as these change the treatment must vary. While in capital seasons good stocks can be easily increased sixfold, and artificial swarms made at the beginning of

August carry in beyond their own requirements, yet, in bad years, with wet cold weather, even old undivided stocks require help if they are to live through the winter. Do not, then, trust to the weather. Increase but moderately and stop in time, at least fourteen days before the pasture is over, because a natural or artificial swarm requires at least this time to make a tolerably complete set of combs and to carry in some provision. The shorter the time in which there is a prospect of abundant pasture the stronger must the swarm be made, and the better must it be furnished, if it is to be capable of passing through the winter.*

THE FURTHER TREATMENT OF SWARMS.

This depends upon the purpose for which they have been set up or made. Young stocks are not always made with a view to their standing the winter; but partly to get rid of the old queens, from which, finally, the advantage can be derived that the artificial swarms, so easily formed by their help, are left to make comb and collect some honey, or that we may have queens in case of need with which to help stocks that have become queenless. With these the chief thing to care for is, that they do not suffer want in long-continued rainy weather; otherwise there is no need to support them. On the contrary, brood-combs can be taken away from them for the support of second swarms which are to be wintered, or for the improvement of stocks that have become queenless.

Greater attention must be given to stocks that are intended to stand the winter. First of all, notice is taken whether the young queens of artificial swarms, second swarms, as well as of parent stocks, have been successfully fertilised. In hives with moveable combs assurance is received as soon as the queen is seen, or the cells of the brood-nest have been looked into. In other hives, especially log-hives, a conclusion can only be arrived at from the activity and behaviour of the bees, as to whether all is right or

---

* This paragraph should be printed in letters of gold and hung up in every apiary, for it points out one of the greatest dangers in amateur bee-keeping. Every one is naturally delighted with the newly acquired power of driving bees and making artificial swarms, but few acquire early enough the judgment in making them that is essential to a successful issue.—C. N. A.

not. In natural and artificial swarms with young queens there cannot well be any doubt on this point. The queen is present if they continue to build diligently and only worker-comb, and if she has been fertilised the fresh-laid eggs will be seen in the ends of the combs without taking them out, and later on by looking higher between the combs a little separated from one another the sealed brood may be seen. It is more difficult to observe early the possible loss of the queen in parent stocks which may have thrown off natural swarms, or had driven swarms taken from them. Opportunity for building is given to them so as to see whether they decline to build, or build drone-comb only, or worker-comb, which last is always a favourable sign. In skeps, a part of one comb on the entrance side, where they soonest build up a gap, can be broken into. In Ständer log-hives or boxes the whole of the combs may be considerably shortened, perhaps, by one-half. The long combs are of no service to the weakened parent stocks, and can but too easily become a prey for the moths. It is of much more advantage to put them in to young swarms, especially to second swarms, whose comb gains thereby in firmness and warmth. The parent stock is not injured by it but rather benefited, because it is excited to greater activity within and developes greater industry outside, and if the queen is present and has been fertilised the bees only build worker-comb, whereas, perhaps, partly drone-comb has been cut away. If the stock should not begin to complete its considerably shortened comb because it is not in a position to cover it, this would be only a proof that it had been weakened too much, and needs strengthening if it is to become capable of wintering.

Stocks, whose comb is tolerably new, should be cut but moderately; but old parent stocks should be cut more freely. The most suitable time for the renewal of comb that has already become too old is just when most of the brood has come out, and the queen should again begin to lay eggs, because the stock provided with a young queen generally builds the same year no drone-comb, which is otherwise so abundantly formed in spring.

Further, notice must be taken whether all, especially all young stocks, are strong enough to make sufficient comb and collect adequate stores. In cold and changeable weather, frequent showers, and storms, quantities of bees often get lost. In old stocks the loss is again made good by the hatching brood, and is scarcely to be noticed in them, but it is the more conspicuous in young stocks that are just beginning to deposit brood, and in which young bees will not be again hatching out before three weeks or later. Even strong swarms may under unfavourable circumstances be unusually weakened after fourteen days. They may be most quickly assisted by the addition of bees brought from a distant stand, though that means always has some danger in it, even though the bees to be added have been left over night to become aware of their queenless condition. These stocks should rather be helped by sealed brood, which soon hatches out. It scarcely needs to be said that here combs should be selected which contain honey at the top and only brood lower down. This causes the bees to transfer the brood-nest continually farther down and to lengthen the rest of the combs so much the more. But it is necessary to guard against extending the brood-nest into which the inserted comb is immediately incorporated too much in proportion to the breadth. For the more combs the comparatively weak stock is made to occupy the more thinly will it cover each separately, and the slower will it be as regards further comb-building.

A swarm builds in its early days exceedingly quickly if it is favoured by the weather. If unfavourable weather sets in, feed; not sparingly, just to keep the bees alive, but liberally, to encourage them to uninterrupted building. In no case is the expenditure of food more richly rewarded than in young swarms. If part of their stores has been taken from stocks full of honey to give them empty space, no more advantageous use can be made of the sweet water that is obtained when the fragments of wax are washed out after being refined, than to give it to the young swarms.

By such thinned food sometimes given to the bees they are powerfully stimulated to building and brood-depositing, but their proper honey-stores will gain no special increase from it; for that purpose honey must be given very abundantly and of greater consistency. The young stocks will be furnished with proper winter stores, if they are not in a position to collect them for themselves, very much quicker, more easily and cheaply, by giving to them full sealed honey-combs, as near as possible to their nest. This can be done in the summer. The bees have then so much the more time to combine everything properly together, and so to arrange as if they had themselves built all the comb and carried in all the honey. In Twin-stocks, when pasture is still abundant and the bees are still building, passages are made by the back wall to the top by partly taking away or shifting the cover-boards, so that the small honey-room will be at least built up over the brood-nest, in which the winter will be spent, and which will be got through all the better for the stores over-head. The completion of the comb is hastened by putting in small combs, or at any rate comb-bars with comb beginnings at the top, because building is now gradually ceasing, and the bees are more ready to fill combs that are in the hive than to begin to build fresh ones. Later on full pieces of comb can be inserted before preparation for the winter, if an addition is still requisite. Pieces of candy can be put in, for whose solution sufficient moisture will form on the top of the hive.

In a Ständer hive, with a higher honey magazine, perhaps consisting of a third part of the whole interior, one would not generally venture to let young natural and artificial swarms go up the first year, because in districts and seasons but moderately good they have enough to do to build up the brood-room. They would put but little for themselves above, and then perhaps would not keep adequate winter stores below.* Early and strong swarms, in excellent seasons and favourable districts, certainly are exceptions to this rule. These may in the first year afford an overplus of honey, and should then be treated as stocks devoted to honey-gathering.

The Management of Honey Stocks.

A honey stock is one that is to yield its profit or produce for the year not in natural or artificial swarms, but especially in full honey-combs. Although when in great strength some thousands of bees are taken away from it, this will do no special damage to

---

* The idea intended to be conveyed, evidently, is, that the bees in endeavouring to fill the 'honey-room' above would consume so much of what otherwise would be store below, that offering them the upper honey space is likely to be mischievous.—C. N. A.

the produce of honey unless the queen be taken from it, because in the latter case comb-building immediately declines in the hive, and does not again progress quickly until the young queen has become fertile. It is best, then, to decide upon those to be treated as honey stocks that do not require a renewal of the queen; therefore, those possessing one of the previous year should be chosen. But if the queen should have been taken from the honey stock either because she was too old or because she was otherwise needed, her speedy replacement must be cared for, either by the insertion of a queen-cell or the introduction of a young queen.

The essential part of the treatment consists in this, that when an abundant honey time sets in, and as long as it lasts, provision is made by continually giving empty space, and, as far as it is possible, empty cells, for the deposit of the honey carried in. All drone-comb that has been carefully removed from the brood-room can be advantageously applied in the honey-room, only care must be taken that the queen does not go aloft and deposit brood there, especially if the season be damp, which will be more favourable to brood deposit than to honey produce. Entrances to the honey-room are then made quite narrow, so that only workers can go up, passages being made only in front by the door and by the back wall,* where the queen seldom comes, because there is no brood there. The application of drone-comb can be delayed till the latter part of summer, when drone-breeding has come to an end. The drone-cells are then chiefly filled with honey, and a drone-comb would then do no harm in the brood-room, but would form a definite boundary (as does every full honey-comb in a hive, for beyond it the queen will not readily extend the deposit of brood†), and all the combs on the outside of it would be filled only with honey, if in other respects the close of the season is favourable.

In space given at the top, bees build in two ways—either by carrying the comb downwards, as usual, from top to bottom, or

---

* Remembering that the ' front by the door ' is at one end of the hive, and the ' back wall ' at the other end, the direction implies that access to the honey-space (equivalent to the English super) should only be given through slits on both sides of the brood-nest.—C. N. A.

† This we have many times noted, and the fact points to the unwisdom of using hives in which the combs are not moveable, which under adverse circumstances cannot be properly regulated.—C. N. A.

by beginning from the bottom and extending the comb upwards. There are many advantages in letting the honey magazine in Ständer hives be built up, according to the latter method. For one thing, bees are accustomed soon to make the combs extended upwards very thick, that they may not bend over, so that there is no chance of the queen occupying them with brood, because the cells are too deep.* The entire mass of honey, therefore, both in the brood-room and honey-room, forms an uninterrupted whole, and at the taking of the surplus from the bees as much can be left as is thought well, as the combs can be cut at whatever height is desired. Rather more than will be used should be left, of course providing for the chance of a long and severe winter, since the surplus even down to the bars can always be cut out in the following spring.

After the cover over the brood-room has been removed, the bees are induced to build regularly by laying on the separate bars strips of comb, if possible extending the entire length, which the bees will speedily lengthen upwards. The larger the pieces of comb put in, the greater is the service rendered to the bees, and the quicker will they be able, in an abundant honey-time, to fill up the space. The largest combs may be put at the back, and can be kept in a perpendicular position by pieces of wax interposed behind, and bars propping in front, till the bees have fastened them. The small strips may be simply laid down. Within a few days, notice should be taken whether they have been fastened by the bees in the same position, that when it is necessary they can be rectified.

Such an arrangement below does not exclude the use of bars in the honey-room at the top. The bees could then in some strength build at the same time from below upwards, and from the top downwards. They would unite the combs in the middle, and, favoured by the weather, would all the sooner complete the building up of the space.

Whether the bees build in the honey-room upwards or downwards, regularly or irregularly, does not much matter. It is of

---

* The advantages appear very small as compared with the disadvantages that arise through the bees being unable to cluster naturally while at work. They can build downward with at least four times the rapidity they can raise their combs from the crown-board.—C. N. A.

far more consequence that the comb in the honey-room and in the brood-room is in communication, so that the bees, which in Ständers do not always keep the necessary winter food over them in the brood-room, may be able to move upwards even in severe cold and live on the stores at the top. To this in autumn a convenient passage must be made, perhaps, by taking out a part of the cover, even if for the summer there had been left but a very narrow passage, or only passages quite at the back and front so as to keep the queen away. To have the honey-room built up in the way mentioned in a bad season, and by weak and even young stocks, would require that the bees should be admitted only into the lower part by inserting the bars at half the height.

A pair of grooves put at half the height of the honey-room, which in its entirety is eight to ten inches high, would afford many advantages. The driving of a swarm to the top would then be far more convenient, because combs five inches from the top are very much more easily put in and taken out than when nearly close under the crown-board. The building-up of the entire honey-room would by this be made much more convenient to the bees by putting the bars below so long as the bees are not very strong and not too widely extended. Later on the bars can be put higher when it is foreseen that the bees will be able to build up and fill the entire honey-room. But if bars were inserted in both pairs of grooves, a double number of combs of half the length would be gained, which would not be without convenience in providing for light stocks. To provide a stock with its necessary winter food, even if it were quite empty of honey, all that would be required would be to take out the upper layer of combs and put them below for the needy stock. Such an interruption and piecemeal arrangement of the combs does no harm in the honey-room, and has the further advantage that a honey-comb cannot readily break off. It is only in the brood-room that such proceeding is injurious and must be avoided.

The Emptying of the Honey-Room.

This can be done at any time we like, when it is entirely or only partly built up, so long as there is not too close an approach to the brood. The operation is best undertaken at the time of best pasture and strongest flight. There is least bother then both from one's own and other people's bees. The process of refining is easily effected because the honey is liquid and the air everywhere warm. The sweet water gained at the same time in washing the wax will be used with advantage as food for swarms. The combs that perhaps contain pollen, or by accident brood, can be put in to the swarms, and the emptying will at the same time keep the bees up to uniform industry, as without it they would notably fall off in carrying in when there began to be a want of room and the temperature had become intolerably hot.

The speedy cutting-out of the honey present, even in the brood-room, is to be especially recommended when the bees have carried in towards the end of May and in June much honey from the pines, which is liable to cause dysentery in winter; but there must be a prospect of abundantly supplying its place with wholesome honey from white clover, lime-trees, buckwheat, and other flowers. The part filled with honey, even of those combs that contain brood lower down, can be cut out with care if the part left is only fast built to the wall of the hive, or it is supported a little from below, so that it will not fall down.

The larger the empty comb-beginnings that are given to the bees in place of the full combs taken out, the quicker will they with continuously abundant sources of honey have again built up and filled the space, and the sooner can the harvest be repeated. If all the empty combs and comb-beginnings have been already used up in a good honey season, help is gained by leaving as much as possible of the middle wall standing when the combs destined for the pot are broken up. The method of emptying entire honey-combs so that they may be put back again to be filled will be spoken of in the description further on of the extracting machine generally called the 'Honey Slinger.'

It is an ordinary practice to give in good seasons collaterals to log-hives when they are built up, but the labour of the bees would be saved, and more would be gained, if more room was given by taking the surplus from the hive itself. It is not only possible to cut out the honey-combs situated beneath a division-board inserted horizontally, but also those at the top from the crown of the hive, although it is rather more troublesome than taking them out of box-hives. This is most easily done when the combs are built across, or only a little oblique, and it is well to induce the swarms by small comb-beginnings to build so. From the first comb a piece as long as possible is cut, from every succeeding one a piece somewhat shorter, so that it can be conveniently taken out. In proportion as the bees have carried in unwholesome honey, have old comb, and the more pasture there is to look forward to, the deeper may the crown-combs be cut into. It is all the better if the gaps that have been made can be partly filled up by empty combs; or, at least, if the bees can have guidance given to them by large comb-beginnings to give the desired direction to the new combs again, or the bees may build up the space so irregularly, carrying the combs partly from below upwards, and partly from the top downwards in different directions, that the taking of the honey would be made extremely difficult in future.

Lager hives, whether they are log, cylinder, or box hives, are especially honey-stocks. The bees limit themselves in them to a number of combs for brood, and fill the rest, situated on one or both sides of the brood-nest, with honey only. Although the honey can be taken in these hives from each side until the brood is met with, it is well to take it where at the same time drone-comb and old comb can be done away with.

In Twinstocks, which may be reckoned among the Lagers, although they have a small honey-room at the top, it is better to leave the honey-room untouched, even when it is built up, excepting it is filled with honey that is not so wholesome and better pasture for the bees is still in prospect. The largest quantity of honey taken with the greatest convenience is gained from the long combs placed near the door and out of the space shut off by the moveable door, to which access is early afforded to the bees. It has been already previously said that in a specially good honey season, in which the brood is only too much limited by the abundance of honey, and requires no checking in any other way, the bees will build up and fill the space sooner if the moveable door (or division-board) is quite removed. Full honeycombs will be taken out in the course of the summer, at one time from one side and at another time from the other, and the door can be put in again at the close of the pasture at the side on which there has casually been left the larger space not built up. In moist summers, favouring the production of brood rather than that of honey, the division-board is left untouched, and the space at the side is made accessible only to the workers. Even if the side space remains not built up, that is better than if the bees spent the little honey collected in increased comb-building and extended brood deposit, and had to meet the winter unsupplied with honey.

The Ständer skep is least of all fitted for a honey stock. It is specially a brood and swarm hive; its round shape and warm material favouring the uniform distribution of heat, and therefore of brood, through its entire extent. There will not be, therefore, much surplus honey gained by cutting from its interior, at most, perhaps, a couple of side combs, which might, besides, be tolerably difficult to get out. If it is wished to treat the skep as a honey stock, and to get a considerable quantity of pure honey from it, it is done by applying a separate super, collateral or eke. To put on a super the skep should be somewhat flat at the top, and provided with a bung-hole. The larger the opening is (which is otherwise closed with a bung), the more readily do the bees take possession of and build in the little skep, box, bowl, or other vessel that is put on the top. They take to it the sooner if in the opening a small comb is put, which they soon extend upwards. Supers, especially small skeps so filled, can again be put on light stocks, which will then be furnished with necessary food in the simplest way.

By an ordinary eke, opportunity would mostly only be given to the bees to extend the deposit of brood, especially of drone-brood, and would in many seasons occasion a lessening rather than an increase of the store of honey. It would only be in exceptional honey years, when the brood business has to retire before the quantity of honey carried in, that the cells built below would be immediately filled with honey, whereas otherwise they would be immediately occupied by the queen with eggs, and so for three or four weeks would be unavailable for honey, within which period the honey yield would perhaps be ended. If the eke—which may either be of straw or may be a wooden box—is wanted to be quickly filled with honey, access must be impossible or very difficult for the queen. The lower space must be divided from the upper by a horizontal board, or the eke must be provided with a thin lid, through which the bees can descend only at the sides and through narrow clefts. The board might be replaced by a set of bars if the spaces between the bars in the middle of the eke were closed by half-inch slips let in.

If we suppose the one side of a box moveable and opened, it is easy to understand how a collateral can be given to it by fitting a similar box to it, so giving to the hive an extension of space at the side. This collateral space can be further increased by putting a box or a skep on the top of it, and it will the sooner be filled if it is already furnished with combs. As communication between the two skeps is brought about by the two nadir boxes adjoining one another, which may form one whole, this may also be done by a channelled board common to the two put underneath. The two skeps to be put in communication could be provided with slits and then be moved close together, but they would be too much damaged by that method. The combination of two skeps could be made by putting underneath low ekes, from which pieces of about three inches long were wanting to complete the full circles. Even if a skep already built up and put in communication with a strong stock in this way is not always filled up with honey, this point at least would be gained, that the combs would not be destroyed by moth.

To get beautiful honey for the table, the best receptacle of all is a little box or skep, but bell-glasses have special advantages as supers in several respects.

Bell-glasses as Supers.

The bell is generally enclosed in a cover when put on the outside of a hive, that the light may be kept off. One advantage of this form of super is, that the cover can be taken off, the progress the bees make in building and carrying in can be always observed, and it can be seen when the bell is built up, and may be taken off. The honey has further a higher value, since a bell-glass so built up is beautiful to the eye, and the honey-comb contained in it is so much the more inviting to the palate, not having been touched by any human hand.

Bell-glasses can, of course, be put into the honey-room of an upright hive, in which case there is no need of any further darkening. For that purpose one large opening, or several little ones, are made in the cover, and the bees are carefully restrained from all access to the space outside the bell, so that they may not build there and leave the bell empty. It is well if the bell is provided with a knob at the top, that it may be conveniently handled. If the knob contains an opening, there can be fastened into it a little pillar, which hangs down in the bell to the floor, enabling the bees to begin and fasten their combs upon it. To facilitate their beginning, and at the same time to induce them to a beautiful regular construction, the pillar can have fastened upon it, in the well-known way along its whole length, six or eight strips of a thin but pure comb. This can be done before it is fitted into the opening mentioned, and fastened into it by wedging with pegs or sealing-wax. The pillar may have as many corners given to it as it is intended, according to the proportion of the size of the bell to have wax strips. The different combs will, therefore, radiate to the different sides, as rows of cells proceed in straight lines from the six sides of a cell. After some time, when the bees have thoroughly got to work, they may be looked after and assisted, where assistance is possible. If the bees have made a comb somewhat askew, it can be straightened. If they have in building one comb come too much in front of another, the former can be pressed in or curtailed. Help of this kind must be given if required, or the longer and broader combs would become thicker than the others, and the symmetry would be destroyed. But if an entirely regular structure is required, care must be taken that all the combs proceed in the different directions, and especially that adjoining combs are kept uniformly of the respective dimensions, and that every inequality is immediately rectified. If the bell-glasses are a little contracted or narrowed below, the taking out of an entire comb is made more difficult, but it cannot then fall out, even if it should have been loosened by shaking on a journey.

If bell-glasses are wanted built up in the ordinary way with combs running parallel to one another, this can be done either from top to bottom or reversed. Vessels that are more flat and dish-shaped the bees will more readily and quickly build in from the bottom upwards. The vessel may first be set up when the bees, guided in the way mentioned above, have already begun to build in the honey-room; that is, have already carried up the combs some inches high. Whatever stands in the way of the vessel can be cut away. Provision might have been made before for the closing of other approaches to the honey-room, and, therefore, the cover of the frames should have been only partially removed. It must not be forgotten that the vessel must be so put in that it will be capable, when built up, of being cut off with a fine wire, or that the separate combs can be detached at the at the bottom by a long thin knife. There must, therefore, be previously put under the vessel a thin rim of wood (or tin), that can be drawn out to give room for the insertion of the knife. The detached vessel, a little raised, is left standing for a short time till the bees have sucked up all the loose honey, and it is then lifted away.

If the bees are to build the vessel from the top downwards, and as regularly as possible, they must have comb-guides attached at the top. For they begin combs of themselves very unwillingly on glass because they cannot well hang from it. And what is begun to be built from below upwards must always be cut away, as the lower comb would, of course, not fit on to the upper one, and would not form a beautiful whole.

Not only beginnings, but larger combs, can be much more conveniently applied at the top; and a building up of the vessel is much more quickly attained by using a glass box instead of a bell-glass. The box may have a moveable cover, and be provided with grooves in front and at the back, so that comb-bars can be put in at the top. According as more or less comb beginnings are put in the honey-comb will be thinner or thicker. If in a box eight inches wide the bees ordinarily make five combs they can be induced in the same space to build only four, or even only three, proportionately far thicker combs, if they have given to them only so many beginnings; because the honey-combs have not, like the brood-combs, a definite thickness, but are made as thick, or the cells as deep, as the space may allow. To prevent the bees from beginning to build another comb between the beginnings put so far apart from one another, a pane of glass could be put either on the top as cover, or under the cover, or strips of glass could be put between the separate bars. The box may have a moveable floor. At first it can be put on without floor, because the bees then take to it the sooner and build the quicker; the larger the access to it is, the more the warmth streams up to it from below. If it is soon built up, the floor provided with passage-holes can be added. All that is necessary then, when it is quite built up, is not to cut it off, but simply to lift it away. When the few bees in the box have discovered their queenlessness and have become disquieted, the box may be put towards evening to the entrance of the hive, when the bees will come out through the openings in the floor and stream into the hive.

Both swarm-stocks and honey-stocks will yield a larger produce the more abundant the pasture and the longer it lasts. In many places all food for the bees is cut off when the comb harvest is reaped, while in other places the most luxuriant pasture for the bees sets in at the blooming of the buckwheat, and later on at the blooming of the heather. To make use of such an abundant pasture, bee-keepers in many districts at certain seasons regularly move their hives to it, or, as they say, travel with them there.

THE SO-CALLED 'TRAVELLING BEE-CULTURE.'

There is nothing peculiar in this; for if a bee-keeper does not travel with his hives because his apiary is so situated that the bees can visit buckwheat and heather without being moved, no one will conclude he follows another method than those who move their stocks to buckwheat fields and to the heath, because by chance they have none in their own neighbourhood. Honey-stocks can be travelled with just as well as swarm-stocks; in this respect there is no difference, only the structure of the hives must be suited for the purpose in districts where the local conditions make it desirable to move them from place to place.

Objection has been made to my hives before this, on the ground that they are not suited for travelling; and with respect to the Manifold, the Presslike Sixfold, the Pavilion, and others, it must certainly be allowed, since they are made of boards, are too heavy to move, and are therefore unsuited for travelling; but the Twin-stock and the double Lager-hive, like two Twin-stocks put one on the top of another, are suited for the travelling culture in a way that no other hives are. To the reasons given for that conclusion in the description of these hives (page 102) many others might be added; meanwhile the following remark may be made here:—

As is well known, the less stores a hive has at the commencement of a honey harvest, the more will it increase in weight. In ordinary skeps, which have hitherto been chiefly used for travelling, the considerable store which they perhaps already hold makes travelling dangerous, and is liable to diminish its success; at the same time, in the boxes the full combs can be left at home, empty ones can be put in their place, and so much the larger increase in the weight of the hives may be reckoned upon. Suffocation of the bees and breaking down of the combs is less to be feared because empty space can be made before travelling by the withdrawal of a part of the stores in hand. The entire set of combs can be made lighter; and even if here and there a comb should break down by accident, the damage with moveable comb hives is easily made good again, whereas every other hive is nearly ruined by such accidents. Further, there is no risk in taking away the honey from the box-hives before travelling, as would be the case with other hives; for as there is no need to press out the honey from the combs, they can be preserved (and this is very advisable), so that they may be partly or entirely put back again, if through unfavourable weather the stocks should be brought home again as light or even lighter than they went. If many of the honey-combs taken out contain brood, they can for a time be put into a hive that is kept at home. In travelling it is not advisable to take all hives without distinction. Those with heavy combs, that are at the same time tender and breakable, should be left at home. For if a large part of the structure breaks down the loss will not be compensated by the additional honey gathered.

The danger is not so great if the road is not too rough and the loading, unloading, and carrying are managed with care and prudence. Laying straw or straw mats under the hives will lessen the effect of the jolting, and the hives must be so arranged that the strongest jolts which come from the side meet the ends of the combs. Twin-stocks would, then, have to be laid along the length of the waggon. Brood-combs or honey-combs of any weight must not be loosened from the side wall immediately before the journey, and the like combs must not be freshly put in, because they would rock backwards and forwards, and thus easily break off. The bees should, at least, have had time over night to fix them up again. There is, on the contrary, no danger with empty combs, of which as many as possible should be put into the hives that are being moved to better pasture.

It is usual to travel with hives in the night, for the sake of coolness. If, however, the heat is not great, it can be done by day. The bees keep perfectly quiet in the rocking motion on the waggon, so long as there is no deficiency of air. Special provision must be made for this, either by incisions in the crown-board, which would be stopped up again by the hives put on the top, by incisions in the back wall, or by slides of perforated zinc put in front of the openings in it. In the Twin-stocks sufficient air can be given to the bees in the simplest way, by providing the moveable door with as many incisions as possible, moving it back before the journey as far as possible, and fastening it there and removing the outer door altogether. The moveable door, which might be fastened by some nails driven in, would be protected against knocks, as it would be at least two inches, or the thickness of the outer door, distant from the outside, and the incisions could never be covered up by the adjoining hives, and therefore the bees would never be deprived of the air that is necessary. Ventilating doors can easily be made from some narrow boards nailed near to one another on to two or three cross-bars.

Ordinary skeps it is usual to reverse or put on their crowns, so as to prevent the combs breaking down. But the unsealed honey will then probably run out of the larger cells, which will cause considerable excitement and heat among the bees. The box-hives are, therefore, put on the waggon in their ordinary position. If it is wished to secure the combs further against breaking down, as the bees never build the combs down to the floor, a piece of wood like a lath can be pushed under each comb, and other pieces packed under it, pressing it upwards against the comb, so that every comb receives due support.

Combs that have served for brood-rearing gain continually in firmness, and do not then so readily break down. If brood-combs fall down they as a rule break above the actual brood-nest near the crown of the hive, near to the bars where only honey has usually been stored. This must be considered in travelling stocks, and the bars should be provided with combs as long and as firm as possible. To lessen the risk of breaking down in the Twin-stocks, when travelling is undertaken with them regularly, the honey-room could, when constructed, have one or two inches added to its height by taking the same away from the brood-room, thus giving to the honey-room a larger share of the weight. In travelling hives, a second pair of grooves can be put at half the height of the lower room, or about six inches from the floor, and bars can be inserted, at any rate, on the outside of the brood-nest, in the neighbourhood of the doors on each side, where the heaviest honey-combs occur; but more hold and support can be given to the brood-combs themselves in the way mentioned on page 115, by applying pieces of wood placed parallel with the length of the hive, to be worked into the brood-nest by the bees, and yet capable of being always easily twisted round and drawn out. By this arrangement, more honey could be taken from the stocks about to travel than in any other way, because the honey could without injury be taken from the upper part of combs whose lower part was still occupied by brood.

The carriage of stocks of bees to great distances always involves a good deal of care and risk, but where the railway can be made use of these do, to a great extent, vanish, and the hives are forwarded quickly and gently, as is already to some extent done on Lüneburg Heath. If the use of the railway should become more general, bee-culture might receive a great impulse. Bee-keepers could then send their stocks away to great distances to rape and bilberry flowers, to buckwheat fields, and finally to the heather; they would gain more, and the proprietors of those rich bee pastures could at the same time derive an income from them. For every bee-keeper would willingly agree to a small payment if at the same time some oversight of the hives was provided for, and if greater facilities were afforded for their carriage from the railway to the heath and back again.

Stocks may be carried just as conveniently, gently, and quickly by water on ship or raft, especially down-stream. On such a raft hundreds of stocks could be ranged in piles, and, without being unloaded, could fly to a field of white clover or buckwheat in bloom, or to a neighbouring lime-tree wood, and when the flowers were over, away again, perhaps, by night, to where a fresh source of honey had opened up. It is, indeed, remarkable that this mode of travelling-culture followed by the ancient Egyptians is now followed so little, or not at all, since now, in the age of steam-engines, it could be used with much greater certainty and convenience.

Whatever road and whatever means of carriage are made use of in travelling, the Twin-stock and the double Lager-hive will prove themselves among all other hives as especially suitable for travelling, because without preparation, and in the easiest possible way, they can be set up everywhere in their previous order, insuring the identification of their hives by the bees and economising space. And a roomy box will easily hold twice as many bees as an ordinary skep, and twice as much profit can be derived from the box-hive for equal trouble and outlay. For it is evident that it does not depend so much on how many stocks are set up, as on how many labourers are sent out into the harvest. If the bees of two weak stocks are shaken together into one box, the carriage, and standing space, and money are saved for one hive when such has to be paid. The doubled number of workers will the more certainly yield twice as much or more when the two small sets of combs can be united into one large one; and in autumn the bees suffer more from too little heat than too much.

What was previously divided, should be therefore united again. Was this then not a beginning without purpose? That is what many would ask. This division was by no means without purpose. It was done to keep the bees up to uniform industry, that they might not be concentrated in too great quantities in one hive, and so fall off in their carrying in through being rendered powerless by the heat. It was done to occasion the rearing of young queens, and the making of fresh comb that more honey-combs should be made, since these are capable of holding more honey; and lastly and chiefly, that the number of workers should be increased by the increased number of queens. For at the beginning of harvest the workers are a capital, yielding an abundant interest; and every bee-keeper who knows what is to his own advantage tries to increase their number, and does not begrudge the honey expended in rearing them, because he will hope to receive it again restored in tenfold measure. After harvest, on the contrary, the workers are unproductive capital, of which the bee-keeper must seek to disencumber himself. The intelligent bee-keeper will, therefore, three or four weeks before the close of the pasture, check or prevent altogether the deposit of brood as much as possible, because the young bees coming out will find nothing more to carry.* If at the time when pasture began, and was ever more opening out, he delighted in seeing the cells full of brood, toward the close of the pasture he is glad to see them full of honey. For it is honey and wax that he will, in the end, harvest as the reward of his trouble. But how often does he not see himself deceived in his expectations. While one stock shows considerable stores, another, which was just as strong, stands nearly empty of honey. The reason of it can only be either that the bees carried in less than they might have carried, or consumed more than was directly necessary. The question therefore arises, what should the bee-keeper do that he may have a good balance at the end of the season?

The Increase of Honey Production.

How can the production of honey be increased, and the profit of the hives be thereby enhanced?

This object is gained by following two chief rules—to induce the bees to carry in as much as possible, and to use up again as little as possible of the honey carried in. The utmost surplus will then be afforded that can be taken from the bees.

It is evident the bee-keeper has to aim at getting his stocks at adequate strength at the beginning of the principal honey harvest, which sets in earlier or later, according to the difference of

---

* English bee-keepers must on no account be misled by this peculiar direction, which cannot and does not apply to the system pursued by advanced bee-keepers in this country. As a rule there is no necessity for checking the production of brood towards the end of the pasture, *i. e.* in autumn, for if there has been a fair yield of honey the quantity stored in the brood-nest will do this sufficiently, and, if the season has not been favourable, the breeding will stop naturally, far too soon for the prospective well-being of the hive. The explanation of the Author's direction may, however, be found further on, where he says, 'For it is honey and wax that he (the intelligent bee-keeper) will in the end harvest as the reward of his trouble,' implying that the production of young bees for wintering is not to be thought of when they are produced at the expense of honey, a policy that must cause many stocks to be unfit for wintering, and render 'uniting' very largely necessary.—C. N. A.

the district, so as to be able to take due advantage of the pasture. In this place, however, we are simply treating of the means to stimulate the bees then already present to the greatest industry.

As it is the case when there is a weak, queenless stock in the neighbourhood to be plundered, generally several of, but not all the stocks, fall on the booty, because they have not all come to know about it, so is it when a new source of honey has opened up for the bees at a great distance. Individual stocks may be carrying industriously from bilberries and rape, while others still remain quiet. The latter should then be fed with honey a little thinned, and some of it should be sprinkled in to them at the entrance, so that they may be induced to play in front and fly after those that are already diligently carrying.

If the bees are only set going, they will carry in as long as there is anything to carry. To keep them at regular work, all that has to be cared for is to see that there is no lack of space, and, when possible, no lack of empty cells in which to deposit the honey that is abundantly carried in. Of course the bees build, but they are not able in a time of extraordinary plenty to prepare as many cells as they could fill, and a part of the bees that would otherwise carry must remain behind at home to prepare wax and build cells. As many empty combs, then, as can be got together are put in both to the strong old stocks as well as to the young ones. They will be filled, and may be taken away in a short time, when the stock can do without them. There is no necessity to leave them for the sealing of all the cells, if the combs are meant for the pot or are soon to be put in to young stocks. Only the combs that it is wished to preserve a longer time, perhaps over the winter, or even to put in to light stocks in the winter, must be for the most part sealed. Only these then, may be taken from a stock when it is wished to give it more space, but those that are unfinished and unsealed must be put back again.

If the hive is too limited for a stock that is too strong, the hive must either be enlarged or the strong stock must be divided, so that the heat may not be burdensome to it, and that its industry may not decline. Temperature may be lowered in hives, and the diligence of bees increased, by shading and ventilation; in Twinstocks, by moving apart and conducting cool air from below, just as well as by extension of space.

To save the honey carried in as much as possible from being consumed again by the bees, care must be taken that the brood does not gain any disproportionate extension, and that the deposit of drone-brood be prevented as much as possible, for the raising of brood costs a good deal of honey. Even if the food of the larvæ is prepared more from pollen than honey, yet the brood costs honey which could be collected while pollen is being collected instead. It costs honey which might be carried in by the bees that nurse the brood, the honey which in an abundant harvest could be deposited in the cells occupied by the brood. Effort must be made to limit even the production of worker-brood, especially when the pasture has already passed its maximum. For it is unreasonable to produce, at the expense of honey,[*] bees that can be of no more use, and which are sulphured, according to the custom of many districts, at a time when they have but just left the cell and are hardly properly come to maturity. The method would have some justification if the skins of the sulphured bees were of any value, if the consumed honey could be extracted or distilled from them again, or if there were no means of preventing the production of these bees. In inaccessible skeps it is certainly more difficult, but easy enough in hives with moveable comb, and therefore it is only these that permit of being really rationally managed. The brood is limited by appropriating to it a definite space—the brood-room so often mentioned—and even this in young stocks, for which it would still be comparatively too large, is further diminished in the way previously indicated, by the insertion of a board. Breeding is stopped altogether when the queen is taken away or caged. A queen that is not to be kept through the winter could be perhaps hindered from brood deposit by crippling her or tying her hind legs. But if this seems repugnant to our feelings, it is not nearly so horrible as suffocating thousands of bees. In doing away with a queen two or three years old in a stock that is to be wintered, it would at the same

---

[*] This direction and the reason given for it somewhat further explains that noted above, neither of which, in our opinion, should have place in a manual of advanced bee-keeping. It is reasonable enough to stay the production of brood when the honey harvest is developing, that every advantage may be taken of it in the way mentioned, but to direct that brood-raising shall be stayed on its decline with the brimstone-pit in view, is, to say the least of it, 'bad form.'—C. N. A.

time occasion the rearing of a young and vigorous one. But this must not be done too late, or the young queen would run the risk of not being mated. There is no need, however, to be in any anxiety about fertilisation, even if the queen should not take her excursions before the middle of September, after the massacre of the drones; for if some quite fine days set in she becomes fertile, for even in healthy stocks drones are tolerated far on into the autumn, and in queenless ones that are left standing by negligent bee-keepers, thousands live on into the winter, though it is well if the young queen produces a generation the same autumn, so that we may be convinced she is not only fertile but not defective in other ways, and does not mix worker-brood with drone-brood, or lay the eggs too irregularly.

To what extent doing away with the queen increases the honey wealth of hives, may be often observed in stocks that have accidentally become queenless. When these are not too weak, and with a little drone-brood still carry tolerably, they accumulate sometimes larger stores of honey than other stocks that are all right. If broodless stocks should fall off in carrying in, which often happens in individual cases, they can be excited again to greater activity by introduction of a comb with young brood.

As a third means of increasing the produce of honey there might be added to the two mentioned the introduction of a variety or race of bees more productive of honey.

Among the ordinary black or grey bees many bee-keepers make a distinction between honey-bees and swarm-bees, because the one kind affords more honey, the other more swarms, and with greater certainty. The distinction is obvious. The habit of repeated swarming has become a second nature with the swarm-bees, originating from the heath. They, therefore, swarm before other bees are beginning to think about it. Even second swarms make preparation for swarming and deposit drone-brood the same year, which other bees never do even under the most favourable circumstances.*

---

* This we are inclined to question, though our objection might be set aside by an assertion that the bees we have known to raise drones in their first year have been so-called 'swarm bees;' nevertheless, through the use of comb-foundation liberally supplied, swarms of Ligurians last year were made so quickly prosperous that drone-brood was freely raised, and drones were duly hatched.—C. N. A.

The difference between the black and the yellow, or Italian, race is more striking still. The latter is especially a honey-bee. In excellent seasons all stocks of any strength become full of honey, but in bad seasons, that are not very rare, even young Italian and hybrid stocks show a good quantity of sealed honey, while others are nearly empty. No other result could be expected from their untiring industry and insatiable greediness to get honey, even after the principal harvest, out of every flower, out of sweet fruits, and even out of other hives. In Italian stocks there is no need to enhance the production of honey by any clever methods; there is no need to limit the brood, for through the quantity of honey carried in, which in the late summer is put more towards the centre of the hive, they diminish the brood-nest only too much, because the outer combs are no longer so thickly covered and warm.*

There is no need to take any trouble about putting away aged, weak queens for the purpose of raising young ones. Because of the extraordinary fertility the Italian queens develope in spring, they appear to reach the end of their life proportionally sooner, and retire to give place to young and vigorous successors. When a change of queens has taken place, it is immediately evident in a stock of yellow bees, since the family of a fresh queen generally appears differently coloured—either lighter or darker, according as the queen has been mated with a yellow or a black drone. There is no need to take any useless trouble with an Italian, as often happens in a stock of black bees from which it is necessary to remove an aged or weak queen, for there a vigorous young one will be found, because an exchange of queens will have been already made unobserved.

Just as the breeder of other animals introduces another breed when what he possesses does not come up to his expectations in respect to fineness or abundance of wool, quantity of milk, or capacity for feeding, or, at any rate, seeks to improve it by crossing, in the same way may the bee-breeder enhance the

* The cells in the outer combs of a hive, when filled with honey, are elongated and sealed, leaving only half the space between the combs that is to be found in the brood-nest; hence the outer combs cannot be so well covered with bees. But we take it the diminishment of the brood is due to the *contraction* of the brood-nest by the filling up of the outer combs with honey.—C. N. A.

honey-produce of his hives by introducing a more industrious race. This is, at the same time, so easy with bees, since by one queen a populous apiary can be improved in one summer.

Meanwhile the bee-keeper, towards the close of the honey-harvest, limits deposit of brood to increase the produce of honey, and lessens the number of queens and the number of the stocks, combining with it the more remote preparation for winter.

Preparation for Winter.

The bee-keeper will now be in a position to judge which stocks are worth wintering and which are to be cashiered. Even the bee-owner that wants to increase his apiary, and, for the time, gives up all pure gain, will, among a large number of stocks, even after a good season, find one or more stocks which for some reason or other will not be worth wintering.* One hive has perhaps a queen that is too old or otherwise not fruitful; another has comb too old, or perhaps short and imperfect; a third is suspected of foul brood; in a fourth there is too great a deficiency of honey for it to stand, and its wintering would be too expensive and uncertain.

These are cashiered without further question. For only good and sound stocks must be wintered, and the chief error that can be made, and which beginners especially often make, is in trying to winter what is not worth wintering. A weak stock consumes nearly as much in the winter as a strong one, because the small population in an incomplete set of combs, possibly in addition exposed to the cold, is obliged to make greater exertion to produce the necessary life-maintaining warmth, and notwithstanding all its efforts, and probably because of them, may come to grief at last. Even if it is fortunately brought through, it hardly pays for the trouble and expense of wintering. The clean comb provided with some stores (in autumn) has a higher value in the following spring than a weak, and maybe, dysentery-afflicted stock, that will probably swarm out and away on the first fine spring day, and may cause great disturbance and damage in the apiary. When some of the stocks are cashiered, especially those having comb too short

---

* As might well be expected after the treatment to which we have called attention.—C. N. A.

and too cold, the means is gained at the same time of making others into more sturdy breeding stocks; while the best queens can be added to some, to others bees can be allotted for strengthening, to others the most useful combs can be inserted for completing their nests. It is better to undertake many of the operations betimes, while the weather is still warm, and not leave them over till winter has set in. If combs are put into a hive, it is better done in time, that the bees may properly attach them and propolise the cover-boards again. A honey-comb must, of course, be inserted where the stock itself already possesses some stores, so that the bees, when they gradually move toward their provisions in winter, may not come upon an empty comb and starve. If much liquid honey has to be given for the completion of winter stores, or if much has to be taken out of combs inserted, this should be done in time, so that the bees can at least partly seal the cells that are filled. If there is any necessity in bad seasons to have recourse to substitutes for honey, as malt syrup, and potato syrup, dissolved sugar, &c., feeding should be done, when possible, while the bees are still carrying something, and still possess the necessary power for the purification of the sweet liquids and the evaporation of superfluous moisture. If there were any obligation, through special circumstances, to winter, contrary to rule, stocks with comb too short and too cold, these should in all cases be made warmer by filling up with pieces of comb the gaps, and the too great distances of the combs from the walls of the hive, and the entire comb can thus in some measure be protected against the penetration of cold air from below. The older the combs, the better are they suited for this purpose. If they contain pollen, they will afford the greater service to young stocks, in which there is generally a deficiency of it. Skeps can be reversed, and the gaps filled up, by pieces of comb applied at pleasure, and the hive can be left overnight in this position. If the pieces contain here and there some honey, or if some had been poured on them, the bees cover them immediately, and attach them so that the hive can be put in its former position again in the morning. In spring, when the bees begin to build again, the pieces of comb that may have been put in irregularly, or in a horizontal position, can be, of course, removed again.

Many of the stocks to be wintered will need reinforcements of bees. A rational and sensitive bee-keeper does not let the bees of the cashiered stocks die, still less does he sulphur them, but allots them to other hives. With proper treatment in other respects there will in autumn be no such quantity of bees present in the hives that there will be any occasion to be afraid of any over-population of the hives for wintering. When the deposit of brood has fallen off, the bees in the hive diminish very much, and if they are to be strong in spring they must in greater strength enter upon the winter, which will destroy a good many. Even if a strengthened stock consume rather more, this can be well afforded, because the larger consumption will be richly repaid in the next spring and summer.

When reinforcement or uniting is to be taken in hand depends on circumstances, chiefly on the construction of the hives. If travelling bee-culture is followed, it would be best to undertake the operation as soon as the hives have been brought home from the heath—before the bees have flown out—so that they may make their first flight immediately from the hive to which they have been added. When time is not available to unite them immediately, or if it could not yet be undertaken because the comb that is required still contains some brood whose coming out is to be waited for, the weak hive must be put as near as possible to the one to which later on it is to be added. The later the driving and the union of the bees come off, the less likely is the comb to be attacked by wax-moths, and the more harmonious are the bees, because they are no longer so excitable in the cool late autumn. One queen, of course the older or otherwise worse, is removed, and the other is caged for some time, if it can easily be done. This is not necessary if the bees to be added have been kept queenless for one or two days. If the bees of the stock to be reinforced are then brought into some confusion by smoking them strongly, and at the same time gently knocking, or if a part of the bees from the combs taken out are shaken down on the floor of the hive, and caused to run and hum in the same way as when a stock is driven, they will not then think of attacking and stinging to death the strange bees shaken to them, especially if they are sprinkled a little with honey. Fighting happens mostly with bees that have been long without queen, and have become drone-breeders, because they probably bring along with them a peculiar odour. In that case the two parts must be brought into yet greater confusion by shaking them all together in a transport-hive, sprinkling them with honey, and as often as fighting is noticed, jolt the vessel violently on the ground, so that the bees fall down and get the more mixed together. Strong smoking with the stupefying methods already often mentioned render good service then. Sprinkling both divisions with the uniting-spirit recommended by many bee writers, can at any rate do no harm. Its principal elements are simply spirit and honey.

The bees are most harmonious at night, therefore the uniting is best undertaken after dark has set in. Of course, the bees to be added must be already in a combless skep or transport-hive in which they have been brought from a distant apiary, or the moveable combs of the stock to be cashiered must already have been loosened by day, and the queen have been caught, because she could very easily be overlooked in the evening by lantern light. When the bees have been shaken in, a piece of touchwood of the size of one's little finger may be allowed to smoulder slowly in the hive when it is shut up again, so as to insure against all hostilities.

If it is with skeps we have to do, the bees can be jerked by a violent blow on to a sheet that is spread out, and they can be allowed to go into the skep that is to be strengthened, which is put upon or near the sheet. Of course, the comb must have been previously taken out, or, if it is wished to preserve it, the bees must have been previously driven into an empty skep. This is done by drumming, as previously described on page 176 and the following pages. Though there is no need to be so particular now as formerly, because the combs have no brood and are light, and so do not easily break down, and if they do the damage is not so great. After the bees have been mostly driven out of their nest by smoking and knocking, it can generally be managed by a violent jerk to throw them down into the empty skep that was put at the top, but of course is now again at the bottom. In skeps that have only half been built up, an empty skep can be quite dispensed with, and during the driving only laying on a cap or a cover. The bees can be immediately jerked on to the cloth spread before the stand, or on the grass, and be allowed to move directly into the skep to be strengthened.

The driving out of bees from hives with moveable combs is very much more convenient. In Twinstocks the bees unite of themselves if one queen is taken away, and after about two days both compartments are put in communication. If a stock is added to a box-hive standing under it, the bees are shaken to them in the way just mentioned, the entrance of the emptied hive is stopped up, but another entrance is made in the other hive in the crown or by the door, as near as possible to the former entrance, after removing the flight-board, so that the added bees which seek an entrance at their next flight may go into the lower hive, to which they have been allotted, until they gradually learn the new flight.

The two stocks which have their flight in the same direction can be united both in Twinstocks and Press-hives, especially if the lower stock is united to the upper one. The hive at the bottom that has been emptied is covered up for some time by a door, shutter, or some pieces of board inclined against it. When the change has been made in the reverse direction, the hive that has been emptied at the top is lifted off and the lower one is in all cases put a little higher. The bees, however, right themselves without this if their new hive is situated close to, and if a quite fine day sets in after the union. They hear then immediately the joyful hum of the other bees, and fly to it. The bees should then be incited by a little food to general play in front of the hive, so that they may fully recognise the change that has been made, that when they fly out at a cooler time they may not have long to seek for it nor get lost.

It is not advisable to unite two stocks which stand in the same apiary but at a considerable distance from one other, because the bees that have been added, at the next play in front of the hive, fly to their old place, and for the most part get lost. The union must then not be undertaken before November, and the united stock must either be put in a cellar or shaded from sunshine, so that the bees may make no flight before they have had a rest for several weeks. They then carefully note their hive, and at any rate will not fly in such large numbers to their previous position.

It is again conspicuous in uniting, as it was in dividing, how very advantageous it is to have two apiaries at a distance from one another. Any stock of one apiary can then be united to any stock in the other, without even one bee getting lost at the next flight, if in other respects a fine day favours it. There is no need to put every weak stock into a separate Transport-hive, but as many bees may be poured together into a larger hive as there is room for. Of course, the queens are taken out. The combs that are most thickly covered with bees can be put in the Transport-hive sloping to the sides and leaning toward one another, with little strips of comb between to prevent crushing the bees, and from the other ones the bees can be shaken off. The few bees still remaining behind in the hive are brushed down into a box held underneath, or smoked and brushed out of the hive, and, after they have collected about the stopped-up entrance, are allowed to move into the Transport-hive. Arrived at the other apiary, as many bees are shaken out to the stocks to be strengthened as are suitable, or as the stocks with the stores they have can sustain. They may either be added immediately or after they have been kept for a day without a queen. Even if the stocks to be wintered were already populous enough, advantage will be derived from the strengthening, because the surplus of bees can be used in the early spring for artificial swarms. The noted bee-keeper Knauff used to drive six or eight stocks together in autumn, and increased the number of his stocks next season to a corresponding number by aid of the empty comb he had preserved. Although such a considerable union is not advisable, yet two moderate stocks can always be driven into one hive, especially with hives with moveable comb, since in these increase can begin sooner next spring with a store of bees and half-full comb, and can be pushed on more certainly and vigorously than was possible to Knauff with his straw skeps.

But it is necessary that there should be no lack of air to a strong stock if it is to winter well. While the entrances of weak stocks are considerably diminished in autumn, because of robbers and cold, it does not do to treat strong ones similarly, even in severe winter. If the combs reach down to the floor they must be shortened a little in autumn, that there may be room for the rubbish and dead bees that fall, that they may not block the passages between the combs and prevent the access of fresh air.

In selecting breeding stocks, the bee-keeper must not only

have his chief attention directed to whether the stock has a queen, but whether she is still vigorous and free from defect in every respect; for what a wintered stock, that is without queen or with an incapable queen, consumes, is just thrown away, because it does not generally live over the winter, and the few bees which perhaps might live to the spring, have no value at all, and a queenless stock, in the disquiet and excitement in which it usually is, generally consumes far more than a sound one. It might, therefore, here be the place to speak more particularly of the signs of the condition of things being right or wrong as respects the queen, which are at this season not so evident and palpable.

Signs of the Possession of a Good Queen and of Queenlessness.

In spring and summer the presence of worker-brood is an infallible sign that a fertile queen is present, or at least was present, a little while ago. Even in autumn, when the queens have left off laying eggs some time, for the most part traces of worker-brood will be noticed in sound stocks, both in the brood-combs as well as on the floor of the hive, because after the close of the pasture the bees draw nearer together, and do not any longer maintain the high degree of heat requisite for brood, and a part of the last young brood dies in the cells, and later on is thrown out. If, on the contrary, in autumn traces of drone-brood are still to be found, either in the cells or on the floor of the hive, it is a sign either of no queen being in the hive or that she is defective. Sound stocks, as is well known, get rid of the drones toward the close of the pasture. Even if exceptionally, some stocks otherwise sound keep their drones unusually long—even into the winter, they do not, however, breed any fresh ones. But if this is the case, and drone-brood is in drone-cells, there is no queen present; if it is in worker-cells, in which case it is called *buckel*-brood, because the cells are then built up high, there is then either an unfertilised or degenerate queen present, although even in this case the eggs may have been laid by simple workers, if there is a lack of drone-comb in the brood-nest.

A queenless stock turns its attention, as is well known, to drone-brood, and builds ordinarily only drone-comb if it continues to build. If, however, it is not building, the drone-comb already present is cleaned and prepared for the reception of brood. This hankering after drone-brood continues for some time after a fertile queen has been added to the stock that has been queenless, so that she often at first occupies some of the drone-cells, so diligently cared for by the bees, before she begins to lay worker eggs. Drone-brood in large cells, while the worker-cells are entirely empty, is always a tolerably sure sign of entire queenlessness, because the case of a worker-bee laying eggs along with an unfruitful queen is extremely rare. The queenless bees form at the same time queen-cells here and there near the drone-brood, and often fill them with several eggs, just as in other cells often entire heaps of eggs are found.* A further infallible sign of queenlessness is when pollen-cells are found extended and transformed into the beginnings of queen-cells. The bees do this when they have no brood at all. They then treat the cells that contain the principal element of brood food as they would otherwise treat brood-cells—they seek to transform them into queen-cells. Mere empty foundations of queen-cells or little cups have no significance. Such are often begun by stocks with a young queen.

A queenless stock possesses no brood, or at most, after long queenlessness, only a little drone-brood; but the bees, still carrying in continually, especially at the beginning, when they cherish the hope of raising a queen, accumulate a quantity of pollen, so that nearly all the cells in the brood-nest are filled with it. From this, and when the pollen cells all appear to be varnished over, because the bees are not using from them, it may be safely concluded that the fertile queen has been absent some time. But whether there is no queen at all, or only a young unfertile one present in the hive, can be determined from the general behaviour of the bees, if the signs mentioned above do not afford any certainty. A queenless stock always shows a certain disquiet, excitement, and confusion, which expresses itself most strongly after the loss of the queen, or rather after this has been noticed by them, and is never entirely pacified. The bees, restless and seek-

* Would it not be nearer the truth to say the bees build the queen-cells around the heaps of eggs already deposited by the fertile worker? We are convinced that bees occasionally remove eggs to queen-cells, but think it 'not proven' that they transfer them in heaps.—C. N. A.

ing, run about in the hive, and principally about the entrance, just as in sound stocks on warm summer evenings they try to keep off the little wax-moths that swarm thereabout. When smoked, or even without this provocation, the bees raise a continuous wail, a genuine cry of distress, that can be heard at some distance from the hive. Early and late, when other bees are quiet single bees keep coming out of queenless stocks, especially when the bees, after long rest, have been able to take a flight again. The ordinary merry playings in front of the hives, which sound stocks so often indulge in, are quite absent in queenless ones; for, as is well known, it is the young bees and the nurse-bees principally that indulge in play, both for sanitary purposes, and at the same time to become acquainted with the position of the hive. But there is no occasion for this playing in front of queenless hives, because no young bees have made their appearance, nor is brood-food prepared by those already existing.

In regard to the disquiet or wailing of the bees just mentioned, it must be observed, to avoid deception, that the signs often occur when the queen, pursued and seized by a hostile party, or only by one bee, gives vent to cries of distress, whereupon part of the other bees afford protection by enclosing her in a thick knot; others begin to run anxiously to and fro complaining. A certain hissing is then to be perceived in the hive, single bees stung to death are seen to fall down on the floor, and among them the queen herself is often found. But quiet is often restored, probably by the hostile bee, which had lost itself in playing in front of another hive, being stung to death before it could give a fatal sting to the queen. But even her own bees often attack their queen in a hostile way, and try to sting her to death, especially when she leaves the hive, perhaps to be mated, or even if she only leaves the brood-nest and returns again, because they probably suspect her of being an intruding stranger, inasmuch as the natural dwelling of the queen is the brood-nest alone. But if a bee has once seized her and sprinkled her with the sting poison, generally more bees attack her, so that often the young queen coming back fertilised is killed in her own hive, on account of a strange odour brought back with her, if the attentive bee-keeper does not rescue her from this danger by caging. Here the convenience of moveable combs is again very manifest, because it is only by such prompt help as can be gained by taking the brood-nest to pieces that thorough conviction can be obtained of the condition of a hive suspected of being queenless. In other hives, on the contrary, one is kept in uncertainty for some time, because the bees keep up a tolerably uniform industry, especially if they are capable of rearing drone-brood, and their condition is only recognised when nothing can be done to help them.

If the queenlessness of a stock otherwise provided with good comb and sufficient stores is noticed in August, it can be put to rights by the simple addition of a fertile queen. With its large store of pollen, the stock will yet raise thousands of young bees, and may become a good hive to stand the winter. Later on, perhaps after the middle of September, there would have to be an entire stock added to a queenless stock, as well as a queen, if it is wished to winter it. The queenless hive could then only be considered as a hive filled with comb, and of course to be set up, or the comb of the queenless stock to be inserted, in the place where the added stock had their flight, if it has not been brought from a distant apiary. No care need be taken of the tolerably old, and therefore enfeebled, bees of the queenless stock; indeed, if they had raised drones, it would be advisable previously to drive them out altogether, and let the new occupiers move in with their queen, letting the bees that have been driven out fly back to the hive individually, by way of addition, so that neither the drone-mother nor yet any unfertile queen that may have been present in the hive should endanger the fertile queen. If the driving of the bees be very difficult, as in Log-hives, in order to prevent a greater evil, we must choose the less, and sulphur the already weak and queenless stock, seeing that the bees that have been long queenless are generally stung to death, causing, at the same time, the death of a part of the others.

In case the comb of a queenless hive is already old, it may be better to make short work of it, and do away with the stock, keeping, however, the combs that have much pollen in them. These may be put in under the brood-nest for consumption to young stocks either immediately, if breeding is still going on, or in the spring. If, however, a part of the comb is left in the crown of the hive, a stock introduced into it next spring will certainly thrive excellently, because there will be much pollen among the honey.

Even the bee-keeper who has no intention of increasing his apiary, and only wishes to derive profit from his stocks in the form of honey-comb, will do well to preserve against possible accidents hives partly furnished with comb, or a store of empty and half-filled combs, so as to be able to put together sets of combs from these. He may then, in the early spring, put into these combs stocks of bees whose combs have been damaged by mice or dirtied through an outbreak of dysentery, or that have been ruined by an unfortunate accident, or have otherwise become unserviceable.

THE REQUISITES OF A STOCK WHICH IS TO STAND THE WINTER.

The things necessary are—

1. *A sound and vigorous queen, not more than two years old.* One that is already three years old we must not winter, and must not shirk the trouble of catching her out of an otherwise strong breeding stock and replacing her with a young one; for since most queens die naturally in about four years, if they have not met with any accident before, the stock does not yield nearly so much as it might if the death of the queen takes place before the next swarming time, quite apart from the fact that an old queen, in consequence of her infirmity, does not lay so many eggs as a young one, and at the same time is especially inclined to the production of drones.

2. *A sufficiently strong stock relatively to the size of the hive.* If 5000 to 8000 bees are sufficient for a small skep in autumn, then for a moderate box-hive 8000 to 12,000, and for a large hive 15,000 to 20,000, will not be too many. For after an unfavourable winter, perhaps there will not be half of them left at the beginning of spring. A stock is relatively strong when its bees can fairly cover the entire comb in the warmer days of autumn.

3. *A clean set of combs, as complete and free from drone-comb as possible, neither too old nor yet too recent.* The bees always winter better in old comb. Other things being equal, a stock for wintering should be chosen with old comb, even if it has to be cut in the spring for the sake of renewal, or even if the stock has to be put into fresh new comb.

4. *A store of honey which will last at least until the following March or April.* It is true that a moderately strong stock consumes, when perfectly quiet, not more than about one pound per month. But as the consumption of honey is greater in colder weather, and, again, in more frequent flight or disturbance of other kind, to be on the safe side it is best to reckon twice as much, that is, about 10 lbs. from about October to March, and about as much again for the time between March and the next honey-season, which would, perhaps, not set in before June. We shall do well when we have a liberty of choice only to winter such stocks as possess the whole amount of honey, which by long experience is found to be required, in sealed cells, because we should afterwards have to give twice as much liquid honey, and, at the same time, disturb the bees inopportunely and to their own detriment. But when there is necessity for feeding after a bad season, not any more liquid honey must be given in autumn than is necessary for wintering, because from experience we know that stocks with too many open honey-cells suffer severely in winter. The deficiency is made up to stocks that are quite too poor in honey with pieces of yellow or brown barley sugar, a part of it especially the little pieces, being dissolved, boiled, and given to them in the liquid form. The barley-sugar has to be so put in that when the bees have consumed their store of honey, they can have access to it immediately, and be able to dissolve and consume it gradually through the heat, as well as by the moisture condensing in the hive, or given to them by means of a sponge or a comb. It should, therefore, be put immediately over or near their winter-nest and winter-store. In box-hives the pieces of barley-sugar are best laid over the bars, after the cover-boards have been partially removed or moved away from the wall. In log-hives they are put into the comb into a cavity made quite at the top in the crown, and the cavity must go down deep enough until either honey or bees are met with. In skeps the pieces must be put into a little box made of thin wood, and put on over the open bung-hole. Skeps without bung-holes must either be provided with them—they can very well be made with a sharp knife—or otherwise the skeps must be reversed, and in the comb over the nest of the bees a cavity must be cut, into which pieces of barley-sugar are put and covered with a comb, and the hive, covered by a board, is left standing in that position.* Sugar can further always be put beneath the bees'-nest. The comb must then be so far shortened that the bees partly hang below it, and the barley-sugar should be put on a board or on a firm comb close under it. In low Lager-hives the sugar can be laid under the bees'-nest on the floor of the hive. Here the bees can the more easily dissolve it because the necessary moisture is deposited on the cool floor. Even if, in severe cold, the bees have to move up and take refuge in their comb, they come down in mild weather and partake again of the sugar put in below, and so save their other stores, of which they must, of course, not be quite deprived. Since bees in Twin-stocks, double hives, and all manifold hives, as a rule take up their quarters for the sake of mutual warmth on their dividing party-wall, and move on this wall up and down, backwards and forwards, the food must always be put in by the wall mentioned, whether it be pieces of comb, barley-sugar, or solid honey wrapped in paper, so that they may always have it at hand.

In choosing stocks to winter and honey for feeding, there is need not only to consider the weight of the honey, but of what it consists. After the season 1858–59 the latest swarms, contrary to ordinary experience, wintered best, because they had only wholesome buckwheat honey, while old stocks and early swarms, which previously had stored much pine honey in the crowns of the hives, were obliged to use from it and suffered terribly from dysentery. The moral of this is obvious, and does not need any further explanation. It does not matter whether the honey is fresh and liquid, or is older and partly candied. If many have come to the conclusion that bees could not dissolve crystallised honey in winter, and must starve with it, this idea has simply been copied by them from others, and has not been arrived at from personal observation; for if the bees can even dissolve the pieces of candy that are much harder, they will be in a position far more easily to consume the crystallised honey if they are not too weak, and are able to produce the requisite heat. Even if by accident many a grain of honey falls down, this is not lost if the bee-keeper does

---

* This seems to be a peculiarly prolix and awkward way of giving bees barley-sugar. In our experience it has generally been sufficient to thrust it *between* the combs close to the main body of the bees, so as to disturb the latter as little as possible.—C. N. A.

not improvidently brush it out. In mild weather individual bees come down on the floor and dissolve the separate crumbs. But if a stock crumbles down an unusual quantity of such honey, either it possesses but a few bees which are incapable of keeping themselves any longer properly warm, or they are so tormented by violent thirst that they seek madly after the small quantity of moisture contained in the honey; for some moisture is just as necessary for the solution of crystallised honey as of sugar. If bees assigned only to such honey can neither obtain this moisture in the hive nor carry in water, it is then the bee-keeper's part to provide it, for which the injection at times of a small spray of water through the entrance will be sufficient.

About the middle of October there was put into a box-hive another comb with candied honey, which the bees immediately began to carry in part into the middle of their comb. On each of the succeeding days single bees out of the hive flew after water, even in rather cold weather, and others, with the greatest eagerness, licked up the dew which had formed in the entrance. When, in consequence of this obvious sign of great thirst, a comb full of water was put in, the bees sucked it dry in a very short time. The same result followed when the comb was filled and put in again, and it was not before a third supply had been given that the bees left half of it as a sign that their thirst was quenched. No bee flew out any more, and the colony that had previously been so much excited soon relapsed into perfect quiet.

From this example it may be seen that bees have need of much moisture, not only for the preparation of brood-food, but for other purposes, and that want of it in autumn and winter, when rest is necessary for the bees, may result in disquiet, uproar, loss of many bees, and at last in the ruin of the entire stock.

But too great dryness is the necessary consequence of the too great retentiveness of heat of the walls of hives, so that a certain moderation must be observed in this respect when bees are left to themselves in winter, as has been already remarked in the description of the hives. For the bees can, in case of need, produce greater heat, but cannot, by any amount of exertion, produce moisture if there is never any deposited in the hive. But the deposit may occur even in the comb of the bees wherever a sharp dividing-wall is formed between a warmer and colder stratum of air. The cover-boards form in box-hives a dividing-wall between the warmer air in the nest of the bees and the cool empty space in another part of the hive. The cooler the under-surface of the cover becomes the more abundantly will the warmer vapour be condensed under it. And this will occur the more abundantly when the upper and lower rooms are carefully divided, so as to prevent all draughts and equalising of temperature. It is evident, then, that the space over the cover must not be filled up, and that, in feeding with candy, a more extensive condensation is obtained when panes of glass, zinc, or slates, are used for covering.*

But moisture will also be deposited sidewards, and in the cells of the comb which separates the cluster of bees, and the warmer stratum of air prevailing in it, from the cooler space that is unoccupied, or not at all built up. As more moisture is found on the window where no second outer window or shutter mitigates the influence of the outer cold air, so the most moisture will be deposited on the comb mentioned when no empty one not covered by bees stands in front of it, and when the warmer stratum of air on that side is parted or isolated from the colder stratum on this side by stopping, or narrowing, as much as possible the distances and passages between the edges of the combs and the walls of the hive. In using frames which stand at a distance from the wall of the hive all round, door and gate are, as it were, left open for the streaming away of warm air out of the winter nest of the bees; and there is no need for surprise that so much complaint is made by possessors of frame-hives of their bees suffering from want of water, thirst, and bad wintering.† The reason lies in the im-

---

* The principle of hive construction implied in this paragraph is diametrically opposed to that we have for years been striving to instil into the minds of British bee-keepers. In England it has been found that the condensation of the vapours from the bees upon the unsealed honey, the pollen, the combs, and the walls of the hive, has been the chief cause of winter diseases, while the use of the quilt, which permits the gentle evaporation of moisture without loss of heat, has been the greatest safeguard against them. We cannot advise our British friends to provide cold surfaces within their hives for the vapours to condense upon unless with a view to preventing their condensation on other parts of the hive, and that the liquid formed may at once be conducted to the outside, away from the bees.—C. N. A.

† In this sentence the author is distinctly in accord with us, and delivers a thrust in our favour against the use of open-ended frames which leave open both door and gate for the streaming away of warm air from

moderate retentiveness of heat of the hives and in the frames, which are just as injurious in the brood-nest and winter-quarters as they are convenient in the honey spaces.

But most bee-hives suffer from the opposite fault. They are altogether too simply and lightly built, afford for winter too little protection to the bees against cold, and produce too much moisture. In summer the light construction has no disadvantage, since the shade of the roofing or of trees standing near affords sufficient protection against hot sunshine; but the damage in winter will be the greater if careful packing up does not supply what is wanting to the hive of heat-retaining power. It is time for this when actual winter threatens to set in, or has already really set in, which in our district is generally about Martinmas. This brings us to the section on wintering.

Wintering.

After the preparations made some time beforehand already mentioned, the wintering itself is quickly provided for, since the bees, if their hives are built sufficiently retentive of heat, require nothing from their owner or attendant but to be left as much as possible in quiet and peace; but he must not only leave them in peace, but insure it for them by keeping off whatever would disturb it, as mice, titmice, wood-peckers, and other peace-breakers. Against mice the entrances are barred by nails or little pieces of wood, so that the bees can go in and out but not their enemies, not even the small shrew-mice which slip through tolerably narrow chinks. Against titmice, that often disturb bees in winter, entice them out on to the snow, and then peck them to pieces and eat them, the simplest means would be to catch them, because those that have once made themselves at home in an apiary come again every day regularly; though in respect to the use of those birds in other ways one would rather protect the hives by boards or straw mats inclined in front of the entrances, by which there is kept off the sunshine, which, when the snow is lying on the ground, entices thousands of bees out to their death.

the brood-nest, causing inordinate consumption of food for the reproduction of the escaping heat, and frequently leading to dysenteric symptoms amongst the bees.—C. N. A.

There must be no forgetfulness in putting aside these things when a day occurs when the bees can, without danger, have a cleansing flight. But if small boards, about four inches broad, were applied to keep off the birds and sunbeams, and the bees could easily get into the hives round the boards, there would be no need to be in any concern about removing them on mild days; for ten times, if not a hundred times, more bees might get lost if the entrance should be forgotten to be darkened again after fresh-fallen snow, followed by bright sunshine, especially in hives that have their entrance high up towards the crown.

It is not advisable to close the entrance with a perforated zinc slide, because the bees, when they have kept quiet a long time in the hive and long for a cleansing flight, enticed by sunshine breaking in on them, might get into general uproar, and try to make an exit for themselves by force, and partly worry themselves to death. It is then better to stop up the entrance entirely, and to open it again in the evening. For old stocks that are mostly built up, and in well-constructed hives, there is no need to take any further precaution against cold. If it is wished to do anything besides, the empty room in Ständer hives can be filled up at the top with straw, and in Lager hives the empty space at the side. In young stocks with tender and incomplete comb, the empty spaces of the brood room may be filled up; moss is especially suited for the purpose, because it accommodates itself to every space, and when it has become damp does not produce any mould or fusty smell. It is, therefore, not at all necessary, nor even well, to let it become quite dry, because a certain moisture in the air appears rather to contribute to the health of the bees than to be injurious. In moss, which like sponge is capable of taking up a good deal of water, the necessary water might be given to the bees for the solution of dry candy and crystallised honey. The filling up should not be undertaken too early, while the weather is still warm, or wax-moths may find a harbour in it and partially attack the comb. In autumn they are glad to move into the crown on account of the warmth, and they may easily be found by and under the cover-boards and be destroyed. A layer of straw, or, better, a straw mat, may be put up on the inside before the door of box-hives if it is not sufficiently retentive of heat; but in that case, when the hive is built up, the first comb at least must be previously taken out, which would without that easily get mouldy from the moisture forming on the door. The wooden door can then sometimes be entirely taken away on mild days with some wind, so as to let more fresh air stream into the hive. The straw mat, that has after some time become a little damp and fusty, may be exchanged for one that is dry and aired. This is, of course, not necessary in hives with doors retentive of heat, especially straw doors.

There is no need of an outer packing for well-constructed hives; it does more harm than good, because it provides hiding-places for mice, which, even if they cannot penetrate into the hives themselves, yet by their biting and nibbling disquiet and disturb the bees in their winter rest. In Manifold hives which stand alone, or are put together with the doors outwards, the large door in common, if such has been made, is now of course put in front of the four or six single doors, more for the purpose of keeping off the driving rain than the cold. If this outer door is fastened by pegs driven in, or more strongly fastened by long wood screws, the hives will be in some measure secured against robbery. But this last object is best attained by the six or eight-fold hives set up in pairs, moved together before winter, and fastened together in the way mentioned where they were described. If the crevices are then stopped up where the two hives do not come quite close together, and they are surrounded at the bottom with forest litter, leaves, or moss, they form with one another, and with the ground an uninterrupted whole, and the most extreme cold cannot exercise on them any special influence. In the Pavilion this is more the case in proportion as its circumference is larger, and the more heat streams up from below as it is withdrawn from the outside.

Other box hives set up one over another are united by stopping up the chinks with moss, wadding, or paper, or by plastering with clay. Under the floor of the lowest hive, which is perhaps only thin, especially of the Twin stock, moss, straw, &c., are put at least at the sides, so that at most nothing more than tempered air from the earth can penetrate to it, and no cold air from the sides. Whether any thing has to be put in at the top under the roof for the protection of the crown-boards of the uppermost compartments depends on the thickness of the crown-boards, and whether the bees have their quarters immediately under them, or whether there is a large space between; though greater protection does no harm, and would only be inappropriate when the bees need moisture for the solution of dry food, and for the preparation of brood-food towards spring.

If there are unoccupied compartments in a pile of Twin-stocks or in a Manifold hive there is no need to fill them up with straw; if each be shut and the entrances stopped up, the immoveable stratum of air enclosed in it forms as warm a wall as a layer of straw equally thick. They can be used for the preservation of empty combs and comb beginnings, which should be taken out of the honey rooms of occupied compartments in autumn, because they might easily decay and become useless there, in consequence of the vapour rising upwards.

It is usual to wrap log-hives round with straw, especially on the side of the door; and when this, as is often the case, consists only of a thinnish board, it is a good thing to do; otherwise this wrapping up is not necessary for hives retentive of heat.

Skeps with thick walls closely sewn together do not need any further protection against cold; whereas skeps, with only thin walls, as they are here and there to be met with, are better covered with an outer wrapping of hay, moss, straw, &c.

Besides, a moderate coldness does no harm to the bees, and strong stocks do not suffer with a temperature of $-10°$ R.($9\frac{1}{2}°$ Fahr.) but if the temperature falls still lower to $-18°$ or $-20°$ R. ($8\frac{1}{2}°$ to $13°$ below zero, Fahr.), and such a degree of cold is maintained for several days or weeks, it penetrates gradually the thickest partitions and walls, the bees hum with a higher tone, and weak stocks suffer considerably. As rooms, if they are to be kept at a certain temperature, have to be more abundantly and frequently supplied with firing, so must the bees put on coals oftener; that is, they must consume more honey, whose essential element is carbon. As the badger in his winter sleep consumes his own fat to maintain the degree of heat indispensable for the maintenance of life, so does the stock of bees consume the fat stored up in summer, that is, its provision of honey; and it consumes it the more abundantly the more it is obliged to exert itself to produce the necessary heat. The badger passes its winter, as is well known, like many other animals, in the earth in its burrow, and escapes thereby those extremes of cold which often prevail over the surface of the earth in our climate. We bury in the ground vegetable produce to which frost is dangerous or destructive, that it may be protected from it. Stocks of bees may be buried, or put into pits in the ground, as is often done in Russia, and thus as mild and equable a winter may be secured as can be desired for bees.

The burying of bees and putting them in cellars has been many times discussed in the *Eichstädt Bienenzeitung*, and Pastor Scholz has lately recommended in it the putting of stocks for the winter into the so-called 'clamps,' as it is now generally usual to winter potatoes; and when provision is made for air, and the hives are secured against mice and wet, the stocks must winter in them healthily and cheaply. This, however, does not secure anything beyond the temperature of the bee-hives. The bee-keeper wishes to visit his charge now and then in winter to convince himself that his pets are alive and well, not in want of anything, and that no mouse has been clever enough to find an entrance, &c. We are able to put potatoes just as well into the cellar for the winter as into the pit or clamp, and then may be free from anxiety as to whether they are rotting or being destroyed because we can look at them every day, and bee-hives may be put much more conveniently for winter into a dry cellar or other kind of cellar-like room. There is no necessity for this to be frost-proof, since the bees, as already mentioned, do not take any harm in $-10°$ R. ($9\frac{1}{2}°$ Fahr.) But much is already gained if they have only to stand 10° instead of 20° of frost, while they are at the same time withdrawn from all other destructive influences and dangers, especially the danger of them dying by thousands on the snow in sunshine. Instead of putting up the clamp every year, it is in all respects better, where no suitable cellar is available, to make such a cellar-like room as a permanency near to the apiary. It may be used for many purposes in summer, as for instance to hinder stocks for a time from flight by putting them in the dark, to preserve them from robber bees, to withdraw honey, that has been taken, quickly from the attack of the bees, or even as a garden-house or milk-cellar. The chief requirement is that the room shall be perfectly dark when the door is shut, so that the bees that are put in remain quiet even when mild weather sets in.

For this mode of wintering the Twin-stocks are especially

convenient, for a quantity of them may be lodged in a small space, since they may be piled right up to the ceiling of the place. They are best put up in piles as they have stood in the open, but in reversed order. In setting out again the topmost boxes come at the bottom as they stood before. The separate hives should only have another position allotted to them after a winter rest of about three months for some special reason, as for the purpose later on of making artificial swarms more conveniently. Otherwise they should have exactly the same place given to them making a note on the hive of its stage and aspect if memory is not to be trusted. In travelling, attention must be paid to this point that the hives are put up again similarly especially at the same height, because the bees which flew at the bottom seek their hive again at a new stand in a similar position, and finding it there become more quickly ' at home.'

There is no need to be over-hasty in housing for the winter even if it should begin to snow and freeze. There might still be genial days coming, and it would not be well to have deprived the bees of another opportunity of cleansing themselves before winter. The later the bees can make a cleansing flight in autumn the longer can they stay in the hives without a flight in spring without injury to their health. If there really comes such a day late in November, or even in December, on which individual stocks begin to play, the others should be incited to play by knocking, breathing into the entrance, syringing honey into the hive, so that the bees may not sleep away the favourable opportunity for flight, which perhaps will not return before March or April.

When the winter at last sets in in earnest, the hives can be housed. Where no cellar or other kind of refuge is available, the plan recommended by Count Stosch in the *Bienenzeitung* may be used. He recommends that the box-hives themselves should enclose such a dark space with somewhat tempered air by placing them one against another with their entrances directed towards the enclosed space, like the four boxes for making a Pavilion (compare ground-plan, page 123), only the fourth side may very suitably be formed by a wall against which three or more hives, or piles of hives, are set up in the way indicated. If the entire room is built up under a roof the erection will involve less trouble.

There have been cases mentioned in the *Bienenzeitung* in which housed hives have not wintered so well as those left standing in the open. This must have had its own special causes, with respect to which nothing more can be done than make suggestions. Either the bees were exposed to disturbances, or they suffered from want of water, which may occur here in hives that keep very dry, sooner than in the open, or there was a lack of fresh air because the entrances perhaps might be too much narrowed, or quite closed, so that the air could not well be renewed in a place free from wind; while in the open, gusts of wind can always drive some fresh air into the hive, even through a small entrance. The entrances must not be closed any longer than while the hives are being taken in, and when this is finished they must be opened again, so that every bee that wants to leave the hive for once, perhaps because it has dysentery, may find a free exit and not disturb more bees through its biting and humming. Even if the floor of the cellar is strewed with bees, there is no need to be disturbed about it. The dead from several hives do not amount to many for each hive, and in the open far more bees would have been lost. As a matter of course, when separate stocks in complete darkness show a greatly disturbed condition, the cause has to be inquired into and removed if possible.

What has to be seen to in Winter.

This has for the most part been touched on in what has gone before. The more the cold increases, the more carefully must every disturbance of the bees—every continuous noise like wood-chopping—be avoided in the neighbourhood of the hive; for individual bees separate themselves then from the cluster and easily chill inside or outside of the hive, before they can get back again into the warm nest. But if the disturbance is greater the bees set up a hum, uncluster, and become more extended. When they contract again, many stragglers are left behind and are lost, as at the setting in of the first severe cold in autumn. The entire stock is excited by repeated disturbances to greater consumption, and since the bees lick up at the same time the moisture deposited on the walls and combs of the hive, they become more gorged, and there is a greater liability to dysentery. Disturbed out of their natural rest, the bees might begin breeding before the time, which is highly injurious, and may lead to the ruin of weak stocks. In stocks that are housed, a too high temperature of the place might have the same result. In that case the room should be cooled by opening it for a time, perhaps for a whole night, when there is no severe cold, and the air will be at the same time renewed. It is better when two degrees of frost are maintained in it than when there is too high a degree of heat. A temperature kept about freezing point might be the most suitable, as those winters in which the quicksilver does not fall much below, or rise much above that point are specially favourable to bees. Very changeable winters, in which severe cold is followed by thaw, and again by frost, are very detrimental. The rime formed in the hives melts, and freezes again into ice, and makes the succeeding cold to be felt more, and may stop up the entrance. This may be also stopped up by the rubbish and dead bees, and many a stock has died from this cause. Inexperienced persons think then that the stock has been frozen to death, whereas it has really been suffocated. The bees may be quite deprived of air by the rime which is formed on cold nights in the entrance, or by the loose snow that is driven in, so that it will then do no harm to make more free access for the air.

Rime is formed not only in the entrance but in the interior of the hive, and generally more in a still frost. It narrows the openings through which warm vapours escape, and so helps to keep the heat together; this does no harm to the bees. Great cold accompanied by cutting wind is felt by them very much more, and is much more dangerous, because there is created in the hive a strong draught which chills every bee that is separated from the cluster. In such frosts bees keep themselves unusually quiet, and are confined to the position once assumed, and may sleep never to awake any more if their honey in that place becomes exhausted, although provision enough may be still present elsewhere. In this critical condition a disturbance to stir up the bees and move them further would be used as a smaller evil to prevent a greater one. The entrances of the hives are then diminished still more than before, especially if they are much exposed to the draught, or those of weak stocks are stopped up altogether. This must be done with snow, so that they will open again of themselves in a milder atmosphere. But if the bees finish their stores under other circumstances, they become extremely excited so long as they are still in full vigour. Single bees search in the hive with great eagerness for honey, and in mild weather they seek it outside as well. They gnaw up, and partially shred to pieces, the wax-cells, and bite holes in the combs to make their way to honey. If it is found in the next passage, they move to it or bring it into their quarters; otherwise they become even more exhausted, and finally die. A sure sign of this dangerous condition is to be found in the unusual quantity of wax-dust falling down. For otherwise there falls but little dust from the broken covers of the honey-cells; but it collects undoubtedly, little by little, beneath the occupied passages into little ridges, so that one may gather from these how many passages the stock has occupied and exhausted, or has begun to exhaust. In that part of the comb under which no wax-dust is observed to have fallen, either on the floor or on a board or comb put in under the comb, the cells have not yet been attacked, and must still be full if they were filled before. Snow affords protection to seed; but the damp arising from it in the thaw is destructive if it does not quickly run off, and so is it with the moisture forming from the abundant rime—it is destructive both to the bees and their comb, because it produces mould and decay. Many are accustomed to open Log-hives when it thaws after severe frost, and to brush and wipe off the dissolving rime from the comb and the walls. But this operation might be too disturbing, and lead to the death of many bees. With skeps and box-hives, however, it might be done without disadvantage in a moderately warm room. In hives that are well constructed and carefully kept there will be little dampness to be found or removed, even after severe cold. At most, there will be a little arising on the door, that can be wiped or turned about to dry. If the next comb shows itself a little damp, it can be taken out noiselessly, when it has been loosened the previous autumn for an occasion of this kind, and may be dried indoors. Dampness in other places the air would immediately remove, if the hive is left opened overnight or only lightly shut up with a straw mat. In most box-hives it would rather be necessary to sprinkle moisture into them, or to throw in snow that the bees may quench their thirst and be able to dissolve food that is too dry. This is especially the case in the middle compartments of triple hives and fourfold Lager hives when all the compartments are occupied, and in which moss that has been put in to fill up empty space is generally found dry enough to rub into dust. Meanwhile moisture is not to be given without obvious necessity. It is better the bees should suffer a little from thirst when severe cold may still be expected, than that they should be incited to breed through a temporary comfortable temperature, and the introduction of moisture which always promotes breeding. The cold would then be to them all the more destructive.

A frost with a cutting wind affects bees far more than a still frost, so does a late frost injure them far more than an early one. In the proper winter, especially in December and January, the bees are stronger, healthier, surrounded on all sides with provision, and can therefore defy even severe frost. But if the store-rooms in their nest and its immediate neighbourhood are emptied, the condition of the bees is a dangerous one, for they are now less numerous and are more clumsy on account of the gorged condition of the abdomen. It is especially dangerous when they have already begun breeding and severe frost sets in so that they cannot uncluster nor bring honey into their nest. They may only too easily starve and chill on the little brood that has been deposited while their owner is thinking how fortunately they have come through the winter. Since it is impossible to know whether there will not be after a few mild February days a longer and more severe second winter, there must be nothing done that would incite the bees to premature breeding. If food has to be given at an extremity it must not be liquid or thinned honey, because the whole stock would be excited by it and would be brought into a kind of comfortable condition that should be avoided. Pieces of candy should be put in for them or solid honey wrapped in paper and only exposed a little on one side or a sealed comb close by or over their nest. They will from these only consume slowly, and hold out far longer with them than if the same supply had been given to them in the liquid form. With some precaution the introduction may be effected so noiselessly that the bees are not in the least disturbed by it.

In the Manifold hives, but especially in the fourfold Lager hives, the stocks, gathered for warmth close to one another, would be obliged to separate if they have used up all their stores in that place; and even if there are still stores elsewhere, food may always be put in for them again at the top till mild weather sets in. For a weak stock might, perhaps, in severe cold rather starve than move away from its warm nest.

It is bad when it has gone so far with a stock that it chills because the food over it, or in its nest, is exhausted; when it is also weak in other respects, and the spring is still distant, it would be best to brush out the bees immediately, and to knock and draw out those sticking in the cells, so that, at any rate, the comb should be kept serviceable for the reception of a fresh stock. But if the stock is not weak, possessing a good queen, and has already got over the greater part of the winter, it is well worth while to attempt to revive it. Bees may, as is well known, be chilled for one, or even two days, and revive again when they are warmed, if the frost has not yet thoroughly penetrated them. But if they are thoroughly frozen they are too far gone to save from death, even if, when warmed, they should still stagger. They do not then take the honey given to them, which is a sign that they are incapable of revival.

How Chilled Bees are to be revived.

This depends partly on the hive, whether it is simple or manifold, light or heavy to carry. To withdraw the bees from the further effect of cold as quickly as possible, either the entire stock is brought into a warmed room, or there is put into the hive a heated stone, brick, or warm bottle wrapped in rags, so that the heat does not affect the bees too suddenly, and that, falling down, they do not burn to death. If the bees after some time show signs of life, quite liquid, lukewarm honey, thinned with water, is given to them, letting it here and there flow in drops between the bees. They will suck it up greedily, licking one another; the heat in the cluster will rise, and they will soon attain their former vigour and liveliness.

With a Box hive it will be best to take the combs apart and put them with the bees sticking in and hanging upon them near the warm stove, and when the bees show signs of life sprinkle them with liquid honey, and then perhaps put them in a Transport hive. The taking of the bees out of a Manifold hive would have to be done in the open, not to disturb the other stocks. But in several degrees of frost this must be done quickly, and the Transport hive in which the covered combs are placed must be previously warmed, and every time be carefully covered up with a warm cushion. At the same time, there must, of course, be special care taken of the queen, that she does not, when the combs are taken apart, fall down on the ice-cold floor, and really freeze to death. To prevent that a sheet of paper may be put underneath, and from this the bees that have fallen down on it may be shaken to the others. Those that have been previously already on the floor will, no doubt, be dead; it is, however, but a little trouble to brush them all out and put them, spread out on a board or a sieve, over the stove. The bees that are still alive will speedily let the fact be known, and they may then be put with the others.

If it is not wished to put the bees that have been brought round into a separate hive for a time until milder weather, but to put them immediately into their former hive standing in the open, this can be done in two ways. Either they are put in again as they were taken out on the combs, after those actually dead have been removed from the cells, or the bees are shaken in the room into a little box put on the window-board; the comb is put into the hive, and the little box with the bees is pushed under it, so that they may be able to move into the comb again immediately.

Not only do entire stocks chill, but hundreds and thousands of individual bees chill on the snow, on the ground, on the roofs of hives, and other places when they begin to play, enticed out by genial sunbeams and when the sun goes in, and cold gusts of wind come at the same time, it is soon all over with them. A feeling bee-keeper does not leave even these to perish. In the shade, in a frost, they are actually dead in a few seconds, whereas chilled in a thaw they all revive again. They are collected, warmed, and when genial sunshine comes again let fly in front of the apiary, or put to a needy stock that wants strengthening. If a little honey is dropped on the chilled bees, and they are pushed into a stock close under the nest, the bees will warm them with licking, bring them to life, and take to them all the more kindly. In sunshine they can be shaken in front of the entrance; when revived by the warmth, they either go in or will fly to the stock to which they belong.

But because so long as the air and ground are cold more bees may get lost in a single flight than in the hive itself in all the winter, effort must be made to keep the stocks quiet as long as possible, and to restrain them from dangerous flight.

STOCKS ARE TO BE KEPT QUIET AS LONG AS POSSIBLE.

Bees lie about chilled not only in the neighbourhood of the hive, where they may, at any rate, be collected, but more still at some distance from the hive, inasmuch as a bee, arriving half chilled, if unable to get into the hive, generally rises again and flies away into the air, or, more probably, lets itself drift with the wind till it falls down completely chilled. Housed stocks should therefore be kept housed as long as possible. If they begin to become restless, cool the place as much as possible by opening it in the night or the cool morning, but otherwise it is kept carefully shut up. In perfect darkness the bees will not fly off, but perhaps will cleanse themselves on the hive and go back again into it. With wholesome flower honey the bees can hold out for four months and longer without a flight, and housed hives longer, because with more equable temperature and greater quiet they consume relatively less. The stocks may be set out to cleanse themselves in February, but it must only be when there is no snow lying and there occurs a specially beautiful day that is cheerful, warm, and free from wind. It will be well to house them again till, perhaps, about the equinox, and to try to keep them again in perfect quiet.

Since bees towards spring are very anxious for a flight, it is often difficult to keep them back in the hives, so that they may not be left lying in numbers on the snow. The danger is not so great when the snow is firm or has a crust at the top, because there has been rain on it or a previous thaw, and if roofs and other elevated objects are free from snow. In bright sunshine and a still air, the bees can even raise themselves again from the snow; they are not so dazzled, because the snow has then a greyer colour, and for the most part get back again safely into the hive after they have cleansed themselves. Those remaining on the snow would very likely have remained in any case, and may be more easily collected than off the dark ground. It is far more dangerous when the freshly fallen snow is quite spongy and of a dazzling white, and roofs, hedges, and trees are draped with it. In that case in a flight the majority of bees are lost because they are dazzled by the snow and fly right into it, thinking they are flying towards the sunlight; they do not recognise their hive, and chill before they can recover a little from their confusion. Effort must then be made by all means to prevent flight. If the hive is spacious, and there is not so much danger of lack of air, the entrance should be closed altogether in the forenoon, so long as the bees are still perfectly quiet, and opened again towards evening when it is becoming colder. If the stock is beginning to get into disorder, and a strong heating of the bees is to be feared, there may be put in front of the entrance a small sieve, a bee hat, or a net, that the bees may be able to cool down in it, vent their rage, and cleanse themselves. To induce them to quiet down the sooner, they may be sprinkled with a little water or have some snow thrown to them, and be smoked at times. On days when there is a raw air, it is sufficient to shade the entrance, surrounding it in all cases with a little snow, that every bee coming out may be induced by the cold to return.

Winters and springs differ very much for bees. While in one year bees not only play merrily in March but carry pollen, in another year March, and even April, may bring snow a foot deep. A truly lamentable time occurs in many a year for the anxious bee-keeper. Scarcely has he shovelled away the snow in front of the hives, brushed it from the roofs, and strewed other places with ashes and earth to take off the dazzling whiteness, but there is a fall over night of fresh snow, and again hundreds of his pets, that have till then come through safely, find their grave in it; or a cold wind rises when the bees are at play in large numbers, and strikes down hundreds and thousands of bees, making them incapable of flight, so that they are found on the ground, on the sills and the roofs of the hives, quite in little heaps, chilled. But if the bees cannot even make the attempt at a cleansing flight because the weather is always too cold, unable any longer to retain the excrement with which their abdomens are distended, they begin at last to dirty the entrance, the walls of the hive near to the entrance, their combs, and one another. The dirt must at least be soon removed from the entrances, so that the bees coming out may not soil their wings with it, and the more certainly fall on the ground and chill. The cleaning of other objects can be done later on at a more favourable time, and a strong stock cares for this itself. It is, however, not only much easier to do when the dirt is not dried on, but a great service is rendered thereby to the bees that love cleanliness so much. Here the great convenience of the Twin-stocks is evident again, for they are capable of being brought one after the other into the warm room to be attended to, and to let the bees cleanse themselves. In this way many a weak stock may be snatched from the destruction to which otherwise it would certainly fall a prey in a long, cold second winter.

The ordinary Spring Cleaning.

For the stocks that are taken indoors, this can be completed at the same time; but for the other hives it is not undertaken before a fine day occurs, and the bees generally are flying out. But because this cannot be so quickly completed with a large number of stocks, the matter can be disposed of quite independently of the weather, if the cleansing of a part of the stocks is undertaken indoors, as opportunity occurs. The dead bees lying on the floor and the wax-dust are removed, but the latter is not thrown away, but parted from the dead bees by a little sieve, and put to the wax that is to be boiled down, because it contains the purest wax. The carrying out of the dead is cared for by the bees; but many living bees are lost through it, by falling into water, or on to the cold ground, and chilling before they can disengage themselves from their burdens. It is preferable then, to save them this labour, which the bee-keeper can do in a few seconds, since by the help of a feather, and a long wire hook, everything that is lying on the floor may be easily drawn out and removed, even from hives closely filled with comb. But since there often hang between the combs, and stick in the cells, many dead bees whose removal by the bees is specially difficult, it is well to subject the comb itself to an examination, and the comb must of course be taken out either entirely, or in the requisite places, for the purpose of removing from the walls and the comb the dirt made by the bees. The dirt may be so completely cleaned from the walls with a damp sponge, or rag, or with a knife, that scarcely a trace of it will remain; while in straw hives it is so firmly attached that it can hardly ever be completely removed.

The combs themselves are cleaned by damping them, so far as they are dirtied, with a wet sponge, or rubbing them with a soft brush. They may then be rinsed, getting the water out of them by striking them sharply against the palm of one's hand, and then drying by the stove or in the air. If no store of other combs is in hand, they may be given back again immediately to the bees a little damp, as these require some moisture as is well known.

If a hive has been very much dirtied, the bees may be put into another clean hive, and perhaps into fresh clean comb. The soiled hive and comb can be cleaned afterwards, so as to be able to put a stock into it again on the same or following day. But if the bees have not yet had opportunity of flying out and cleansing themselves generally, and if such an opportunity is not very soon to be expected, the opportunity must be made for them in the room, because they would otherwise dirty hive and comb over again. A room with only one window, with a southern aspect, is best suited for the purpose. If there are more windows, the others may be darkened so that the bees only fly to one, and may the more easily be put back again into the hive. The bees of course sprinkle everything near to the window with their brown excrement, which may, however, be completely wiped off the floor and window, where most falls, with a wet sponge. Other objects near, that are not easy to move, must be covered with paper, or cloths if they are to be kept from being soiled.

The bees can cleanse themselves partly in flight, and partly at rest. If they are to do it in flight, the hive is set up on a chair at some distance from the window. If the sun is shining brightly through the window, the hive may be put directly in the sunshine. The bees incited by a little thinned food, which may be put or sprinkled into them, will soon begin to play, and all the more strongly if they have previously become a little heated, through their entrance being stopped up. Most of the bees circling around will have relieved themselves before they reach the window, but many may not do so, and the window must be continually wiped clean, so that the bees may not dirty themselves. This continuous wiping may be spared if a lattice or fine net is applied over the window-board, so that the dirt may fall through and the bees may not come in contact with it. Combs otherwise unserviceable, especially drone-combs, render the same service by taking up the excrement of the bees in the cells. The bees fluttering at the window, sunning themselves, or gathering in little clusters here and there, will stream again joyfully into the hive, when its entrance or open side is after some time put to the window. Combs may be put on the window-board continually, and the bees immediately collecting on them in quantities, may be put again either into the hive or by it, when they will joyfully move in. The remainder is brushed together off the window and the floor into a vessel and shaken to the same hive or to another one, which may be strengthened at this opportunity if it needs it, and the other has a surplus of bees.

The bees are accustomed to cleanse themselves immediately, when they settle separately, especially on a cold surface. One comb after the other may be taken out of a box-hive, and the bees be shaken on to the floor and spread out with a feather. By the little clusters forming separately, pieces of board or other objects may be put on which the bees soon mount individually. Most of the bees as soon as they are separated from the cluster will cleanse themselves, and then either fly to the window, or collect again in little clusters, which may be brushed on to a sheet of paper and put into the hive again. The floor must of course have been previously cleaned, and the spots of dirt must be continually removed so that the half-chilled bees may be brushed together with a feather without dirtying them. If the collected chilled bees are put close over or under the nest of the stock, they will be immediately revived by the heat proceeding from it; otherwise they may be so far revived in a glass by the heat of the hand, that they may be able to betake themselves to their cluster when put in front of the entrance, or shaken into the hive.

If the bees had dirtied their hive and their comb very much, especially if they were already half chilled in severe cold, they have often dirtied one another so much, as to be entirely incapable of flight. They must then be sprinkled with a little water, when they are sunning themselves in the room on the combs by the window, so that they may be then able to clean their wings again, or they may be bathed and washed all over, letting them be revived and dried again in the sun or by the stove. It is unnecessary to say that the queen, who is generally quite clean, is not to be subjected to this bathing.*

The hives that are subjected to cleaning are of course on this occasion provided with the food that may be requisite, or in box-hives the full honey-combs are put near to the nest of the bees, if severe frost may still be expected. Since even in the beginning of spring cold nights always are prevalent, the hives must be still carefully protected, because they need warmth now more than in winter because of breeding.

Breeding begins, in moderately strong stocks ordinarily after the first considerable flight, if not before. Many bee-keepers believe it to be of great advantage to feed their stocks repeatedly with liquid honey for the purpose of promoting breeding. Spring is, moreover, the time when feeding is most usual; this, then, may be the right place to say something on the feeding of bees, how it must be done, and what is attained by it.

FEEDING.

Since this costs a part of the honey that has been gained, it is generally looked upon as a necessary evil which has to be limited to the most urgent cases, when some important object is to be attained by it. The behaviour of the bees when left to themselves corresponds always to the present position of vegetation, and the condition of the weather. If vegetation is at a standstill, or cold, unfavourable weather is prevailing, activity in the bee-hive is at a standstill too, or becomes ever more limited. But if nature affords nectar and pollen, the bees will carry in and breed without

* The Author paints a vivid picture of the miserable condition of dysenteric stocks in winter, and of the trouble necessary to be taken in remedying it. We well remember similar experiences to those indicated, when we used hives with space above the frames, and a tightly fitting crown-board, but, thanks to modern improvements in hives, and careful attention to the preparation of the winter nest for the bees, we are now able to prevent dysentery almost entirely, and so save the laborious and disagreeable proceedings above described. It is peculiar that in the countries where it is claimed that bee-culture is best understood, notably in Germany and America, the scourge of winter, dysentery, runs riot each year and decimates their apiaries, while here in England, by attending to the wants of the bees, as by them expressed, the disease has been practically rendered impossible in well-managed apiaries.—C. N. A.

being otherwise stimulated to it. If liquid and thinned honey is now given to the bees to supply artificially the lack of natural pasture, and to revive the courage of the bees, they are led into error which always, more or less, brings its own punishment. The bees are misled into making flights, which not only are of no use, but cost the lives of many bees. They are incited to deposit brood, which later on they will perhaps not have the power adequately to nourish and bring to perfection. The food that is given to stocks for the purpose of increasing brood in early spring, is in most cases not only pure loss, but often brings about actual damage, since more bees get lost on the one hand than are added by breeding on the other.*

In some cases, nevertheless, a stimulus from food may be very advantageous. For instance, an artificial swarm may have been made from which it is wished to obtain a good many queen-cells, but quite unfavourable weather sets in directly, efforts must then be made to keep the bees active and on the alert by means of food, because, otherwise, they would fall into a condition of despondency and inactivity, in which they would form either few queen-cells or none at all. But what they omit at the beginning they would not be able to make up later on, because meanwhile the brood would have become too old. Or an Italian stock has already begun drone-breeding and unfavourable weather set in again; or

---

* This at first sight appears to militate against our (Abbotts') system of gentle, continuous, stimulative feeding in spring and autumn, for the production of young bees early and late, to take the places of the aged and the worn. But the system of hive construction recommended by our Author does not permit the use of the bottle and the perforated disc, and we may presume that he is not therefore acquainted with the system of feeding, of which they form so essential a part. He, however, recognises the principle of gentle, continuous feeding under certain conditions, 'to keep the bees on the alert continuously;' as the reader will find further on. We have the greatest faith in stimulative feeding when sensibly conducted; *i.e.* when made to eke out the natural supplies of honey and pollen in the spring when they begin, and in the autumn when they are declining. In the former case the supply of syrup should begin (as a rule, with the crocus blossoms) with as much as the bees can take through one pinhole, gradually increasing to three or four, and peaflour should be given proportionately, beginning by partly filling the crocus cups and increasing the supply by sprinkling it largely in inverted skeps filled with shavings; and in the latter case by doing the same thing inversely. Nevertheless it will usually be found that bees will take very little, if any, of the artificial pollen in autumn.—C. N. A.

a stock has been too much weakened by changing its position when making an artificial swarm, and there is reason to fear the destruction of a part of the brood. In these and similar cases feeding is strongly to be recommended, so that no injury may grow out of the passing depression and weakness.

If there is much brood to nourish, feed abundantly, otherwise let the food be given oftener and in smaller portions, or it may be given in that way that the bees can only take it into the brood-nest in smaller portions, and will so keep on the alert continuously, or for a longer time; that is, it is given to them in a condition not easily soluble, as sugar, or solid or candied honey, either in a comb or a piece cut out of the barrel. The water necessary for solution may be put in for the bees near it in a comb, when the weather does not permit of their fetching it. A supply of water may always be kept in a vessel in a quiet warm place near to the apiary, because the bees use a great deal of it for the preparation of brood-food, especially in spring. The surface should be covered by moss, straw, reeds, or the like, to prevent the bees falling in. They fetch it gladly out of combs, especially if it is warmed a little by the sun. The bees may easily be enticed to it by a little honey water. When other people's apiaries are not in the neighbourhood the bees may later on, at a warmer season, be fed altogether in open vessels with honey, sugar, syrup, boiled beer, wort, and other sweet liquids. But in apiaries with numerous stocks there arises then such a great hum, that a weak stock may be moved to swarm out by the sound of the swarm note. It has been already mentioned that when pollen is not to be found, wheat-flour, rye, or oatmeal, may be put out for the bees in combs, and they immediately industriously form pellets and carry it in. In respect to meal-feeding, what has been said on honey-feeding also holds good, yet it might not be useless, and might save honey when used immediately before the first pollen carrying, when the weather is already warm, since the bees if they only receive sugar and water with it, convert the starch in the meal into sugar, and are in a position to extract all the nutritive elements contained in it. Meal-feeding in the first fine days of spring has at least this advantage, that a queenless stock may be more easily discovered because it remains inactive while the others are carrying meal, although there is also a tolerably sure sign of queenlessness in the continuous disquiet that lasts until evening on the day of the first flight.

Through the so-called speculative feeding, the object of increasing deposit of brood in stocks is only certainly and completely attained when the bees are excited thereby, not only to greater activity, but when there is afforded to them in the food all the necessary material for the preparation of brood-food, especially the nitrogenous element which is not contained in honey, and for the sake of which they always so diligently, and often at the risk of life, collect and carry in pollen, especially in spring. Old stocks certainly as a rule possess a considerable store of pollen; but a part of it will have perhaps gone bad and become useless, and the well-preserved is often used up before the district and the weather permit of carrying in fresh in sufficient quantity. Breeding must then suffer a check when its continuously larger extension is most desirable, since the season of best pasture is ever drawing nearer.

It is, therefore, a principal problem of rational bee-culture to discover a substitute for pollen just as a wholesome substitute for honey that often runs short has been discovered in sugar, especially in candy. This problem has been brilliantly solved by Mr. Emil Hilbert, landed proprietor, Maciejewo, by manifold experiments in egg and milk feeding, the results of which have been communicated most disinterestedly to the meetings at Halle, Strasburg, and Breslau, and which have earned for him the gratitude of all bee-keepers. The materials mentioned being the means of the nutrition of young animals, contain in especially concentrated form the necessary nitrogen for the building up of animal bodies, and thereby, and from their comparative cheapness, recommend themselves peculiarly for bee food. This is especially the case with milk, because it affords bees at the same time the indispensable water for the preparation of brood-food. The new milk is boiled and strongly sweetened with sugar. Honey is less applicable for it, because it easily curdles the milk and induces robbery. The contents of the egg are thoroughly mixed with honey or dissolved sugar and used as food in that way, though the bees only take as much of it as they can use at the time for the preparation of brood-food; whereas of the sugared milk they deposit a certain store in the cells. At isolated apiaries the sugared milk may be given to the bees in the open in combs, little troughs, or other vessels.

The treatment is quite different when stocks are fed for the purpose of completing stores for the winter that are not quite adequate. Pure honey, or pure syrup boiled to the consistence of honey, is then used. With moveable combs the simplest, most natural, and cheapest way of feeding is by putting in sealed honey-combs. The bees have no need to suck up and carry away the honey given in combs and they are not excited and induced to useless flights by it. Through such an increase of the honey stores of a stock, let it be at what time it may, none of those disadvantages therefore are introduced which ordinary feeding at a wrong season causes. Besides, as has been already said, bees manage far more economically with sealed honey, and hold out much longer with it than with liquid honey.

If no sealed honey-combs are available, or if the bee-hive does not allow of their insertion, the bees must not be fed so as to excite them to activity and breeding. They must not be fed as a speculation, as it is called, but as a necessity for the increase and completion of the necessary stores. The food is given in as large portions as possible, and only a little, or not at all thinned, so that repeated excitement and extensive breeding may be avoided, and that the honey given, which is to be laid up for the future, may not, partially or entirely, be immediately used up by the bees.

Feeding must be avoided as much as possible as a necessary evil, and be limited to the most urgent cases, because, if it is not managed with proper precaution, it may easily introduce another evil, viz., robbing, especially with weak stocks.

Robbing.

As was previously remarked, towards the end of the theoretical part, page 39, bees are much disposed to rob one another and to penetrate every place where they discover traces of honey and find no resistance, or only a weak one. This is especially so at the beginning of spring, before they find fresh nectar in nature, and before they are accustomed to seek their food more at a distance. At the beginning individual foragers shyly circle about the entrance, trying to penetrate, quickly retreating when a bee on guard seizes them, or they try to release themselves by force when they have been caught. But if a number of bees succeed in penetrating, escape loaded with booty, and return home, they come again with strong reinforcements, attack the hive in greater strength, and put its inhabitants into such confusion that it gives up all further resistance and lets things go as they will, so that the strangers plunder from early morning till late in the evening, so long as there is anything to carry. Not only is the honey carried away, but at last the comb is partially shred to pieces, because a whole cluster of bees is crowding at every cell. In consequence of the mutual thrusting, and the consequent heating, the slipping into deep honey-cells, and penetrating into narrow chinks and crevices, and the pulling and biting on the part of their opponents, and the licking on the part of the bees of their own hive, the robber bees become of a brilliant black colour and are easily recognisable by this, although some bees may gain it by accident, especially if they are shut up and have become heated. These black bees that were formerly so much talked about have no validity beyond this. Bees become more disposed to robbing the oftener they have been engaged in it successfully. Habit becomes here a second nature. If one hive has been plundered, another is attacked, and it may have the same fate if it does not courageously resist the attack, or if the bee-keeper does not come to its assistance.

Once used to robbing, they lay aside labour even in the time of most abundant pasture, and carry away from even populous stocks many a pound of honey at the time of most abundant flight. They become then half at home in the strange hive, and go at last in and out quite unmolested. Many an one at last does not take the trouble even to go into the hive, but attacks one bee after the other of those lying in front, causes it to give up its last drop of honey, and then flies away laden with booty to its hive.

It is at last hard to break bees of a habit so deeply rooted. For it is well known to every one who is in any measure acquainted with their nature that it is simply an acquired habit, and that there is no variety of bees which has been destined by nature to robbery, and that bees are not to be induced to robbery. Robber bees do not, therefore, permit of being made.* Bees may

---

* There seems to have been an idea among some of the old-fashioned bee-keepers of Germany that the advanced teachers of a better system, amongst other clever things that they did with bees, had the power of

certainly be made excited, spirited, and then perhaps pertinacious, by feeding, especially if spirituous liquor is mixed with the feeding-honey, but no bee-keeper is in a position to induce his bees directly to robbery. To do this would be extremely unreasonable, because robbing bees put themselves entirely into the power of the man whose apiary they come to rob. He could kill or poison them, and destroy the whole stock. Robbing is unpleasant for both sides, and on that account the so-called speculative feeding even of strong stocks, in which it, at any rate, might be of advantage, is by preference to be omitted, because it excites the bees before the time, makes them obtrusive, and may easily beguile them into robbing. But since we cannot blame any bee-keeper for trying to make his stocks strong and spirited early, he cannot be made responsible for robbing. For with all the courage and pertinacity of robbers, stocks that are found in order, and that are treated with precaution, will take no harm. When robbing is destructive, it is a proof of want of knowledge or attention on the part of the bee-keeper whose apiary is robbed.

Robbing is more easily prevented and avoided than cured when it has once taken a start. If apiaries numerously occupied are in the neighbourhood, the bee-keeper has double reason to avoid everything which might give occasion for robbing. He should leave no weak stocks standing, least of all those that are queenless, and should thoroughly investigate every one immediately which gives any ground whatever for suspicion of queenlessness, especially any that is much troubled by robbers. If he is not able immediately to strengthen the weak by adding bees, or to restore order in the queenless by introducing a queen, let it be cashiered or put into the dark, only letting it fly at times, perhaps shortly before sunset, until it can be effectually assisted. But if the bees offer a tolerable resistance to the strangers, the entrance has only to be narrowed, so as only to allow of the passage of one bee at a time. The entrance may be rubbed with strongly-smelling things, such as garlic, onion, wormwood, and the like. The robbers especially retreat before the pungent odour of their own sting-poison. The

training them to plunder other people's hives. We can well imagine that when the neighbours found the Doctor's yellow Italians in such numbers come prowling around, that they were looked upon not only as born robbers, but as robbers who had enjoyed a higher education in the art. —S. S.

sting of a bee that has been stung may therefore be rubbed down in the entrance, or the bee itself may be fixed there, so that the intruding bees must brush past the sting dripping with poison. Since the acid or poison of ants has the same effect, some ants, of which bees have always a certain shyness, may be rubbed down in the same place or be fixed there. The entrance may further be darkened by a little board, whose inner surface has been rubbed with some of the things mentioned, so that the robber bees may not be able to get into the hive in a straight course and nearly on the wing, as they generally do. Feeding by day must be carefully avoided, not attending to it before evening, and what the bees have not carried up, and, indeed, even the empty vessel smelling of honey, must be removed. Robber bees fall most greedily upon thinned food, being able to load themselves quickly with it, and robbing may in that way be quickly set going. For this reason solid honey, put in at the crown, is more advisable for food than the more liquid. There is less danger if it can be given at the top by putting it for skeps into a glass covered with muslin or perforated paper, and reversed over the bung-hole.

It sometimes occurs that a strong stock, that is otherwise all right, lets strangers go in and out freely and carry away its honey. Sometimes two stocks mutually rob one another in this way. This tolerance, which can only have its ground in similar odour or other like conditions, may be done away with by taking away the queen from the one stock, or by giving to it a very different odour, perhaps with musk. The means recommended by many of exchanging the position of the two stocks is only available when they belong to one owner, or when the two have agreed to the exchange.

If the habit of robbing is too deeply rooted in the bees, it is hard to break them of it. It is best, at any rate, to send them to a distant apiary, where they must accustom themselves to a new activity, as well as a new flight. Either the entire robbing hive must be sent there, or only the bees that are accustomed to rob. These must then, of course, be caught. This is most easily done in the hive they have been robbing. If its entrance is shut up till the robbing bees have gathered in considerable number in front, and is then opened, they will all stream in together, and for the most part be caught if the hive is again closed. The robbers can be still more completely captured by the aid of a tube put into the entrance, projecting a little into the interior of the hive, and provided with a lid that can easily be moved. Placing the hive then for some time in the dark, till the bees have become more accustomed to one another, and putting them then on a stand at a distance, they will continue to work unitedly if the stock was otherwise all right.

But if the robbed hive is not removeable because it forms one of the compartments of a Manifold hive, the robber bees must be caught in another way. The robbed hive must be closed before the beginning of the flight, and the robbers be caught altogether in a Transport hive in the following way. As often as the robbers have collected on a comb containing, or that has been smeared with, somewhat firm honey, they are quickly shaken into the Transport hive through a small side opening till they are nearly all together. Brought to a distant apiary, and allowed in the dark to move into a stock needing strengthening, they would mostly remain with it. If we wanted to add them by day, they would immediately fly apart out of the opened hive in all directions, and for the most part get lost, except they had been shut up with some brood, honey, and water, until they had determined to build queen-cells. In spring and summer robber bees may be used for making an artificial swarm, or, at least, for strengthening it, by shutting them up with a comb of brood and honey, and then removing them to a distant stand. By this a wrong is certainly committed against the owner of the robbing stock, who is quite innocent of the robbery. Meanwhile he is already undoubtedly in some degree recompensed for the weakening of his stock by the honey carried in by it, and it is still open for him to be recompensed in other ways. The man who would poison robbing bees, and perhaps thereby destroy the whole stock, would be guilty of much greater injustice, and if he had made use of a poison for the bees that would also be dangerous to men, he would deserve not only to be mulcted in damages, but he would be liable to criminal proceedings as a poisoner. An apiary in which robbing is evidently going on ought to be capable of being always examined without previous notice, in order to ascertain and remove the cause, and to be able to have proof of poisoning if any were going on.

To mark robbing-bees with meal, as is often done for the purpose of ascertaining the robbing stock, is a trouble quite useless, since not the least blame attaches to the owner. Whoever complains often of robbing proves thereby that he does not know how to manage bees, and should not meddle with their culture. The proverb is rightly applied to him, 'Whoever by his own folly injures himself, must not mind being laughed at.'

We come now to an operation that deserves to be the more particularly spoken of, as it is the only one that many bee-keepers undertake, especially many owners of Log hives, and as there have been brought forward the most different and contradictory views from different sides upon it.

THE SPRING COMB-CUTTING.

Many look upon the cutting, that is, the shortening of the comb, as being equivalent to the ruin of the stock, and a notable bee-keeper went so far as to give the advice that not even a single cell should be cut away from a log-hive filled with comb and full of honey.

Such a view deserves at most to be laughed at, but not refuted. To follow such advice is equivalent to renouncing all profit. We certainly give up a little gain willingly enough, when by so doing there is the hope of harvesting the more in the future. But this hope does not occur here, as, by experience, uncut stocks do not compare with cut ones more favourably but less so, so that the loss is a double one. The opponents of spring comb-cutting argue from the view that a stock, when it is not cut and does not need to build, can breed the more, and, having that advantage over the cut one, will therefore yield more produce.

But it is firmly established that a stock, the more it has given it to do, developes also in every other respect the greater activity both inside as well as outside. The stock that is cut tolerably close will generally deposit not less, but more, brood than the uncut. For when the workers have been excited to greater activity, the queen will lay eggs more industriously. Even a closely shortened set of combs will at the beginning contain the necessary brood-cells in sufficient number; and later on, when the brood requires a larger space, the bees will provide for the necessary extension, since they have at that time a special inclination for building. But the bee-keeper must know how to take advantage of every inclination of the bees. He must, therefore, let them build new combs industriously in spring, that he may be able to take away the more full ones later on. The bees have sufficient time for this in spring, so long as the pasture has not completely opened out, and can, with the care of the brood, at the same time conveniently construct the cells which the ever-extending deposit of brood requires. The heat which comb-building requires, and even in turn produces, is at the same time to the advantage of brood. In this sense comb-building may be called a bye-product, as Pastor Köhler called it a short time ago in the *Bienenzeitung*. But if this designation is taken in the sense of the bees producing wax itself only as a bye-product, without special expenditure of honey, we should be found manifestly in error. The production of wax in all cases costs honey, but the additional consumption of the workers for building is carried in again in abundance by reason of the greater industry they develope outside, so that also from this side the fresh comb which the cut stock has been caused to build in spring is to be looked upon as pure gain. Through comb-cutting even the bee-keeper will have some profit who does not know how to make any better use of the combs than to melt them up for the sake of the wax, but he will have a far greater profit who preserves them carefully for hanging in the honey-room of strong stocks at the time of most abundant pasture, or for the purpose of furnishing young swarms with them. There is no way of gaining a large store of empty combs so easily as by the spring comb-cutting.

Many important secondary objects are further attained by it. The cleaning stocks from litter and dead bees, which not only lie on the floor but often hang in great numbers between the combs or stick in the cells, is notably facilitated both for ourselves and for the bees; feeding that may be necessary is made more convenient; queenlessness, foul-brood, or any other defect, is much more easily and certainly recognised; and, at the same time, the comb of the stock is renewed. Even in Lager-hives the combs may, at least, be so far shortened that all refuse may be removed at the bottom; and that a comb with honey for feeding, with pollen or water, may be pushed in underneath. Since the lower edges of the combs have the most wax, there is at least something gained from them, even if larger combs can only be acquired from the upright hives.

After a long discussion carried on in the *Bienenzeitung* on the matter, the most weighty authorities have agreed that cutting combs close in spring is advantageous, and to be recommended without hesitation. But if it is wished to gain profit, and not to bring about loss from it, we must know when the comb-cutting should take place.

The Time for Comb-cutting.

Since the bees are excited to activity by the cutting, it is evident it must not be before a time has arrived when the bees can be active outside the hive continuously, and at which their flights bring the stock more profit than loss. The bees must already be able to carry pollen abundantly; and since they cannot prepare any wax from the pollen alone, they must either still have honey in store, or be provided with it, or be able to find some honey in nature. So long as these conditions are not present it is better not to disturb the bees, but to let them await better times in quiet. Moreover, it is better to undertake the cutting of the stock later, rather than too early. Even if a little brood is cut then, this has but little significance, and there is better assurance of the stocks being all right. The time does not allow of being accurately fixed, as it may be different according to the difference of the district and of the spring. But in most districts of Germany the fittest time for cutting might be about the middle of April, or at the time the willow blooms.

How and Where should We Cut?

The lower part of the comb—perhaps about half—is evenly cut off; but in high hives, like log-hives, more, so that only the upper part is left about a foot long. Whether an inch more or less is cut off is of no consequence, since a strong stock provided with stores is in a position to make again hundreds—indeed, thousands—of cells in the night if it needs them. The older the comb is, and the more drone-comb it contains extending far up, the more honey the stock can spare, or the more accurately we wish to know its condition,—the closer do we cut, that is, the higher up is the section made. Sometimes it is a matter of getting as much honey as possible at the cutting, because the stock is over-provided with it, or because it is needed for the feeding of poor stocks. In that case, when the honey-combs stand in front they may be cut much shorter; and as the bees'-nest is approached the knife may gradually be directed lower for the sake of sparing the brood. Making gaps in the crown by taking away whole combs is to be avoided as much as possible; because the brood-nest is cooled by it, and the stock's development is delayed. Should a gap be made in the crown, the space should be immediately shut off, or at least filled again immediately with empty combs. Rather than make gaps in the crown, it is better to shorten all the combs a good deal, but equally; even if it should involve the taking of the brood partially or entirely from the stock. There is no need to let this die; but it may be suspended in a box-hive, or used for making an artificial swarm, which made about the middle of April might succeed very well. The bees necessary for it may easily be taken away from the strong stocks at the cutting, and after it when they hang in quantities under the shortened comb.

The opponents of close cutting especially object that the bees may make much drone-comb where worker-comb has been cut away. But it is just as possible that the bees may form worker-comb where drone-comb had been cut away, if this had only been cut away until worker-comb was arrived at. Very strong stocks are much disposed to build drone-comb in spring, in warm, damp weather favourable for brood and swarming, and pass on several worker-combs immediately to the building of drone-cells. But the bees allow themselves easily to be turned from their purpose if only one of the drone-combs—perhaps the furthest—is left for them; but removing betimes the drone-cells that have been begun on the other combs. It is for that purpose the bee-keeper is there to overlook; and he has the means, in the empty combs gained at the cutting, of making the construction of a drone-comb absolutely impossible if a comb with worker-cells is put in for them at the requisite place, which may be done easily enough even in log-hives and skeps.

Whoever, on the contrary, has neither time nor desire to look after his stocks at the right time, but leaves them altogether to themselves, may certainly treat the comb with more forbearance, if it for the most part contains worker-cells and is not old; especially if he has no need of any empty combs. And in districts in which the full pasture sets in all at once with the blooming of the rape, without gradual passage, it will be necessary to cut less; partly because there might then be a deficiency of cells for the deposit of the abundant flow of honey and the bees would be obliged to take up valuable time in building fresh, and partly because at that time it would be difficult to restrain them from building many drone-cells in the overpowering swarm impulse arising. Moreover, the deposit of drone-brood ceases when the old queen has gone away with a natural or artificial swarm, or when the bees have given up swarming in a time that is too dry, although otherwise yielding abundance of honey. For it is just the drone-cells which, before others, are filled with honey; so that much drone-comb may turn out not a disadvantage, but an advantage. There is no need, therefore, in this respect to be too anxious.

FURTHER MATTERS THAT HAVE TO BE ATTENDED TO.

These may be put into short compass. Many a stock, which was all right at the spring cleaning and examination or at the spring comb-cutting, may later on become queenless. Notice should be taken whether all the stocks carry pollen with equal diligence, complete the comb that has been cut down, deposit brood, and have young bees playing in front on fine days, and careful examination should be made of any suspected from one or other of these causes; so that robbery, to which the bees are especially disposed in spring, may not be introduced by allowing stocks to remain standing queenless for some time. In hives with moveable comb a queenless stock may be preserved, if it is worth it, by putting in a brood-comb from time to time, and letting a queen, that has perhaps been raised earlier, be removed about the middle of April for a new one to be raised, which may then very well become fertile. But log-hives and skeps which have lost their queens before April are best cashiered. Their cure is very uncertain and especially difficult, because it is not easy to know whether they are entirely queenless or have an unfertile queen. In the latter case, a good queen added is not accepted, but is sacrificed unnecessarily, and no new one is raised from brood put in. To catch the flighty, unfertile queen is very difficult, because it is not easy to get her out of the comb in the cool season by drumming. Success would be most simply attained by putting a quantity of bees from a sound stock to the one suspected; or by putting the stock, if it is in a skep, in combination with another strong one for some time, so that the one is reversed and the other put on the top of it. The unfertile queen is then either immediately killed by the strange bees, or she is enclosed in a knot of bees so that she may be caught or flung out by a sharp jerk. If no queen at all is present the strange bees will become uneasy and begin to run apart from one another. The stock will then be more sure to raise a young queen from brood put in and suitable for the purpose, because the craving for a queen, which gradually dies out in a stock that has been queenless for some time, has been again awakened. A Twin-stock that might get into such a condition would be put into communication with its neighbour, the bees being allowed to work in common; and they would be parted again at the approach of the swarming-time.

The chief care of the bee-keeper in spring must be to see that his stocks do not suffer want in any way. If they have used up their stores and nature does not yet afford any fresh honey, or if they are hindered by the weather from carrying in, he must assist them with food. Although inciting bees to breed prematurely in spring is to be deprecated, yet when it has once been begun— which, even in weak stocks, will be by the time of the equinox— he must see to it that it is carried on without interruption, and even more extensively; and that the stocks never have any necessity to pause in it, and even to suck out again and eject brood already deposited. Every gap or pause arising in the natural bee-pasture, especially the gap occurring in many districts after the fruit-blossom, must be filled up artificially; trying always to keep the bees in good spirits. Since bees use much water for their brood the food may be much diluted; and may, in case of need, consist only of quite sweet beer-wort. At isolated apiaries the bees may be fed in the open, but where other stocks are in the neighbourhood feeding must be managed with greater precaution. If the food is poured into the cells of empty combs indoors, and they are then hung up in the hives quickly for the bees, or pushed under their nest, there will be but little cause to be alarmed about strange bees.

If the bee-keeper has not omitted any attention and care for his pets in the cold season that yields no food, he will be the better able to rejoice over their industry, when the bee-pasture has at last developed in its entire fulness, the stocks carrying in heavily from early morning till late in the evening, and increasing in strength and weight from day to day, and at last beginning to throw off swarms. This begins for the bee-keeper a fresh business, but such an one as is at the same time the greatest pleasure to him, namely, the hiving of the young swarms that appear, or the increasing of the number of his stocks by artificial swarms. Since we started from this point, and have already treated of it more particularly, we break off here our descripton of the different doings and occupations of the bee-keeper that repeat themselves every year in the same order, that we may touch upon some of the subjects relating to bee-culture that are not connected with any particular time or order, especially the different ailments of bees, their different enemies, and some of the implements necessary, or at any rate useful, in carrying on bee-culture.

The Ailments of Bees.

These are not numerous, and science has lately absolutely erased from the catalogue of peculiar diseases some kinds that figured in early bee-books.

Dysentery

Even the dysentery that often occurs towards the end of winter, and which has therefore been already mentioned in its place, is perhaps an evil only in countries with long and severe winters; it consists in the inability of the bees to retain their excrement beyond a certain time and measure, but is no special disease because the evil is removed as soon as the bees have been able to relieve themselves. The causes of dysentery are long and severe winters, honey that is unwholesome, or that has been carried in or given too late, and which is therefore mostly left unsealed, coldness of the hive and comb, frequent disturbances, too much dampness, as well as too little, for then the bees often become disquieted, premature breeding, and further, every circumstance that incites the bees to a larger consumption of food, whereby more excrement accumulates in their abdomens, and whereby they are longer deprived of the opportunity of ejecting it in their play in front of the hive. But because in some circumstances, the abdomens of many bees are so swollen by the accumulated excrement, that they can only drag themselves outside the entrance, and have no longer the power to fly off and cleanse themselves, then the evil certainly developes into a disease of which many bees die both in the hive as well as outside. Indeed the evil sometimes appears to become contagious, as we have learned by experience that when a stock that has been diminished by dysentery has been strengthened with healthy bees, the mortality continued and the stock became soon just as weak as it was before. It is only the queen that is not subject to dysentery, and keeps in good health, although the larger part of the stock may have died of the disease. The reason is that she partakes only of the most refined honey and food, and ejects her excrement, which consists of a watery and somewhat turbid liquid, always naturally in the hive and with ease. Whoever wants to winter his stocks wholesomely and cleanly must avoid the causes specified above, so far as it is in human power to do so, and must moreover only winter stocks that are capable of wintering well. The method of management when dysentery begins to show itself and how in extreme cases bees may be furnished with an opportunity of cleansing themselves in a warm room, has been given in former chapters.

An evil that is incomparably greater, indeed indisputably the greatest, which can visit the bee-keeper, is foul brood.

Foul Brood.

As the designation already indicates this does not have to do with the bees already developed, but with the young brood, which dies and goes to decay partly in the larval condition, and partly as nymphs. The danger of the disease consists in this, that not only the brood attacked by it is lost, but the brood-cells are tainted and are made useless for further deposit of brood, that the evil does not simply extend in the same stock to an ever-increasing number of cells, but by contagion spreads, to an ever-increasing number of stocks of the same apiary, and of the same locality, if efficient means for its removal are not immediately used. Even the bee-hives become useless on account of the germs of contagion attaching to them for a long time, even if not for ever.

Symptoms of Foul Brood.

An infallible symptom of the presence of foul brood is the discovery of dead, dried-up, shrivelled larvæ or nymphs in separate cells amongst healthy brood. These dead larvæ have passed into a pap-like or tough mass, and later on into a greyish-brown or quite black crust on the floor or the lower surface of cells. If the majority of the cells is in that condition, the infection has taken place some time ago, and the evil has already become very great. Because a stock with foul brood generally ventilates considerably, the evil may be recognised in hives with immoveable comb by an unpleasant smell proceeding from the entrance; the smell is similar to that of putrid glue or meat. As the bees take the trouble to bring out separate larvæ that have not yet entirely rotted, such will be found sometimes on the floor of hives affected. The bees take the trouble partially to remove to the outside the blackish brown crust forming finally from the rotten matter. There are, therefore, found on the floor a dark-coloured dust and entire skins torn off, which, when rubbed down between the fingers, give off the same unpleasant smell. In spring, when other stocks are already diligently building, the foul-broody do not generally make any preparation for it; at most they will only do so when they are still fairly strong, and unusually good pasture sets in. If the combs are examined the sealed brood is never found *en masse*, but standing in isolated, irregular patches. To be thoroughly satisfied, a piece of brood comb must be cut or torn out, and if it shows cells with the matter described above foul brood is certainly present.

Foul Brood is of two kinds.

There is one kind that is mild and curable, and another kind malignant and incurable; both kinds are, however, contagious.

The curable occurs in this way: more of the larvæ die still unsealed, while they are still curled up at the bottom of the cell, rotting and drying up to a grey crust, that may be removed with tolerable ease. The brood which does not die before sealing mostly attains to perfection, and it is only exceptionally that individual foul-brood cells are met with sealed.

This is exactly reversed in the malignant kind of foul brood. In this the larvæ do not generally die before they have raised themselves from the bottom of the cell, have been sealed and begun to change into nymphs. The rotten matter is, therefore, not found on the cell floor, but on the lower cell wall; it is brownish and tough, and dries up to a firm black crust, both in consequence of the heat prevailing in the hive, and of a small opening bitten in the depressed cover. This matter the bees are not able to remove; and when they are in some strength, they can at most get rid of it by entirely biting down the tainted cells and making fresh ones.

This dangerous disease may most readily be caught from honey which is given to the bees, or which the bees themselves fetch out of foul-broody stocks. If this happens at the time the stocks have brood, the disease will be fairly certain to break out, even if not immediately. It is only in autumn, when breeding has ceased, that in case of necessity honey from affected stocks may be used for feeding.* This must be done without putting in whole combs, because the hive might only too easily be infected with the germs of the disease, if the combs were left in it till spring. Although examples have occurred when one stock in a double hive was in the highest degree foul-broody, but the other was perfectly healthy and remained so, yet it is perfectly certain the disease can pass from one stock to another through their mere nearness one to another; indeed, the bee-keeper himself may very easily carry the contagion if he had undertaken an operation in a foul-broody stock, and without having thoroughly washed his

* But never, under any conditions, without it has first been boiled.—C. N. A.

hands immediately after, had undertaken another in a healthy one. In case we are visited by the evil, the great advantage of two apiaries at a distance from one another is evident, since the stocks remaining sound cannot be more surely withdrawn from the risk of contagion, than by separating them from those affected, and setting them up at a distance from one another.

The Method of Treatment of Diseased Stocks.

This will be different according to the time of year at which the disease shows itself or is discovered, and according as it is of a bad type or not. The curable kind may occur of itself, under certain conditions of ingathering, especially when the bees are working on bilberries and pines, and sometimes disappears again of itself when the conditions have changed. But the bee-keeper cannot with safety reckon on this favourable issue. The disease may quickly increase continuously and ruin the stocks. To put a stop to the evil immediately, catch the queen without delay as soon as any foul-brood cells have been observed. In spring and early summer she may be advantageously used for making an artificial swarm. If bees are added to her from healthy stocks we may be sure of having a healthy stock. But if bees were given to her out of her own, or other foul-broody stocks, the swarm must be left in a Transport hive, sieve, or the like, twenty-four to forty-eight hours before it is put into its hive, and the queen must be kept caged here for some days so that brood may not be deposited nor brood-food prepared before the bees have used up all the honey and food taken with them out of their parent stocks, and have expended it in comb-building.

Because there is now no more brood deposited in the stocks robbed of their queens, none can die and go bad, and till a young queen is reared, fertilised, and has begun to lay eggs again, the bees will have gained time, if they are still tolerably strong, to completely purify the brood-nest. They may be assisted in it by cutting the comb so close that the bees are able to cover it thickly. The new generation will then generally thrive quite well, and the stock be brought back to health again. There would be a greater certainty of this if the entire previous comb, as soon as it has become empty of brood, were cut out, and the entire stock were driven into a new hive. The question would have to be considered whether the stock is still strong enough, and if there is yet so much pasture in prospect that it might still form new comb and collect adequate stores.

A fresh outbreak of the disease will be more certainly prevented if the stock is again robbed of its young queen as soon as she is fertilised and has fairly occupied the brood-nest, if any advantageous use may be otherwise made of her. Further, no greater advantage can be derived from the foul-broody stock than when it is used for the production of fertile queens for making artificial swarms, or for helping stocks that have accidentally become queenless. Only there is this risk connected with it, that the rearing of a young queen by it, especially after the taking away of the first queen, does not always succeed, because all the queen-cells may easily become foul, and they have then to be assisted, either by a sound cell from another artificial swarm, or with a surplus young queen. In this way diseased stocks may bring in greater profit than sound ones equally strong. They afford in the course of the summer several queens, and at the end some pounds of honey besides. The bee-keeper, visited by foul-brood, will very well be able to make use of the queens; for he must work at the raising of young stocks so as to be able in autumn to cashier all that are not thoroughly sound. It would be folly to want to winter stocks suspected of foul-brood, when hundreds and thousands of sound stocks are sulphured in autumn simply for the honey and wax. It would be especially inexcusable if any one in autumn wanted to let a stock stand that suffered from the worse kind of foul-brood, as is often unfortunately done in districts where log-hives are kept, because the hive affected was a good stock before. Even in spring such a stock should be immediately cashiered, but in such a way that other bees may not taste the honey, because they may easily carry the disease home. The emptied hive should be singed with straw so that other bees may not be infected by it, or suck up the honey that may still be found about it, or gnaw off the propolis and carry it into their hive.* But the infected hive is

---

* This is a fruitful source of contamination by foul-brood, and one against which bee-keepers in England cannot be too carefully cautioned. When a stock dies, or is 'cashiered,' as our worthy Author puts it,

not made useable by this means, it must rather be put out for two years, open to the air, to be sure of the disease not breaking out again. We have at present no reliable evidence that the germs of infection are destroyed by sulphuring, as Hübler concludes, although it is quite possible. Though even the boiling of an infected hive in a large brewing-pan has shown itself to be ineffectual. When the hive was dried and occupied the foul-brood reappeared.* With a disease of such obstinacy it is easy enough to see what is the value of certain receipts recommended by many writers on apiculture. At most the mild form of the disease might be cured by such means, but never the malignant form. With this the question at most is how the owner of the diseased stocks may come off with least loss. First of all the queen may be made use of, and must be caught as quickly as possible, if the whole stock is not immediately cashiered. But if the bees that have been made queenless should have removed all the foulness up to the time of the hatching and fertilising of the young queens, it does not do to rely on having obtained a healthy stock. The disease would soon appear in greater severity than before, because in the meanwhile the poison has probably permeated the accumulated brood-food all the more completely. We hasten, therefore, to take out the queen again as soon as she has become fertile, and after some time we put a queen-cell into the stock again, or clear out the hive using the honey for any purpose

through 'foul-brood,' it is usually thought sufficiently protective if the combs, honey, and brood are done away with; but it often happens, through carelessness or want of consideration, that the quilt, the frames, or the floor-board are left exposed, and on the first sunny day the propolis on them, or either of them, becomes softened and gives off its attractive odour, and the bees pack it on their thighs, and carry it home, thus introducing the germs of the disease, which it will be sure to contain, into their own hives. A quarter of an ounce of propolis exposed may thus be the means of causing disease in a hundred hives.—C. N. A.

* But was it quite certain that it reappeared through an insanitary condition of the hive? The germs of all diseases that affect humans are said, on the best authority, to be destroyed by heat not greater than that of boiling water, and that surely ought to be sufficient to destroy the germs of foul-brood, which are almost, if not quite, of the same nature as are the typhoid germs, which, under ascertained conditions, are so destructive to human life. Singeing infected hives with straw may be beneficial, but leaving them exposed for two years in the open air without boiling or baking, is, we submit, a likely way of spreading the disorder.—C. N. A.

other than bee food, which might be given to brood. But if still we cherish the hope of recovering a healthy stock from the bees they must be subjected to a similar but longer treatment than that already described. After they have been kept two or three days shut up in an airy vessel without food, or with food given very sparingly, they are put into a new hive, the queen being kept caged for some time, partly for the sake of preventing the laying of eggs, partly to hinder their going off, to which such a stock is much disposed. But to put in a comb of brood, or even larger combs, is not advisable, because the stock ought to work up all the nutritive material it has by it as much as possible into wax, so as not to deposit it in the cells. Notwithstanding that, the stock may at last be up and away, or show itself again foul-broody, so that all the time and care bestowed on its cure is lost. It is better, therefore, to make short work of it, break out the contents of the diseased hives, make the best you can of them, and buy in their place healthy breeding stocks. Since it is easier to prevent the disease than to remove it when it has broken out, we should be cautious in buying foreign honey, especially for spring feeding. Instead of buying American, especially barrel honey, rather use sugar or malt syrup for bee food. Candy is in all respects the best substitute for honey after bad seasons, but grape sugar is the cheapest.

Foul-brood, which was formerly so much feared, and was justly looked upon as the greatest evil in apiculture, has, in recent times, lost its terrors. Through the studies of Dr. Preuss, Sanitary Councillor of Dirschau, Pastor Schönfeld of Tentschel, and other investigators, it has been established that the cause of the disease lies in certain fungoid growths, only to be seen through the microscope, in certain conditions multiplying enormously and destructive to the tender organism of the bee grubs. Since that, fortunately at about the same time, Professor Kolbe of Leipsic, has discovered in salicylic acid a means of killing those growths without injuring the bees or the brood that is still healthy, if the remedy is applied with precaution and properly diluted. Mr. Hilbert, the landed proprietor, who has been already mentioned in connexion with feeding with milk, has also made many experiments on the treatment to be observed for the cure of foul-broody stocks, and has most disinterestedly given their results for the advantage of all bee-keepers in the papers communicated to the meetings of German bee-keepers at Strasburg and Breslau.

Whoever wants to make himself thoroughly acquainted with the treatment may read the reports of the discussions in the last numbers of the *Eichstädt Bienenzeitung*, the organ of the German bee-keepers, for the years 1875 and 1876.

The essential part of the treatment is that the salicylic acid, which is not easily soluble in water, is dissolved in eight to ten times its weight of alcohol, and is kept well corked for future use. For use it is considerably diluted with water, which we must try to keep at a heat of 20° R. (77° Fahr.). Thirty parts of water are added to one part of the solution for spraying the brood-combs of the diseased hive; and the spraying is repeated about every five days till the cure is completed. But for the disinfection of emptied bee-hives, and of combs no longer containing brood, a stronger mixture may be used. The cheaper carbolic acid may also be used for the cleansing of the hive.

The remedy may be also administered internally, and a little tincture of salicylic acid may be mixed with the bees' food. On account of its property of preventing decomposition, a little tincture of salicylic acid may always be mixed with the sugared milk.

Although in applying this remedy the removal of the queen of the diseased hive would not be absolutely necessary, it is at any rate safer to use this preventive at the same time, and so to insure that the combs are not reoccupied with brood afresh before they have been completely cleansed from all impurity. Combs too much dirtied are best entirely removed if they contain but little or no healthy brood. In log-hives, skeps, and other hives with immoveable comb, the part of the comb that is altogether too foul is best cut out; leaving only so much for the bees as they can completely cover, and are in a position to properly cleanse.

Vertigo (*Tollkrankheit*).

This is indicated by individual bees falling down on the ground, either in the hive or outside of it, struggling, till at last they die. It is not so much a disease as a poisoning, which may be introduced either by malicious men or by Nature herself.

Towards the close of the fruit bloom, which is a very critical time for bees, Nature appears often to prepare a poison for bees; so that in this district we often find at this time in the hive and outside a quantity of bees, mostly young, struggling with death. Whether the bees carry in the poison from the mountain-ash, from the crow-foot, or the apple, which bloom at this time, or whether it is a consequence of some night frosts occurring, has not yet been established. Wherever bees can work on rape—which they prefer to every other flower, and to which a frost does no harm—there is none of this poisoning to be noticed.

The Horn or Tuft, formerly reckoned among bee-diseases, which appears after the fruit bloom, deserves at least to be mentioned as a curiosity. Professor Theodor von Siebold has proved, in the *Bienenzeitung*, that the yellow or dark-green tufts showing on the foreheads of individual bees are not a fungoid growth at all, but are the very elastic pollinia of certain orchids that are left sticking to the bees.

There is a peculiar phenomenon connected with bees when they are working on buckwheat; they lose the power of flying when the sky becomes overclouded, falling down in numbers in front of the hives, and falling to the ground in other places as well, and this they do even in a temperature of 15° R. ($65\frac{3}{4}$° Fahr.) in which they are otherwise able to fly continuously. But as soon as the sun shines and temperature becomes warmer they rise immediately and fly briskly into the hive. It is doubtful whether this weakness is a consequence of over-excitement, or intoxication, or of actual weariness; and whether it has to do only with the organs of flight or the entire organism. Another phenomenon, which has only been observed here once, appears to be similar to this, but, on the other hand, to be essentially different from it. In this, it is not the bees arriving loaded, but the bees coming out of the hive, that fell; and instead of remaining quietly settled, fluttered about impatiently, so that they were collected in depressions in front of the apiary in considerable numbers. They do not appear to have regained ability to fly. Perhaps it is the same evil of which bee-keepers on the heath so frequently complain, which they call *Füssgangerei* (footing it instead of flying); and whereby the hives, under certain conditions of weather, are sometimes considerably depopulated. If the cause of it is in the weather, the bee-keeper will hardly be able to meet it in any other way than by keeping his stocks strong, and so making up abundantly for the decline in population by breeding largely.

If naturalists, who lately have made bees quite specially the objects of careful investigation, have discovered in the stomachs of bees many kinds of fungoid growths, this might but little interest the practical bee-keepers for whom these lines are more particularly written, because nothing remarkable could be noticed in the stocks from which the bees subjected to investigation were taken.

The Enemies of Bees.

These are considerably numerous, and are found among mammals, birds, amphibia, and insects. There is no need to think of animals which, from their nature, have a certain aversion to bees and persecute them; on the contrary, their enemies are altogether too fond of them—some, of the bees themselves; others, of the honey; and others, again, of the wax—so that they consume bees and honey and eat up the comb, and so are injurious to the stocks and often destroy them entirely.

Among mammals, the bear is notoriously fond of honey, although no longer to be feared generally by bee-keepers, or at most only in some districts of Russia. In all countries the tiny mouse causes bee-keepers much greater anxiety. But it is only dangerous in the cold season, when the bees are in a closely crowded cluster from which they cannot separate themselves. The mice then slip into the hives, first of all eating up the dead bees, then the honey and pollen, and gnawing and nibbling the comb to pieces. We must make their entrance impossible, either by nails or a slide, without depriving the bees of a free exit. To catch them all is the best way of preventing their disturbing the quiet of the hives; and they do disturb the bees, even on the outside, and may injure straw hives in many ways.

Martens do not appear to despise honey altogether, for there are instances in which they have repeatedly visited stocks of bees living in hollow forest-trees, and called the attention of people to the stocks by so doing.

The enemies of bees are more numerous among birds.

Amongst these, the most destructive are those that visit the hives in winter; because the bees are least numerous then and cannot soon make up their loss, and because they are considerably disturbed when they require the greatest quiet.

The woodpecker, especially the green woodpecker, entirely destroys many a stock of bees living in a forest tree, by visiting it every day after it has been once discovered, and, by pecking at the entrance, brings the bees into the greatest disquietude, eating up many of them, and occasioning the death of still more by their chilling in the severe cold both inside and outside of the hive. If such an enemy and disturber of the peace has been repeatedly seen in an apiary, we must try to shoot or trap it. The great titmouse is not content with picking to pieces the bees left chilled on the snow, and partially eating them, but disturbs the bees in their winter rest, and entices them out by pecking. On account of their use in other ways, one would rather try to make them uninjurious by covering the entrances than by catching the tits. It is scarcely credible that the stork, when it is stalking about in the meadows, is snapping up bees, though that must be so, because a quantity of bees have been repeatedly found in the crops of storks that have been shot. Whether the redstart, the swallow, the fly-catcher, and other insectivorous birds, catch and eat bees, has been many times a subject of dispute among bee-keepers. If they do in case of necessity catch some bees, the damage done by them to the stocks is hardly worth talking about, and certainly affords no reason for persecuting and destroying these birds that are otherwise so useful in ridding us of so many troublesome and injurious insects. They do not appear with us before the bee-hives already contain a quantity of brood, and the loss of some hundreds of bees is not so much felt.

Among the amphibia, toads especially eat up many bees that accidentally fall to the ground, especially when they are left lying over night. They even jump up on the hives and snap up individual bees.

But their small, but more numerous enemies that are among the insects do the most harm to bees. The different stocks of bees themselves are mutually dangerous enemies, because they are so inclined to attack, rob, and destroy one another. When in autumn the number of the bees so sensibly diminishes in the hive, this has principally its cause in the bees being specially disposed when the pasture is ended to intrude into other stocks, and are then seized and stung to death. Stocks of the same apiary do not molest one another so much, but stocks of different apiaries situated at a little distance from one another are more inclined to it. In addition to the abundant pasture, this is one reason why bees thrive so well in isolated apiaries in forests or their neighbourhood. Bees that live in different compartments of the same Manifold hive, may get fighting if their entrances are either put too near, or are not properly divided, or if chinks open in the doors situated on the same side, or even in the mutual party wall. This must be carefully avoided. Although in a time of great heat, the bees clustering in front get along together very well, a queen, especially a young one, may easily be stung to death or injured.

Among other insects, the fly-wolf (*Philanthus apivorus*) may be the greatest destroyer of bees. It is a good deal like the ordinary wasp, and is mainly distinguished from it by a thicker head, a stronger thorax, and a longer and thinner abdomen. It knows how to seize the bees on the flowers very adroitly, throttling them with its claw-shaped mandibles, and at the same time stinging them with its short sting, and then flying off to its burrow with them, pressing them with its legs so firmly to its own body, that they seem only to form one body with it. Its burrow is dug, like that of the fox, anywhere in the ground, especially in the banks of ditches and other slopes. If these enemies of bees in dry summers greatly multiply, they bring great loss on the bee-hives in autumn, and depopulate them very much. The hornets, also, are notoriously genuine bee-wolves, which catch bees sometimes on flowers, sometimes in front of the entrances of the hives themselves, and after they have torn off the offal, fly away with the abdomen to their nest to lay it before their brood as food. In autumn, when they have become more numerous, they do great injury to bees in the neighbourhood of forests, especially oak forests, where they find abundant opportunity of building their nests in hollow trunks. Effort must be made, therefore, especially in spring, to kill the large queens because thereby just as many nests are prevented as they would have formed.

The small wasps are less injurious which only seize a bee, and perhaps also kill it, when they are attacked by it, and obliged to resist; otherwise they only go after the honey, and in autumn carry many a drop away when the bees have drawn more together, and the entrance is no longer so well guarded. When they multiply very much in some specially dry seasons, they give the bees a good deal of trouble, and may cause them great loss, so that it perhaps might be well to protect the entrance into the hive by the aid of a perforated slide, without, however, omitting to open it again when the bees desire to fly out in warmer weather.

The so-called bee-louse, a small insect of nut-brown colour that takes up its quarters mostly on the back of the bees, occurs only in isolated cases, and most frequently on queens. Other small insects and larvæ of insects have been observed on bees; but since they only occur exceptionally and rarely, and the bee-keeper can do little or nothing against them, we will not delay by enumerating and describing them. Whoever is more interested in the subject may find in several places in the *Bienenzeitung* explicit information upon it.

In the journal named, the death's-head moth has been often mentioned recently, sometimes being declared to be a mischievous enemy of bees, at other times spoken of as entirely uninjurious. It is certain that it intrudes into the bee-hives, attracted by the smell of honey. It has been met with here in box and log hives, but generally dead. But since it occurs here very rarely, the damage which it does to the apiaries is hardly worth mentioning. But if it should occur in numbers in various districts, it might well cause considerable injury to the stocks of bees, since it not only very much disturbs them with its fluttering, but being so large it is quite capable of stealing from them a teaspoonful of honey at a time.

Far more dangerous to the bees and their comb is another much smaller one, the so-called wax-moth, and especially its larva or grub, which is only too well known to every bee-keeper, so there is no need to describe it more particularly. There are, as is well known, two species of them; one small, the other, considerably larger, is the one that is peculiarly dangerous. The smaller is generally present in far greater numbers, but causes the bees but little loss. The small species whose silver-grey wings lie in the form of a roof-ridge swarm about in great numbers in the evening twilight on warm days in front of the entrances of the hives, so that the bees, for the purpose of keeping them off, run about anxiously and fretfully, as if they had lost their queen and were looking for her. The small grubs are mostly found on the floor, and feed on the wax *débris*. They do certainly gnaw the combs to pieces as well, but without spinning such a tissue all through them as the larva of the larger species of wax-moth does, which, when it is grown up, attains pretty nearly the size of a quill. It makes the storing of combs and entire sets of combs extremely difficult. For they are only too easily gnawed to pieces and perforated by the spun tubes, and made partially or entirely unserviceable. Single combs are, indeed, more easily secured against them, because they may be carefully looked through from time to time. The eggs, which are somewhat thicker than bees' eggs, may be removed with a pointed instrument, and so may the small grubs hatched from them, which close with web the entrance of the cell in which they lie hidden. As long as it is cold the eggs remain unhatched, and the larvæ that have come out remain torpid and cannot carry on their destructive work. But as soon as the weather has become warmer the combs must be more often seen to, if there is no sufficiently cool place available for their further preservation. Set up, or hung up, separately in an airy place, they are, at any rate, not easily attacked; wax-moths cannot bear a draught.* They do the greatest mischief when they get into brood-combs. Biting the dividing-walls to pieces, they march out of one cell into another, entangling the young bees in their web, or making them incapable of flight. Often the larvæ of the small moths burrow under the covers of the brood, and as a consequence the bees carry the covers away entirely, so that often entire rows of unsealed nymphs are to be seen. The moth-larvæ may be best scared out of the sealed

* It would be much nearer the mark to say that wax-moth cannot bear a *dry atmosphere*. Where there is a draught there will of necessity be means of access for the moth, and a moist condition of weather will help them in their work of destruction, but in a warm, dry cupboard, the combs being separate, they will not be likely to invasion, and moth-eggs that may have been previously deposited in them will dry up and come to nought.—C. N. A.

brood-combs by shaking. If the noxious insects have got the upper hand too much in a hive, which may be seen from the many nymphs encumbered by their webs and thrown down by the bees, taking away the queen is the best means, as in cases of foul-brood, of helping the hive into a better condition. If it should be too weak as well, it must be strengthened by adding bees. When the entire brood has hatched out, and the moths no longer find any hiding-places in the brood-combs, they can be thoroughly cleaned by the bees, and repaired where they have been damaged, and the new generation originating from the queen that has been raised or introduced thrives again quite well. The moth-larvæ must be removed from other parts of the hive as well, stopping all small chinks where the bees cannot get at them properly, lifting the cover-boards and destroying the grubs found under them. It is best to lay these boards a little hollow, so that the bees can get everywhere, and the moths will not be able to conceal themselves. If a board is laid on the floor of the hive the moths go under it, and the bees are not able to reach them. They may be very conveniently destroyed there. The moths have a similarly comfortable and safe lodging under the drawer that many recommend and put into the hive. This should be thrown away altogether, because it will be propolised by the bees and can only be drawn out with difficulty.

The wax-moths are not so very dangerous as they are generally thought to be. At most they affect weak stocks to some degree, and may destroy them in the end or impel them to leave their hives. Such small stocks as are kept, perhaps, only for the sake of the queens may therefore often be assisted in the warm season of the year by cleansing the hive from *débris*, particularly examining the combs that the bees are not in a position to cover, killing or drawing away the moth-larvæ contained in them, perhaps by the method of putting them out in the hot sunshine, without, however, letting it go so far as to melt the wax. Strong stocks know very well how to defend themselves from these enemies of theirs, if they have not altogether too much empty comb left with them.

Ants are further known as enemies of bees, and the large brown species, if it has its nest near, may give the bees a good deal to do; though it has this use, that it drives away the smaller species of ants, which often become troublesome by making their nests in or between the bee-hives themselves. They may easily be driven away by ashes, and may be kept at a distance from honey-vessels by the same means. The material with which the hollow space of the double walls of hives is filled up it is, therefore, well to mix a little with ashes, especially at the bottom, whether it may be moss, fine shavings, flax-waste, or the like. If box hives are piled one on the top of another, ashes may be strewed between them. This makes an accurate junction between the two, and the harbouring of ants as well as of wax-moths is made impossible there.

Every bee-keeper recognises spiders as enemies of bees, because spiders catch, entangle, and suck the blood of many of them in their webs, which are often put directly in front of the entrances. We brush these webs away and kill the spiders when we can, which is best done in the evening twilight, when they come out of their hiding-places. Meanwhile, the loss that house-spiders cause bees is but insignificant and scarcely worth mentioning. The injury is greater which the field-spiders, which are so numerous in autumn, prepare for them by covering the heather-bloom especially with their web, making it very difficult, or absolutely impossible, for them to work on it. Only heavy dashes of rain can remove this hindrance, after which a more abundant flight into the heather is always to be noticed.

Destructive Influences of Weather.

Many kinds of destructive influences of weather, both in winter and summer, occasion greater loss to the bees than all the enemies named here, to which in many districts many others might be added. In winter, in a few minutes, thousands of bees may be lost when they begin to play, enticed out by the sunbeams, and then either find their grave in the snow, or, struck down by cold gusts of wind, are left lying on the ground chilled. But severe and long-continuing frost is no less destructive. Not only do many individual bees then chill in the hives, but entire stocks fall asleep for ever when their honey-stores overhead or in their nests are exhausted, through severe cold having entailed greater consumption.

In spring many bees die from being so eager to carry in pollen that they venture on distant flights, and chill when the sun is concealed behind dark clouds. In summer the weather is destructive to bees when it is of the kind that Nature either affords no honey, or the bees are hindered from carrying in by cold winds or continuous rain, or when the bees are surprised and destroyed when foraging by frequent heavy rain.

Although bees generally like warmth, yet an excess of it may be injurious, and indeed destructive, to them. In altogether too great heat, not only does their zeal languish, but their comb becomes a great deal too soft and may partially or entirely fall down, and the entire ruin of a stock be brought on by it.

Although man generally can do but little against the forces of Nature, yet the bee-keeper may keep off, or at least alleviate, many of their injurious influences. He incites the bees in late autumn and winter to a general play and cleansing flight when a really favourable day occurs for it, that they may not want to fly out on less favourable days. In the winter, when the ground is covered with snow, he either houses his hives or keeps off the sunbeams carefully from them. To save the bees in spring from distant flights for water, he provides them with it, or keeps a supply of it continually at hand, in a place free from wind. He gives to his apiary a position as much as possible sheltered from winds and storms, and protects his hives from excessive heat, by shading and ventilation. Against other destructive accidents of weather, to which many bees succumb when foraging at a distance from the apiary, the bee-keeper can do nothing further than keep his stocks thoroughly strong, whereby a considerable falling off in population may be abundantly replaced by brood.

APICULTURAL IMPLEMENTS.

In what has gone before, mention has here and there been made of different operations which are performed in bee-culture. Although in these the bare hand does most, yet certain tools cannot well be done without. We will, therefore, proceed to speak of the different implements which are, some of them necessary, and others useful and convenient, in the pursuit of bee-culture.

The Smoking Apparatus.

This is the most necessary implement in the pursuit of bee-culture, and without it no long operation in the interior of the hive can be undertaken. Bees are smoked for the double purpose of protection against their stings, and to drive bees away from certain places where it is wished to operate.

The stinging of bees is certainly unpleasant, though after all their desire in that respect is not very great. The bee does not sting without a cause. Bees are excited to anger and stinging, by knocking, by noises, by every shaking of their hive or other objects covered with bees, by every quick or violent movement, and especially by breathing on them or submitting them to any other animal exhalation. Especially do they settle on animals, and other objects of a dark colour, and if these present a hairy or otherwise rough surface, they remain hanging on it with their claws, entangling themselves, and by their hissing and humming exciting an ever-increasing number to anger until gradually the whole stock will become furious.

In simply opening a hive, a part of the bees will be excited to anger; smoke must therefore be blown on them without delay, whereby an immediate change of tone will be produced. Instead of anger, a kind of embarrassment takes possession of the bees, in which they have no thought of stinging. For the purpose of pacifying them, and making them submissive, tobacco-smoke is sufficient, and is peculiarly suited for quieting them. The smoke may be blown out of the pipe itself, especially if its cover is provided with a cone-shaped tube directed forwards, but it is best to take a mouthful of smoke and blow it on to the bees. But if it is wanted for driving bees away and setting them in motion, the smoke of smouldering touch-wood is better. If this is in large clumps that may easily be split or sawn to pieces, there is no need of any further vessel for smoking. A piece is taken of about the thickness of one's thumb, or a thin billet, lighting it at one end, and letting the smoke rise, or blowing it with the mouth to the place where we want to direct it. If sometimes a wax-comb is touched with the burning brand, or a little wax is put on it, the vapour causes the bees to move and run more freely. The effect of the smoke is increased by dipping the touchwood in saltpetre, and of course well drying it again. Instead of touchwood, a glimmering match may be made use of for smoking. If touchwood is only to be had in small pieces, or if it is of such nature that it will only burn when put on hot coals, then of course some kind of vessel must be used, the simplest kind is a small pot into which are put first of all some glowing coals, and on them from time to time some pieces of touchwood. Failing touchwood, other glimmering material such as dried cow-dung, may be used for smoking.

In driving, and other operations in which it is often necessary to force the smoke deeply into the hive, a proper smoker like the one here represented affords excellent service. It consists of a small bellows and a tin capsule which may be either put on the nozzle in front as in the figure, or be put on the air-valve. It is provided in front and behind with a piece of perforated zinc, so that sparks may neither be drawn in

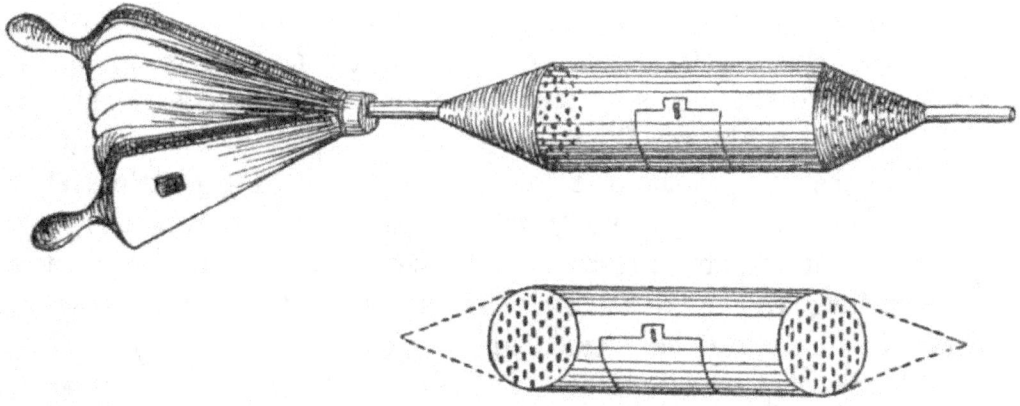

nor ejected. A simple pair of bellows is capable of being used by holding a glimmering piece of touchwood by the air-valve, and the bees are in driving put in motion by the mere draught. If a pointed nozzle is put in front of the machine, the smoke can be directed with greater accuracy to the part aimed at. Another less necessary, and therefore less honourable piece of defensive armour against the stings of bees, is the so-called

Bee-Cap or Bee-Hood.

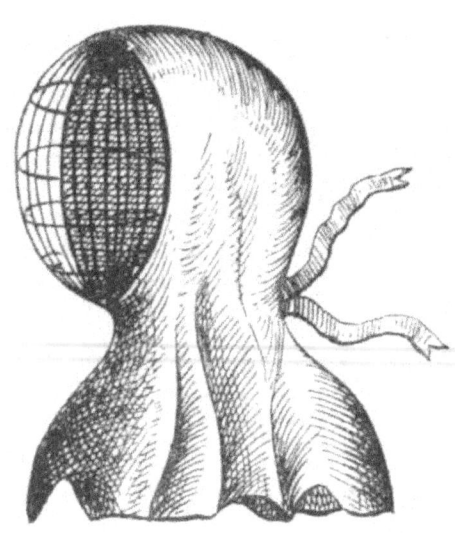

In its usual shape, it is a small oval sieve with a bag attached to it of corresponding width, and in use it is put over the head, but it is extremely uncomfortable, and very oppressive in great heat. The form represented here is far more comfortable, in which the place of the sieve is occupied by a wire mask.

But a proper bee-keeper rarely or never makes use of a bee-hood. Penetrating as the pain of a bee-sting is, it quickly goes off again; and if the sting is only removed at the moment and the place is immediately forcibly wiped with a wet cloth, whereby some of the poison that has entered the wound is removed again, the sting has no further result, especially after getting used to it. A wet cloth should always be kept at hand in long operations. But whoever is much disposed to swell, if he is a clergyman, schoolmaster, or other official who has to appear at stated times, he should always protect himself more carefully from bee-stings. But the best defence against them is the introduction of the gentle Italian bees, which sting extremely rarely, and only when they are greatly irritated or pressed. In operating in the hive, it is best to turn up the coat-sleeves if they do not fit quite close, so that bees may not be pressed and obliged to sting if by accident they have crept up the sleeves. It is, of all things, not necessary to protect the hands with gloves, partly because a sting there has so little significance, and partly because there is much that could, only under these conditions, be performed with difficulty and clumsiness.

The Sieve.

A small sieve of the size generally used for a bee-hood is, however, very useful in bee-culture, as for instance, for separating the dead bees from the wax-dust at the spring cleaning, for preventing the bees from flight in snow and at other times; or for catching a little swarm that is coming off, by fitting the sieve on the hive. A somewhat larger sieve may be very conveniently used for catching and hiving the swarms, as well as for sending them away to an apiary at a distance, if the sieve is bound at the edges with a cloth of the requisite size.

Transport Hives.

But far more convenient are the so-called Transport hives, small boxes with a wire-gauze lid that easily shuts up; these are used for carrying natural swarms, driven swarms, or bees for strengthening. According to the difference of the methods of culture, and difference of purpose, they may be had of different sizes. The smallest may be just wide enough to permit of its being pushed directly into the hive, so as, for instance, to be able to insert, perhaps at night, added bees for strengthening immediately with their Transport hive, or to be able to drive an artificial swarm at once into the hive pushed in at the top reversed, and so on. But it may, for instance, be made of the size of a Twinstock on the outside, so as to be capable of carrying, at any

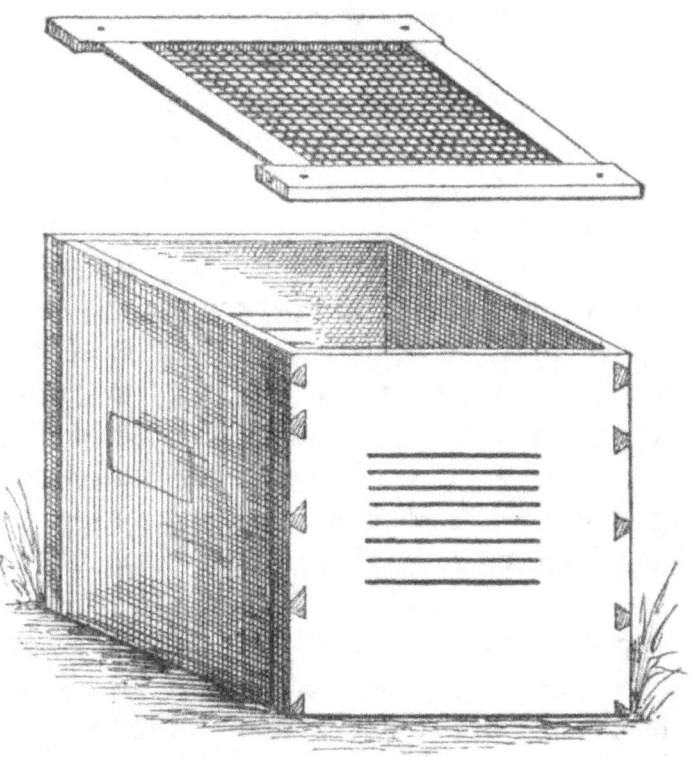

rate four swarms altogether, or bees for four swarms. The moveable lid, consisting of wire gauze in a thin frame, may either be placed at the top or the side. We may at the beginning put the Transport hive with the open side towards the top, and when we have shaken in so many bees that they begin to crowd toward the open top, we lay it down and put in the remainder at the side. The great point is, not only that the lid should be as airy as possible, therefore consisting of wire gauze or other similar material, but that the other side should have openings to permit of currents of air, so that even in great heat the bees may take no harm. It is evident that the Transport hive must be made of boards as thin as possible, so that we do not have to carry an unnecessary weight backwards and forwards. For the lid, firm material must be selected, if the somewhat costly wire gauze is not used, because a weak material like ordinary fly-gauze is in a short time bitten into holes by the bees, and so allows them to escape. And this would not do as the bees are often left in the Transport hive for whole nights, and indeed for some days perhaps, without queen or only with some brood to make them, for instance, more disposed to receive a young queen, or to make them more harmonious in uniting with another stock. There should be either no rabbet at all, or only a very small one, put in the hive for the lid, because otherwise hundreds of bees would have their heads trapped through incautious closing. Whoever has a skilful basket-maker in his neighbourhood may have quite convenient Transport hives woven for him out of osiers or willow roots. If they are only close enough, and allow of being shut up so securely as not to let any bees out, they are excellent for transport and for sending swarms away. But regular boxes have this advantage, that they allow of the bees being far more easily detached from the smooth walls and shaken out on to a cloth, and they may be also used as honey-boxes, so that fine combs may be carried home in them from a distant apiary. The hive must be waterproof at the bottom, so that the honey which may drip out will not be lost, and the air-holes must not be put at less than a couple of inches from the bottom. Otherwise a sheet of paper must be laid at the bottom with its margins turned upwards, through which even limpid honey does not quickly penetrate. Transport hives having the requisite depth are very good to use for hanging up combs while an examination of a hive or any other operation is going on; and they are the more convenient since in the robbing season the combs may be immediately with-

The Comb-horse. 293

drawn from the attacks of strange bees by covering them with a cloth. Whoever has Twin-stocks with a brood-room ten to twelve inches high must have his Transport hives made of a like height, so that combs completely built may be hung up in them. The length, too, might amount to as much, or more, so that entire honey or brood-combs might be moved in them, in which case they could not be hung up, but must be put leaning with a side-edge on the floor of the hive. The breadth of the Transport box may correspond to the breadth of the hive. Meanwhile, when the Box hive has too great a width, the case may be met by putting a strip across the open top and hanging the comb-bars by one end on the strip, by the other on the one side wall.

But it may be wished to look on both sides of a comb, that has been taken out, more particularly to catch a queen on it, to cut out a queen-cell situated on it, or the like, which are not possible when it is hanging down in a box. To have them before us with free scope to look at them, they are put on the comb-horse.

The Comb-horse.

This consists of one shorter piece of wood and two longer ones connected together, and three legs directed a little outwards fixed in them. The two longer and parallel pieces of wood stand as far apart from one another as the hives and the combs are broad. As far as the comb-bars extend with their two ends into the grooves, so far do they extend over the pieces of wood which may have a corresponding groove of about a quarter of an inch, so that the combs may lie firmly. The figure on preceding page indicates the shape and appearance of the comb-horse.

The Pliers.

For taking the combs out and putting them in again, it is best to use the appliances that every one always has with him, and that every one can manage most skilfully, namely, his own hand and fingers. But whoever shrieks and retreats as soon as a bee creeps on his hand may avail himself of a pair of pliers. An ordinary

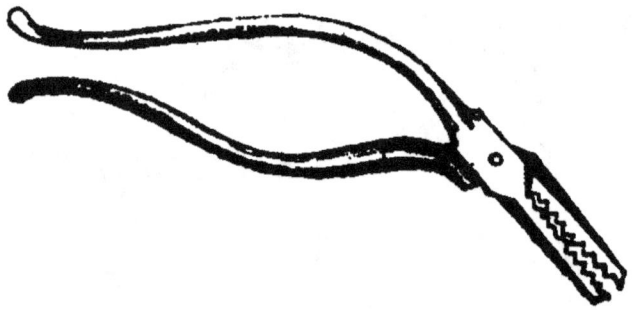

pair of wire pliers is quite suited for it. They may be bent down a little, as indicated in the figure, so that the comb-bars may be more conveniently gripped with them close under the crown-board of the hive.

The Fork.

A fork of suitable length may be serviceable at the cutting-time to a bee-keeper who is shy of being stung. Even owners of Log hives avail themselves of such an article sometimes for the purpose of sticking into a honey-comb that is to be cut, and by this means forking it out from the recesses of the hive. In hives with moveable comb the fork is, of course, still more capable of being used. For if a comb is pierced immediately under the comb-bar it lies on the fork firmly, and may not only be taken out without spoiling it much, but may be put in again in the same way. The fork may be still better managed if each of the two tines has an inch long depression not far from the point, so that when the one-inch bar falls into it the comb gains thereby a firm position on the fork. The points should not at most project

more than half an inch beyond the depression, or otherwise the next comb would be unnecessarily injured.

The Knife.

For loosening the combs which are intended to be taken out and that are here and there built to the hive-wall, any ordinary knife will do, but a table-knife, rounded off at the point, is especially suitable for the purpose. The thinner, sharper, and more pliable it is the more conveniently will it detach the combs from the walls. In cool weather, when the wax is harder, the knife may be heated at the same fire used for smoking; it then slips through the wax a great deal better.

The Feather.

For brushing bees from combs, walls, &c., there is nothing better than a single strong feather. When an entire goose wing is used some of the bees get their claws entangled in the down upon it, and, becoming angry, are the means of exasperating others. In cleaning the hive a single feather is sufficient, since the principal part of the *débris* and the dead bees may be previously raked out with a bar or a wire hook.

The Hook.

A hook may be used for loosening and pulling out the comb bars, which are sometimes considerably propolised. This we may easily make for ourselves by bending the point of a nail with a large head. A door-nail does very well for the purpose. Such

a hook is especially necessary when we want to draw out combs from close under the crown-board, which the finger cannot grasp, or close under honey-combs, which will thus be left uninjured.

Queen-cages.

For caging the queen, partly for the sake of having her in our power, partly to protect her from bee-stings, or to prevent her laying eggs, we need a so-called queen-cage.

The shape is of no consequence if it only corresponds with its purpose—allowing the queen to move freely in it, to be fed by the bees, and easily to be let in and out again. A queen-cage may be easily made by taking a piece of a willow branch, about two inches long and a little more than an inch thick, making a notch not far from each end, a little more than half an inch deep, removing the middle part of the wood, and wiring over the space left so closely that no bee can creep through, and if possible, not even thrust its head through. At one end an opening must be bored for the introduction and release of the queen. Instead of being rounded off at the top it may be quite flat and low. The opening for letting the queen in may then be situated on the one side, where instead of putting wires a moveable wedge is fixed between the projecting parts. Into such a cage a queen-cell may be inserted, and may be left in it for some time by itself, and may be put into any hive without any fear of its being destroyed. Such queen-cages, quite flat, are convenient on this account—that they may always be put between the combs, or may be laid under the cover. Because the opening is larger in the latter cage than in the former, the caged queen is more certain of being liberated after some time by the bees supposing it (the opening) to have been closed with wax.

The Drone-trap.

This serves for catching drones, which is a useful thing to do when they are present in too great numbers, and especially so when we want to keep Italian bees pure and to propagate them. Its essential points have been given in the description on page 197, and the figure we hope will make it sufficiently plain.

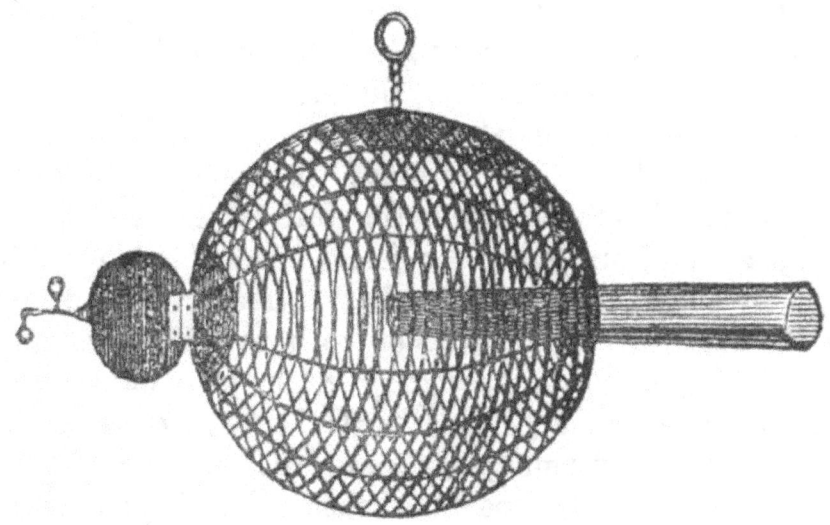

It does not matter whether it is round or square, or whether it is entirely or only partially of wire. Whoever may not be a skilled wire-worker would be able to make it much more easily square, and if he makes of wood the side through which the tube goes as well as the side opposite to it, which may contain the door for shaking out the drones, which may be killed in hot water, the construction will be still further facilitated. The wires might run from one board to the other as in the queen-cages, either turning down the ends, clamp fashion, and hammering them into the edges of the boards, or drawing them through small holes bored close to the margin. It might be more quickly prepared by putting four pillars at the four corners from board to board, and winding wire round the frame-work thus made, continuously, without break, in a spiral direction. If the wires have been either placed too narrowly or too widely, the error might then be more easily corrected. If the wires are placed a trifle narrower than a worker-cell is wide the distance will be correct. The workers can then slip through directly, but the drones are obliged to stay behind. In a trap found to be correctly constructed twenty wires with nineteen spaces would be reckoned to the space of $3\frac{3}{8}$ inches, which is equal to about the width of sixteen worker-cells. In this case it must be noticed that the wires were only very fine. The wires that form the tube must have the same distance, and the shape of this must be fitted to the entrance. Even if the entrance were not quite filled up it would not matter. It would be even well if such openings remained at the side, that workers might get into the hive by these directly without first going through the tube.

Since the queen is of about the same thickness as the drones, and makes her trips at about the same time as they do, we must guard against using the trap in a hive that has a young queen still unfertile, for fear of catching and killing her. To use the trap in a hive that is ready to swarm for the sake of catching the queen, or to prevent the escape of the swarm during a casual absence from home, would also be dangerous, for there would be a risk of the tube choking with bees in their hasty swarming out, and the stock might then easily be suffocated. The tube must then be of about the width of the entrance through its entire length, whereas otherwise it may gradually diminish in size.

The Swarm Net.

The swarm-net is used to prevent a swarm entirely flying off, and to catch it immediately it comes out. It may be mentioned here simply for the sake of completeness. There is no need to describe it more particularly, as what is necessary has already been said upon it on page 158, and as the rational bee-keeper, who carries on the increase of his stocks systematically, will rarely get so many swarms as to be obliged to have recourse to the swarm-net to hinder their flying together. To be at hand with the net immediately the swarm begins to come off would involve such unintermittent, anxious attention that the bee-keeper would be obliged entirely to overlook other things perhaps of greater importance.

The Scooping Vessel.

For detaching and scooping out bees we may certainly use the first vessel that comes to hand, as a box, a scoop, or the like, but we may have specially made for it a wooden or tin scooping vessel provided with a stick as handle. It is more convenient rounded off for round skeps and often for round log hives, but for box hives it is better square. If it is round at one end and square at the other, it may be used in all kinds of hives. It must then be fastened on a handle so as to allow of its turning on the nail with which it is riveted, and so turning at one time the

round end to the front, and at another time the square end.

The Feeding-box.

Such a box, without handle, may at the same time be used as a feeding-box for giving the bees liquid honey, though it is better then if it has directly perpendicular sides, so that the reeds, straw, or pieces of wood, with which the honey is covered to prevent the bees drowning in it, may, as the liquid lowers, be let down to the bottom without hindrance. In hives with moveable combs we may very well do without special feeding vessels, since we can pour the feeding honey into the cells of an empty comb indoors, and may then suspend them in the hive when we have no more full combs overhead. Pieces of solid honey may be put in the honey-room, and in warmer weather be laid on the floor of the hive, or be put underneath in the first vessel that will answer the purpose.

The Tin-pan.

It has been already said that either a tin or earthenware pan is needed for melting the wax, in order to fasten the comb beginnings

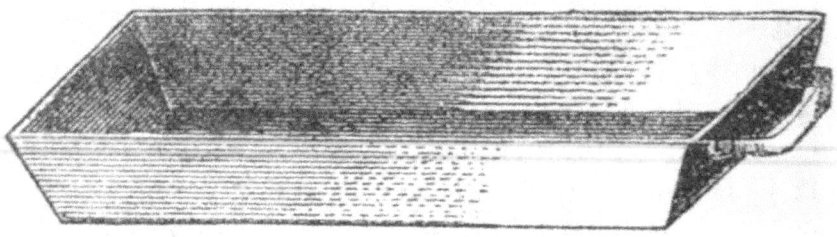

on the bars with it. A pan about 12 inches long and 2 inches wide at the bottom will do very well for the purpose. At other times it may be used as a scooping vessel and feeding trough.

The Syringe.

A syringe with a long point and a fine opening, therefore somewhat like a clyster-syringe, may often be very serviceable in bee-culture. A swarm may be conveniently sprinkled with it, the bees of a swarm may be induced to settle more quickly, if we use the syringe under them; water for a stock may be syringed into the combs through the entrance, or from the door; a stock may quickly be induced to play in front, by exciting it with thinned honey, and the reverse effect may be obtained, and the bees be driven to rest, when they want to begin to play at an unfavourable time, by syringing cold water against them, causing them to suppose it is raining.

The Screw.

An ordinary pocket-knife is generally sufficient for opening the door of the hive. Meanwhile in winter, wooden doors may swell so much that we have to apply some force to get them out. In this case we may make use of a screw on which a large ring is fastened. After we have screwed it into the door, and put a somewhat long stick through it as lever, the door is bound to come out, let it be swollen as much as it may.

The Honey-Slinger.

One of the most useful inventions recently made in practical apiculture is the machine by which the honey may be extracted from the comb without destroying it. Previous attempts had been made to get the honey out of the cells turned downwards by the simple force of gravitation, but this only succeeded imperfectly, because the force of adhesion with which the honey is retained in the cells is for the most part stronger than that of gravitation. Major Von Hruschka, of Polo, near Venice, hit upon the happy idea of applying another stronger force, the socalled centrifugal force which is produced and may be increased at pleasure when bodies are set into quickly rotating motion. According to the first law of motion, every body moves in a straight line, unless acted on by some external force, but when put into quick rotation its direction is altered continually, and everything not firmly attached flies off from its surface, as the stone out of the sling, the water from the sprinkling brush, the mud from the carriage-wheel. So does the honey of a comb fly out of the open cells directed outwards, when the comb is put into quick rotation. The honey is slung out, thus the designation of honey-slinger, has been generally adopted for the sake of brevity. It consists of two principal parts, namely, a kind of winch on which and by which the combs are put into a quick rotatory motion, and the vessel enclosing this, which catches and collects the honey slung out. If we imagine a small square-shaped table with its four legs directed upwards, with a perpendicular axle in the middle through the table-top, and the ends of the legs connected by arms, we have something like the form of such a winch for four combs, as they are generally made. The honey-combs to be emptied are put in or hung up between each of the two adjoining legs, and when the side of the comb put outwards is emptied, it is turned about, and the machine is again put in motion. To prevent the combs themselves from breaking while they are still heavy, they must be inclined against a kind of net, which the honey can spurt through, or against several stretched wires or threads.

For the outer vessel, any ordinary large boiler has been recommended in the *Bienenzeitung* by a practical bee-keeper, but a wooden tub-like vessel with perpendicular wall is certainly better adapted for it. The pan in which the perpendicular shaft works is in the middle of the bottom. At the top the shaft goes through a bar, that is fastened across. A principal point is the mechanism by which the winch may be put into the

quickest possible motion. The simplest is merely a somewhat strong cord, which is repeatedly pulled forcibly and quickly after it has been wound round the shaft at the top, and the further winding up it does itself when set in motion. Mr. Schmidt, beekeeper of Ingolstadt, considered this apparatus as sufficient, and made many machines on this principle. But the motion resulting from it is jerky. An equable motion obtained by turning a wheel with a handle is much more suitable. Whether the incomparably quicker motion of the winch is obtained by a cogwheel, or whether it is from a cord, or a driving-belt, probably does not very much matter. In the last case, the wheel to be turned by hand, may be put anywhere outside the machine since the driving strap or cord may have any length we like given to it. Of course, the honey only comes out of open cells into the

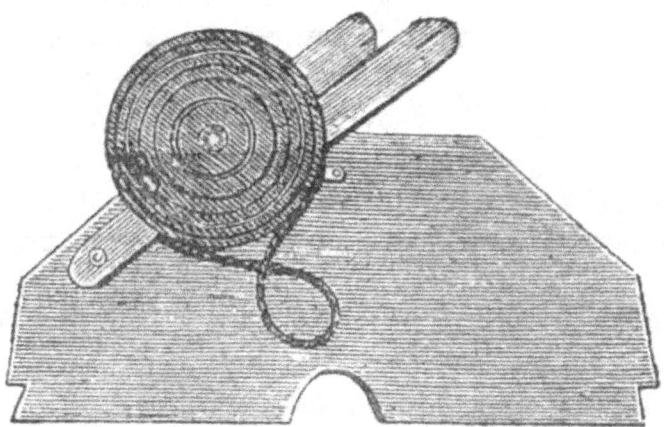

machine. If the honey-combs are entirely or partially sealed, they must be unsealed before they are put into the machine. For this, a peculiarly shaped knife is used,

The Knife for unsealing Comb.

Another instrument has been recommended for the same purpose, the comb-roller, by which the sealing is broken into

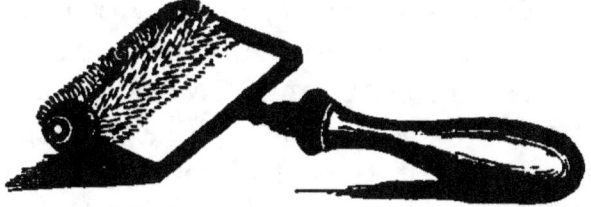

small pieces so that these are slung out at the same time with the honey. But the knife is to be preferred as simpler.

The Refining of Honey and Wax.

Strictly considered, this is not a matter for the bee-keeper who sells his honey-combs and wax-combs in the raw, and leaves to others the extraction of the honey and refining of the wax. But since bee-keepers generally attend to it themselves, some hints may be given upon it.

The extraction, or straining of the honey, may either be done cold or warm. Cold strained, the honey preserves its beautiful light colour, and its pure natural flavour. This succeeds readily so long as it is limpid. The honey-combs are thrown into a basket, made of osiers, they are broken to pieces, so that every cell is broken up, letting the honey trickle down into a vessel put underneath. The warmer the temperature is, the more quickly and completely does the honey flow off. The remainder is extracted by repeatedly pouring on warm water and is used either for feeding bees, or for vinegar or mead, or for boiling down to more solid honey. Honey partly crystallised must first be made liquid by moderate heat, and then be treated in the way mentioned. But because honey only too readily assumes a darker colour, and a somewhat burnt flavour from the fire, it is best managed by putting the vessel with the honey into a boiler, filling it partially with water and letting it boil till the honey is properly dissolved.

Honey and wax are separated by heat, by putting the vessel full of honey, perhaps, into the baking oven, after the bread has been taken out of it. The melting wax rises to the top, and it is found on the following day in the form of a firm cake, which may be either taken off, or left for the better preservation of the honey underneath.

If the honey-combs had previously contained no pollen nor brood the cake forming over the honey would be composed of quite pure wax. Most honey-combs, but especially the empty wax-combs coming to the melting-pot, contain pollen-cells here and there, and a quantity of nymph cocoons, which the young bees, reared in them, have left behind. The refining of the wax consists in the separation of these two foreign elements. It is evident this cannot be done without dissolving the wax, that is, it is boiled up in the boiler or pot with a suitable quantity of water till all the wax particles have become liquid. All that has to be done now is to make a clean separation of the wax from the other elements. The wax separates itself from the water since it rises in it as a fatty substance of lighter specific gravity. The separation of the other substances, which likewise swim on the water, is more difficult. The only thing to do is to hold them down without preventing the rising of the liquid wax. If a perforated lid is put into the vessel half-full, and hot water is poured to it, the wax will rise over the lid with continuous moving, and may be drawn off, skimmed, or be taken off later on as a fine cake. With a large quantity the end might be more quickly gained by pressing out than by a simple boiling out. The principal thing in this is to press out the boiled mass as hot as possible, and therefore as quickly as possible. For this a bag, quite strong, but not too close, is necessary, and a press with which a strong pressure may be used. It does not matter whether it is a stamping press or a screw press. A crushing press, with long arms, is a simple but quite practical apparatus for it. A strong pressure may be brought to bear with it, at the same time allowing of being relaxed every moment, to shake up the bag a little. To prevent the wax squirting to waste, combs may be arranged to catch the liquid wax, which can then be boiled again. The mere flowing down into a vessel, partly filled with water, separates the wax immediately by its congealing. It is collected and boiled up again with water to purify it completely. After the scum has been taken off, and the wax has cleared itself, it is poured into a vessel. A pan with tolerably high walls is best, so that the impurity still to be found in it may fall to the bottom, and not partially on the margin, which would give to the wax-cake a dark appearance. The vessel may be rubbed with a little honey water, that the wax may be more easily removed, and as soon as a thin cake has formed it may be carefully loosened with the point of a knife where it sticks to the vessel and begins to crack. The impure wax that has been skimmed off the top, and which has been taken off the bottom of the cake, is stored and boiled up again at the next melting.

The further Use and Manufacture of Wax and Honey.

This scarcely belongs to a bee-book. But as a better knowledge of how to convert bee produce into money may indirectly contribute to the promotion of bee-culture a few hints may be given upon it here. To allude to the manifold uses of wax might well be superfluous. The frequent inquiry after it is proof enough that it is a much-esteemed article that cannot be done without, and which always will find ready buyers.

There might rather in excellent seasons for bees be a want of a market for honey, although its uses too are many. But the consumption of honey might be very much larger and more general than it is. As the bee-keeper, failing honey, uses sugar for bee-food, so honey might, when there was a surplus, take the place of sugar. Honey does excellently for preserving and sweetening fruit. Melons taste much better with honey than with sugar. For sweetening tea honey is quite well adapted, as it also possesses a sudorific character.

On account of its near affinity with grape-sugar honey might be used for the preparation of mead and honey-wine, or for the improvement of wines which are wanting in saccharine matter, it might be used in far greater quantity than it is. What splendid beverages may be made from honey, Mr. Blume, apothecary of Berlin, has shown, who brought to the meeting of German bee-keepers at Hanover different kinds of mead and honey-wine of his own manufacture, which met with the warmest approval. Large sums, which now go abroad for partially adulterated wines, might be kept in the country if the consumption of mead and honey-wine were more general. The different associations that have sprung up for the promotion of bee-culture should make an effort in this direction, and themselves lead the way in this respect, especially at their meetings.

There is no need, therefore, to be anxious about what is to be done with the honey if the pursuit of bee-culture becomes more general and good bee seasons set in. At a moderate price there will even then be no want of a market. Fine honey, especially honey-comb, will perhaps always be worth 6*d*. per lb., and if larger quantities are harvested the profit will always be remunerative. If we can take from a good stock, in a good honey season, 60 lbs., this would give a profit of 30*s*., with which we may be very well satisfied. Moreover, honey keeps for many years, and does not require to be given away at a ridiculous sacrifice; and a careful bee-keeper always preserves a part of his harvest to be safe in case the next season should be bad. He selects for this, of course, the honey that is least saleable, containing pollen, occurring in old black combs, or having a dark colour as buckwheat honey has, but which is a specially acceptable and wholesome food for bees because of its peculiar strength. If honey should sink still lower in price, as in certain districts of America it is said to be nearly valueless, it would still be in the power of the rational bee-keeper to have more wax produced than honey.* This would be attained by cutting as much as possible in spring and after the swarming time, and by the largest possible multiplication of stocks in summer and reduction in autumn.

But if the profit of bee-culture is to be remunerative, it is not simply a matter of increasing the number of stocks, or always managing them as should be, but attention must be paid to this point, that the bees are never allowed to suffer from want of abundant food. For at most the bees cannot do more than collect and refine the honey which Nature affords them, they have no power of producing it themselves. If the bee-keepers of a district only thought of increasing the number of their stocks an overpopulation might easily occur after good swarming seasons. It may, therefore, not be superfluous, in conclusion, to say something on bee-pasture, from which every bee-keeper may gather at what he has to aim, so that his bees and himself may succeed.

BEE-PASTURE.

Bees certainly visit most flowers, though particular ones are specially acceptable to them, partly because they yield pollen and honey in especially abundant measure, and partly because they are easily accessible for them. Many have special value for bees by affording food to them very early, as the hazel-nut, from which they often carry pollen and prepare brood-food as early as February, but as a rule in March. Then there are the alder, the aspen, or trembling poplar, but especially the different species of willows which flower one after another from March to May; and among which the male willows are of great value to bees, because they not only afford honey but pollen as well in abundance, and promote breeding very much. The palm-willow, *Salix caprea*, is notoriously valuable for bees, and it may be

---

* Considering that every pound of wax produced will have been at the expense of about 20 lbs. of honey, we can scarcely conceive of wax being looked upon as a remunerative product of bee-keeping. An ordinary swarm of bees, in one year, will not produce an average of more than, say, 2 lbs. of wax in the form of comb, and to keep them to this work it will be necessary to continually remove what they make, and the result at 1s. 6d. to 1s. 8d. per lb. cannot be considered highly remunerative.— C. N. A.

budded on other willows, as it does not root so readily as the other species, and in this way large trees may be quickly obtained, at about the same time the sycamore and elm bloom and afford pollen and honey in abundance. Soon the different kinds of fruit-trees follow, whose bloom forms the principal spring pasture in cultivated districts, whereas in wood-lands the bilberry or whortle-berry, blooming at the same time, offers an abundance of honey, which perhaps exceeds that of any other plant, so that it pays to travel there with the bee-hives, where it is extended over a considerable area of forest. Unfortunately there occurs an interruption of the pasture and of the flight of the bees in most districts after the fruit-bloom is over and when the hawthorn is in bloom; and since there is now a good deal of brood in the hives feeding must be attended to if the bees have no old stores. This pause is less noticeable in districts where oil-producing plants, turnips and rape, are cultivated, the latter especially remaining in bloom for some time till the ground ivy, white clover, and corn-flower begin to bloom. As regards shrubs the blooming of the bird-cherry occasions a more lively flight towards the end of May or beginning of June. This shrub then deserves at least to be protected even if not specially cultivated. The buckthorn (*Licium*), since it affords honey abundantly from spring to autumn, and is suited for quickset hedges and for shading arbours, deserves to be very frequently planted since this can be done so easily by cuttings. Among trees deserving of mention there are the horse-chestnut and the maple, of which all the species are abundantly visited by bees, and especially the common acacia and lime. They afford not only a quantity of honey, but it is of an extremely agreeable flavour, and readily commands a higher price. Their cultivation, therefore, cannot be too urgently recommended, not only in the interest of bee-culture, but for the beautifying of the country, since they have such beautiful foliage affording abundance of shade. Among the conifers, the firs and pines (especially the pine scale insect living on them) often afford, as is well known, large quantities of honey. But, unfortunately, when this honey forms the only food of the bees, and they are long hindered from flight, it usually occasions an outbreak of dysentery. The same holds good, more or less, of all aphis-honey, which, in warm and continuously dry weather, often appears on the leaves of many trees, and is known by the unsuitable name of honey-dew. Although the bees do not despise this aphis-honey, but collect it very industriously, and carry it in when it is dissolved by dew or a fine warm rain, yet the flower nectar forms their natural, and therefore their most wholesome food. Among the so-called weeds the field radish, *Raphanus raphanistrum*, the blue corn-flower, *Centaurea cyanus*, and the eyebright, *Euphrasia odontides*, would take perhaps the first place as honey plants. Near Carlsmarkt, the corn-flower forms nearly the only summer pasture, and without it bee-culture could hardly be carried on here.

Among cultivated plants perhaps the white clover, *Trifolium repens*, affords the most abundant pasture, because the flower is not only very rich in honey but lasts a long time. It is a pity that the bees only visit red clover under certain circumstances, because it has such a deep corolla. Among the species of clovers that especially deserve to be cultivated there are the Incarnate clover, *Trifolium incarnatum*, which, although properly only an annual, may be sown in the winter, or perhaps in August, and then begins to flower as soon as the fruit-blossom is coming to an end; and the bright-coloured Swedish clover, *Trifolium hybridum*, which comes into flower over again later on, and still affords food to the bees even in August. Sainfoin is well known as an excellent honey-plant, as well as the heath-corn or buckwheat, which thrives especially on light soil, and is sometimes sown as an early crop; but not before the end of May or in June, on account of its sensitiveness to frost; and sometimes it is sown as an after-crop in rye-stubble. Horsebeans and vetches exude much honey, but, very remarkably, not in the flowers, and are under favourable conditions of weather much visited by bees. The lupins, that are now always being more extensively cultivated, are certainly not entirely despised by the bees, but yet are not so frequently visited as might be desirable. Heather is generally well known as affording abundant autumn pasture. The flowers of tobacco, when it is extensively cultivated, yield much honey in September, and even in October, if it is not previously destroyed by frosts, as well as the ground-ivy blooming in considerable abundance on precipitous slopes, and in rape and turnip fields. There are still many plants whose cultivation would much improve and prolong the bee-pasture, among which only one must be mentioned here. A zealous lover of bees from Russia, Mr. John Kalinski, strongly recommended it to the Author, and left some seed of it with him. It is a summer plant, growing quickly and blooming quickly, even if it is not sown before July, perhaps on the ground which has previously yielded a crop of peas. The plant becomes a large shrub, the leaf is similar to that of the cypress, the blue flower to that of the viper's bugloss, *Echium vulgare*, and is worked on by the bees in October, and even in November. He called it *Phacelia congesta*.

But the pasture of the bees may be improved and prolonged by sending them to the place where they may have a better pasture, or where it still lasts after it has come to an end in the neighbourhood of their previous stand. Certainly bees fly a very long way, as much as five English miles, when they can no longer find anything nearer; but, of course, they then cannot make so many flights, and may die on their distant journey, and are left lying on the ground exhausted in rough winds, or become a prey to their numerous enemies, so that from such a distant flight in the end the loss is greater than the profit. Even in districts in which proper travelling culture is not pursued, it may in some circumstances be very advantageous to set up a part of the hives at some distance when the rape, clover, and buckwheat are by chance at some distance off. By removing a number of the hives a better pasture is made for those that are left, because now there falls to their share what the others would have carried off. All those advantages are at the same time gained which present themselves in parting and reuniting two apiaries at a distance from one another, and which have been many times mentioned at the places relating to them. From this it may be seen, that just as there are advantages and conveniences when the individual combs of a hive are moveable and can be put in another place at pleasure, it is not less advantageous when individual hives are moveable and may easily be brought to any apiary we like.

In conclusion, the Author cannot, therefore, too strongly recommend the Twin-stocks, as well as the double Lager-hives. An additional motive for him to do so is the possibility of wintering bees excellently in Twin-stocks, however destructive to bees the winter may prove. Wintering bees well is the masterpiece of the bee-keeper, as Von Ehrenfels has called it. Although the Author frequently receives most bitter complaints of the devastations caused by the continuance of intensely cold weather, and although he himself has lost many a colony, even when populous, in press-shaped hives, and in hives made of logs, he has always found bees to winter well in Twin-stocks, however small a colony may be, their combs remain quite clean, and they consume very little food, while the consumption in other hives is unusually large in cold weather. When the winter shows signs of becoming severe, the Twin-stocks are put into some place for shelter, so that the cold in the interior of the hives scarcely reaches the freezing point, while the temperature inside the hives that are allowed to remain in the open air is often as low as — 20° R. (13° F. below zero), or even lower. Straw-hives, too, can easily be taken indoors, but cannot be piled up as conveniently as Twin-stocks. Where space is limited the latter may be placed one upon another, like logs of wood. Should the hives happen to touch, the entrances may be closed, and the doors having all been turned towards a passage, which should remain quite free, air may be admitted either by easing the doors a little or removing them entirely in case the colonies are large and require more air. In the absence of any suitable place to keep the bees in during the time when the cold is most intense, a bee-keeper might put them into a spare room, or even remove them into a cool corner of his sitting-room, and by doing so he would render the bees a great service, for but a few unusually cold days are sufficient to give a colony the death-blow.

But in the apiary in the open bees always winter better in Twin-stocks than in other hives, especially when we see to it that the young swarms put into them build fewer combs in the brood-room, but build up and fill immediately the small room, at any rate, over the brood-nest and winter-quarters. They are then surrounded on all sides by stores, can more conveniently move backwards and forwards, and it will not be their lot, as it is that of many colonies in other hives, to chill on empty combs while adequate stores are still to be found in the adjoining passages.

We have allowed ourselves to make many kinds of alterations in the Twin-stocks, but an old expert, the Forester Krüger, rightly designated them as anything but improvements. The only alteration recommended which has a certain justification relates to the setting up or piling not one across another but parallel, so that the entrances come to stand only towards two directions, and the bee-keeper, in operating, is less inconvenienced by the bees. A pile of six simple Twin-stocks, or three double-hives, would then appear as indicated in fig. on page 107. Every human work is capable of improvement and perfecting, and so also the bee-hives and the entire method described by the Author. Every improvement and new experience he will, as previously, bring before the attention of bee-keepers through the organ of the Association of German Bee-keepers, the *Eichstädt Bienenzeitung*.

It is usual in an introduction to bee-culture, especially for beginners, to give a compilation of the different operations to be undertaken in an apiary in the course of the year, according to the succession of the months. Although the Author follows the custom in this guide to rational bee-culture, he would not wish to indicate more by it than the orderly succession of the bee-keeper's work, and by no means establishing fixed times for his operations. Life in nature awakes according to the difference of the climate of the district, and according to the season; and upon this the bee-hive operations depend, being at one time earlier and at another time considerably later. The further development of the stocks, and the time for the bee-keeper to attend to certain work, depend on the season. This specification of time, then, is only to be considered as an average one, and by no means as appropriate for all seasons.

BEE-CALENDAR.

JANUARY.

The bee year may very properly be begun with the civil or calendar year. For with the setting in of the cold season the bees fall asleep for a long rest; to awaken, when the sun gets higher again, to a new life and a new daily labour. This awakening to fresh increased activity, and renewed breeding, often happens as early as January; but it is then only to the injury of the bees themselves as well as of their owner, inasmuch as stores at least are consumed which later on might either be taken away or used more advantageously by the bees. If the bees have already set about breeding, severe cold following after, it is all the more dangerous. Having begun to breed, the bees are now tied to a definite part of their comb, and can no longer move farther on, or they do it at least with extreme unwillingness; and may starve to death, with ever so large stores, if they have used up the stores in their nest. In that case, severe cold continuing a long time renders it impossible for them to uncluster, and to bring in honey from the combs situated more at the side. Although, therefore, the rule holds good generally that the hives must now be disturbed as little as possible, it might be, in certain circumstances, useful or quite necessary to insert a few full sealed honey-combs close to the nest of the bees instead of those used up, so as to ward off the danger indicated above. To bring fresh stores near to the nest of the bees is good also on this account—that in double and manifold hives the colonies may stay in proximity to one another by the mutual party wall, and not be reduced to the necessity of separating from one another to move after their stores. Even weak stocks—if they have always food at their side or over them, and thus can remain quietly for the whole winter in the warm nest they have once taken possession of—for the most part get through the winter, and then thrive excellently;

whereas strong stocks often suffer considerably, or may be quite ruined, when they have used up everything in their nest, and, in severe cold, are obliged to move farther on, in which case a large number of bees is generally left behind chilled. Fresh food is most conveniently given to the bees from the top, even in the middle of winter. In order that the bees may be able to mount up from all the passages into the honey-room, a slit is made by the wall by which the stock has its quarters—in double hives, therefore, by the wall common to the two compartments. By this wall and in this angle pieces of honey are put, a piece of board is inclined over it, and the rest of the empty space is filled up with straw. Because heat always rises the bees can betake themselves upwards even in great cold, and can bring honey down into their nest. Through this slit the bees in the different passages keep in better communication with one another; and even the further moving on into other passages is much facilitated by it. But this small passage into the honey-room must have been seen to in the autumn, and the cover-boards have been a little moved away, because a forcible removing of them in winter would cause too great a disturbance, which is to be avoided as much as possible. For a while the slit may be covered up with a board—perhaps with a comb-bar, not opening it before honey for feeding is put in at the top. This is best put in in sealed combs. It does not matter whether they are put in lengthwise or across, or whether they lie on the flat side or on the edges. The bees will know how to get at the honey, even if they should be obliged partially to gnaw down the cells. If the emptied combs have not been taken out before for the purpose of putting in fresh full ones, they must, at any rate, be taken out at the spring cleaning to put everything in its previous order.

Solid or crystallised honey is very convenient for introducing in this way. Wrapped in paper, it may be pressed into any shape that the space to be filled may require, and is gradually dissolved and consumed by the bees if a little of it is uncovered by the passage. If we are short of honey and there is reason for being economical of it, pieces of candy may be kneaded among it. The sugar becomes milder and more soluble with the honey. In case of need candy alone may be put in in pieces; but when the colony is somewhat weak and when the hive keeps very dry the bees may be deficient in the necessary dampness and power to dissolve the sugar, and they may at last starve or die of thirst even on the pieces of candy.

Besides food in their immediate neighbourhood—therefore near or over their nest—the bees need warmth, in winter especially. The best protection against frost is afforded to the hives by putting them into a room which, if not entirely frost-proof, is yet in some measure protected. In this the bees are left as long as possible, not allowing ourselves to be beguiled into putting out the hives by a few genial days, because February is not to be trusted nor yet March, and severe cold may yet ensue. Whoever has neglected to house his hives, or perhaps has no opportunity of doing it, protects them on the stand in every possible way against the cold, if the hives are not already built so retentive of heat that the severest cold of winter cannot be destructive to the bees. They cannot well be too warm now. The more protected they are against the influence of cold the less do the bees themselves need to be at the trouble of producing the necessary heat, and the more are they able to economise their strength. With too much wrapping up, or too much narrowing of the entrances, there might very easily occur a deficiency of fresh air, the access of which should never be quite shut off. Both warmth and fresh air are promoted by removing straw covers, comb, &c., that have become tarnished and damp, and replacing them by fresh ones, or by opening the doors for a time on mild days, or, better still, on windy nights, and letting the moisture be removed by the draught.*

It is well known how dangerous mice are to bee-hives at this time. Although a free exit must always be left for the bees, the mice must be carefully debarred from using it either by nails or slide or by catching them. The small shrew-mice especially are difficult to keep off. Where these occur plentifully the hives must be the more carefully protected. It is true they do not gnaw the comb, but they eat up bees; and not only the dead ones, but, with great probability, the living ones too. If the bee-keeper keeps off all other kinds of disturbance by birds, domestic animals, and sunbeams, from his hives, and provides for some

* We should be sorry to be included amongst those who approve this latter kind of treatment.—C. N. A.

protection against robbery and theft, he has done what, in the main, is incumbent on him to do in this month.

FEBRUARY.

In this month the days become noticeably longer, and the effect of the sunbeams ever more powerful. All Nature gradually awakes out of her winter sleep. Just so does there awake in the bee-hive a life ever more active. Strong stocks at least generally begin breeding now, even in cold winters. If it cannot be entirely prevented, it should, at any rate, not have any impetus given to it, because it is as yet more injurious than useful. In the flights which stocks undertake more frequently after breeding has begun, far more bees get lost than are added by breeding, and the stores of honey and pollen are consumed; and, even in the most favourable case, the injury from it is greater than the benefit. But if severe cold should follow, or an aged and infirm queen exhaust her fertility in consequence of premature egglaying, and succumb at a time when no substitute for her is yet possible, the entire ruin of the stock may be the result. Whoever wants to bring his stocks through the winter, and out of it, as cheaply, surely, and healthily as possible, must try to keep them in perfect rest and restrain them as long as possible from breeding. A cleansing flight may, however, be very well afforded to them this month, because the bees have by this time a strong desire for it, if they have not been able to play perhaps for a quarter of a year or more. The bees may undertake this flight—the weather being otherwise suited for it, the air warm, still, and genial—when the ground is for the most part covered with snow, if only roofs, hedges, and trees are free from it. In a warm air and cheerful sunshine the bees may even fly up again from the snow, if it is only firm or covered with a crust. Around the hives, when a flight of the bees is to be expected, the snow may be either shovelled away or stamped firm and strewed with ashes, chaff, &c., so that the bees are not dazzled by the brilliant whiteness, and can more quickly and easily find their bearings and return to their hives.

But if the bees themselves show no need of a cleansing flight let them rather be left at rest, and do not put housed stocks out.

For stocks which have had a flight often crave to fly out again sooner than those which have not flown out since the autumn. For these, after the flight they have had, begin to breed, to clean the combs, to suck up the moisture found on them, and this may in a short time result in an urgent necessity to fly out again for cleansing, of which those stocks know nothing which kept themselves quiet in compact clusters. If a fresh position is given to a part of the hives it must be done before the first flight. If the bees do not entirely forget their previous position after a four months' winter rest, they accustom themselves to any other without difficulty and without great loss if the day on which their first flight from the fresh place occurs is quite genial, and if they immediately begin to play in large numbers at the fresh place, thereby mutually attracting one another. This is attained by giving the bees somewhat diluted food, squirting through a tube into the entrance, or even by only breathing into the entrance. If the entrance is stopped up for some minutes, and is opened again when the bees crowd very much about it, the whole stock will soon be playing vigorously in front of the hive. If we wish to give some stocks out of a Manifold hive a fresh position, in which case comb and bees must be taken out and put into a new compartment, this could only be done in the same bee-garden before the first cleansing flight, and would succeed best just at the time when the bees, scarcely put into their new compartment, could immediately begin to play. In hives easily moveable, like the double hives, the operation may be conveniently performed indoors, so that, even in cold weather, not a bee is lost. At the same time, the bees may be provided with fresh food, and may, perhaps, be put into fresh comb previously fitted up for them in the new hive, at most not needing to give more out of the previous one than those combs that contain brood. If a quantity of fine combs, partly filled with honey, have been preserved from the autumn, or if surplus stores have been left in old stocks, convenient and advantageous use can be made of them in the next months for putting together fresh sets of combs, for improving and completing those sets that are damaged and dirty, and for providing needy stocks in the easiest way with food.

It is evident that if old stocks are bought in the place they must be brought to one's own apiary before the first spring flight, so as not to lose many bees. If, however, the distance between the two apiaries were considerable, say a thousand paces, the removal might succeed very well after, since bees rarely go very far in their first flights, and a small loss is again replaced by the brood which the stock has certainly begun to rear after its first flight. If, however, stocks are bought from another place, the removal may be made at any time. But the removal is combined with less danger so long as the stocks have not built any fresh comb and have not much brood, and the comb is not as yet much softened by the heat. It would, however, be quite desirable that the bees should have already cleansed themselves at their previous position, because then there would be less danger of their dirtying their comb during the removal. The bees can then sustain a longer imprisonment without injury. The removal of bee-hives may be effected most conveniently and gently on a sledge, especially after a fresh fall of snow.

Whoever wants to set up an apiary should select a quiet position, protected from the prevailing storms, therefore with a west or north-west aspect. The protection may be afforded by buildings, shrubs, a wall, or a hedge. For setting up hives such as the Author has recommended and described there is no need of any costly building. Two beams or sills laid parallel for double or fourfold hives, of about the length of ordinary logs, are the whole apparatus on which the hives are put to stand one over or adjoining another. For sixfold hives the logs must be of about double the length. Although, in case of need, a considerable quantity may be set up on a small space, yet the bee-keeper does well who has a larger area at command to set up his hives more in groups, the individual piles of box-hives and the pavilions therefore more separate, and, where possible, so that the hives enjoy the shade of a neighbouring tree, at least in the hottest part of the day, although this may also be afforded by the roof projecting a little more, or by shutters. In the choice of position the sun need only be so far considered that the hives are not too much exposed to its beams, for they are injurious to them in winter, and are extremely oppressive, if not equally injurious, in the heat of summer.

Since February often does not come behind its predecessor in cold, what has been mentioned for January is applicable, and is to be heeded now, and certainly what relates to the supply of food is more applicable now, because the bees may have used up their stores in their winter nest more completely.

MARCH.

March will certainly afford the bees their first cleansing flight if February has not already done so. Notice is taken whether one or other of the hives appears to be queenless, which may easily be observed, especially on the day of the first flight, by a certain unrest, by the coming out and flying off of individual bees till late evening, and by the plaintive tone mentioned before, which the bees sometimes strike up, especially when the hive is opened and the bees are smoked.

Since the rearing of a young queen at this time leads to nothing, either a weak stock with a queen is united to the queenless one or it is united with its nearest neighbour, so that there may be no inducement to robbery by letting it stand any longer, for the bees are especially disposed to rob in the first warm days of spring. For weak stocks narrow the entrance, rubbing it with garlic, onion, wormwood, or other things of pungent odour, before which the robbers retreat. But the smell of their own poison, or formic acid, is very effectual in frightening them away. One or more bee-stings or ants may therefore be rubbed on the entrance, or a board or sheet of paper rubbed with them may be put in front as a blind; it is best placed over the entrance. In warm weather the bees begin to extend more widely in the hive, to clean the combs from all impurity, and to carry out the dead bees lying on the floor. Because many a living bee loses its life in this business we should relieve them of this trouble by brushing out all the rubbish lying on the floor and taking out or cutting off the mouldy, decayed, or very dirty combs. When the winter cold does not last unusually long even weak stocks begin to breed, and the stronger ones are continually extending their brood further. Of course we cannot check or prevent it any more than we can prevent the sprouting of the trees, but we do well as yet to do nothing to further it, just as no prudent gardener would have the wish to see his trees bloom earlier than the season of the year would warrant. The so-called speculative feeding for promoting breeding is at this time generally useless; not only are the food and trouble expended pure loss, but there is generally actual injury from it, as the stocks, repeatedly fed and incited to premature flight, lose more bees than they add by breeding. They become, by this means, out of health, and, at the beginning of somewhat remunerative pasture, are far behind stocks which were not incited to breeding, but were at most provided with food in a way as little disturbing and exciting as possible at a time when there might be a risk of actual want. Feeding with meal, which may take place on genial days of this month, by shaking wheat or rye-meal into combs, putting them in a quiet, sunny place in the garden, and enticing the bees thither by a little honey-water, although it is certainly quite uninjurious, excites to breeding considerably, and affords a very enjoyable spectacle to the bee-keeper, but yet, for the reason given above, has no specially successful result. The bees are excited both in the hive and outside to a premature activity, which is not in unison with the first awakening life of nature, and which, therefore, does not produce the fruit that it might if developed at a right time. Certain objects may undoubtedly be attained by artificial forcing in the bee-hive. By meal-feeding or pushing under a comb with pollen, combined with repeated supply of diluted honey, we may in this month so augment breeding that the queen occupies with eggs drone-combs that are situated in the brood-nest or are now purposely inserted. But if we want with certainty to rear drones from these we must either deprive the stock of its queen or put the drone-combs into other sufficiently strong, queenless stocks or artificial swarms, because, otherwise, the bees, noticing their premature commencement, and thinking to improve upon it, would begin for the most part to throw out the drone-brood that is being reared. But otherwise practical results are scarcely attained by speculative feeding. If the weather is genial the bees may carry in pollen, and even some fresh honey from the hazel, alder, and perhaps from the aspen and willow, and other spring flowers, and so they will soon have a quantity of brood without any artificial stimulus. If, on the contrary, the weather is raw cold, it is better if the bees keep quiet and await in patience the beginning of the actual spring.

April.

This month affords the bees a better, and indeed a luxurious, pasture in districts where there are many osiers and other kinds of willow-bushes. When the bees begin to carry in pollen from these in abundance as the fine weather sets in, it is time to cut the combs in the hives. Many hold every shortening of the comb to be injurious, and therefore will not let even an inch be cut from it. Others cut the comb, especially in log-hives and in skeps, unusually close. Pastor Scholz brings forward, in No. 7 of the *Bienenzeitung*, a striking proof that even closely cut stocks are by no means ruined, but in certain circumstances thrive splendidly, and may even surpass those that are not cut. But because we must never trust to the weather, especially in this month, whose changeableness is proverbial (even May may yet bring raw, cold days), so in cutting the golden mean may be the best. We cut then only moderately, so that in a warmer season the bees may be able completely to cover their comb, whereas on colder days they may retire between the combs. Where the comb is recent and clean, containing few drone-cells, it should be dealt with more leniently. In low hives that are more of the Lager type it is sufficient only to shorten the combs so much as to allow of the hive being thoroughly cleaned. Stocks which already show considerable industry do not therefore need any stimulus, and need not be so closely cut as those which appear indolent. When, in cutting close, no trace of brood is discovered, or when the stock makes no preparation for completing its comb in favourable weather, any possible defect may at that time be far sooner noticed and remedied.

If we do not want to cut the combs, where else are we to get the empty combs for artificial swarms, or for realising the profit of an abundant honey-pasture? The smallest bee-keepers have the opportunity of getting combs in requisite quantity by purchase, but then the method may be justly met by the objection that it does not stand on its own legs, and requires somebody else to help it out. For the combs that are gained in the autumn reuniting might not be sufficient, and are not so cheaply come by, while those obtained at the spring-clip are pure gain, inasmuch as the stocks, if they had not been cut, would not have been a whit better.

Through the clip many another secondary purpose is attained, and much convenience is afforded by it. Food that may, perhaps, be necessary which is now better given from below, may be afforded to the bees far more conveniently by putting a honey-comb under the nest of the bees, either on the floor or on two pieces of wood propped across. Since the brood becomes now continually more numerous, and the stores are perhaps coming to an end, we must not now omit to feed the stocks, especially when cold or rainy days occur. This is done not so much for the purpose of exciting them to more extensive breeding, but to prevent their sucking out or pulling out, through want of food, the brood that is already being reared. With stores of honey still in hand this might be done from want of water, of which the bees require much for the preparation of brood-food, and after which they fly out even at the risk of life in cold and rainy weather. A great service is rendered to the stocks if they are provided with it on cold days in too dry weather by pouring a couple of spoonfuls of it into the cells of a comb, which may be pushed in for them covered with some pieces of solid honey. It is easier for the bees to get rid of the possible surplus in flight than it is for them to fetch in what they require of it. The moisture that is so necessary is produced for the bees in the hive itself if any wall, door, crown-board, or floor, afford a cool surface, on which the vapour in the hive may condense into drops. It is especially convenient for the bees to suck up the moisture precipitated on the crown-board or floor. The grains of honey that have fallen down, or honey or pieces of candy laid on the floor, are thereby partially dissolved and made more easy for the bees to enjoy. For this purpose we should have to insert a wedge between Box hives put one on the top of another. In hives furnished with dry candy this might also be done for a time in winter. It is possible for the bees to raise the heat that is withdrawn by this, but it is not always possible to supply the want of moisture indispensably necessary for the solution of sugar.*

Weak stocks, of which there will always be some to be found in a large apiary in spring, we should strengthen, otherwise they

* The principle of providing for condensation of the hive's moisture within the hive itself, may hold good in Germany, but in England, perhaps through greater humidity of atmosphere, the chief difficulty in winter is to keep the hives sufficiently dry to prevent mouldiness of pollen-combs, and fermentation and sweating of the honey in open cells. We do not see

might swarm out and cause some injury in the hives on which they might settle. With only one apiary this can only be done by the insertion of a comb with brood, for the most part sealed, and at the point of coming out. Later on two of such combs might be put in. The bees on them may be given with them, inasmuch as the young ones will remain with the hive. There is in that case less risk of the brood put in taking a chill. With two apiaries at a distance, strengthening with bees is to be preferred, because strong stocks feel the loss of some thousands of bees much less than the loss of a comb of brood. Bees are collected into a Transport hive by shaking them off the combs, removing them by means of the box from under the shortened comb, enticing them on to honey-combs, and the like. When possible, the bees must be collected from several hives, to put them into mutual confusion. Bring them to the other apiary and let them move into the weak hive in the dark, when the bees have given over flying; or, if several have to be strengthened, shake out a portion to each. We must not omit, before shutting the box, to provide them with honey, so that the bees do not become exhausted, and may be the more willingly accepted. If, besides the strengthening with bees, some brood be added too, the weakling is helped all the more effectually. We must, however, be convinced that the hive has a fertile queen, free from defect. For if the reason of the weakness lies in the unfertility or worthlessness of the queen from any other cause, strengthening does not help at all. The operation that is otherwise so useful is then quite thrown away, and only loss comes out of it.

When the clipped stocks begin to complete their comb again, which happens mostly at the time of the fruit-blooming, we must take care that they do not build too much drone-comb. At the side in front of the door a comb may perhaps be left to them, where it may later on be conveniently removed. But, further, in the brood-nest we must cut off the drone-comb up to the worker-cells, or hinder them from carrying the drone-comb any farther by pushing a bar underneath. Since only stocks that are already

the Author's difficulty in providing for the solution of sugar or candy, as, when either are given to bees, water could also be given in small pieces of comb, or the whole difficulty could be got over by feeding with barley-sugar, which of itself will absorb the moisture of the hive's atmosphere, and become excellent bee-food.—C. N. A.

very strong are much disposed to build drone-comb, the removal of a part of the bees from them for strengthening weaklings is not only no disadvantage, but may turn out to be an actual advantage, because then, instead of drone-comb, they build more worker-comb, and at the decisive honey-harvest have more workers and fewer useless consumers.

MAY.

The month of bliss is not only one of the most joyful to the lover of nature generally, but is so to the lover of bees. The stocks become continually stronger in population, and continually develope a greater industry. On exceptionally warm days in the second half of the month probably drones will be seen among the bees at play, and in districts with early pasture, where rape is cultivated, towards the end of the month the first swarms will come off. Unfortunately there are districts in which, after the first bloom, nearly all supply ceases, and the bee-keeper has in many seasons to feed till the proper summer pasture begins with the blooming of the ground-ivy, the white clover, or the blue corn-flower. In such districts swarms are hardly to be expected before midsummer, inasmuch as the bees, dispirited by the long gap which cannot be completely filled up by feeding, first have to make a fresh start in preparation for swarming. Feeding may be very conveniently managed when it is carried on in the open in large troughs, feeding-boxes, or large drone-combs, into which the much-diluted food is poured, since the weather now mostly permits of the bees flying out. But if other people keep bees in the same place, these have to be fed at the same time, and much honey or sugar has to be consumed, if there is to be anything like an imitation of a moderate bee-pasture, and if we want to force early swarms by it. The end is gained more certainly and cheaply by making swarms artificially and expending the honey for feeding on these. We may begin to make artificial swarms as soon as drone-brood is seen in the strong stocks. By the time the young queens are reared and undertake the pairing-trips, drones will become out and playing in front of the hives. To make artificial swarms altogether too early would certainly not be well. But if a part of the hives (by preference those that want to rear too many drones) are driven fairly early, or if artificial swarms are formed by another method from brood and bees, after ten to fourteen days a quantity of young queens or queen-cells are gained, which may be used with the greatest advantage in the artificial swarms to be made, or in the hives to be driven later on. A fertile queen has at this time great value, because she can produce many thousands of workers, which appear at the best time of harvest. The value of many a weak stock, wintered with trouble, in spring does not amount to much more than that of its queen. It scarcely needs to be remarked what an advantage it is to gain quite early fertile queens, which are still more valuable than those of the year before. A weak stock, provided with such a queen, and having one or two brood-combs put into it, whose removal a strong stock scarcely feels, may develope into a very fine stock if the bees have the prospect of a two to three months' pasture before them. It is not every queen that is reared that becomes a fertile mother. Many come to grief on their flights. Care should be taken to replace the loss soon by another surplus queen, or a queen-cell, or to put in a brood-comb, which may be done in any case as a precaution. The hives, therefore, with the young queens are to be observed especially on any day on which they have played very much. In strengthening artificial swarms that have young queens, precaution must be used, and it is best effected by putting in sealed brood, which at the present warm time of the year does not so easily chill, even if it should not be completely covered by the bees. If the colony is twice or three times as strong, we may always add more brood-combs, and so raise the hive in a short time to any strength we like.

June.

This month affords to the bees, if the weather is not otherwise unfavourable, the most extensive pasture, which increases later on only in districts where buckwheat is largely cultivated. At the beginning of the month the white clover, the blue corn-flower, and the field-radish, come into flower, and afford just now the most nectar. The days are now at their longest, and allow of the bees carrying in the more. In most districts the swarming time begins now, and the swarms coming off at this time are the best, because they are stronger than those appearing in May, and find immediately abundant pasture, which is often wanting to those that have come off too early. Now is the time when the bee-keeper is most full of work, which, however, every one does very willingly. From nine o'clock in the morning till three in the afternoon, and even later still, he must have his apiary watched so as not to lose swarms, and must keep empty hives and the necessary implements for catching swarms, in readiness. The bars with comb-beginnings must be prepared beforehand in the hives to be occupied by swarms, so as to be able to put in the swarm when it comes off without loss of time. A greater haste is especially necessary when the apiary is large and other hives give indications of swarming. But if the swarms are delayed, or if for special reasons there is no wish to let them swarm, we proceed, when the stocks are populous and have begun already to cluster in front in large numbers, to driving or artificial division, or go on with it if a beginning has been already made in the previous month. We shall have no cause to repent of it later on unless it has been pushed too far. Especially do we drive those hives which have either too old comb or a queen that is too old. When all the brood has come out, after about three weeks, we may either entirely or partially cut out the comb that is too old, and either put fresh into the hive, or cashier it altogether, according as the one or the other may appear advantageous from the condition of the hive, the pasture still in prospect for the bees, and the strength of the stock. If with continuous good honey-pasture the bees begin to suffer from want of room and of empty cells for depositing honey, we open the honey magazine, which may be situated either at the top or at the side, if the brood-room should not yet be built down to the bottom. The bees are more disposed to build up space that is afforded them now than later on, when the pasture has passed its maximum, and when in addition somewhat too dry weather may have set in. But on that account it is sufficient to provide the space to be built up with mere comb-beginnings when we have no surplus of empty combs, because the bees themselves push on rapidly with the building of comb at this time. Since the bees later on very well fill combs that are put in if there is still pasture for them, but do not willingly build fresh comb, so it is well to preserve what store there is of it for the later summer, so as to be able quickly to furnish the later swarms with comb as extensively as possible. The preservation of empty comb has now in the heat of summer its special difficulties, inasmuch as the wax-moth may quickly get the upper hand in it, destroying and making it useless. We should, therefore, frequently look through the separate combs and immediately do away with the little moths that may be seen in it. The bees themselves best preserve this comb from destruction. We may, therefore, put it in, fastened on bars, to such colonies like recent swarms which have surplus empty space in their hives. The scout bees, which often are seen now, when the bees have the desire to swarm, in hollow trees, in cracks, in buildings, but especially in empty hives with comb in them, protect the comb not only from further destruction, but clean it with great zeal when it has been already attacked by wax-moths. The stored empty combs may therefore be put into such hives that are often visited by scout bees, and they will be preserved from being moth-eaten. Sometimes a swarm coming off immediately take possession of a hive so prepared, especially if it is rubbed with balm, the smell of which the bees are very fond of. The swarms that come off settle willingly on places that have been rubbed over with it.

The later a swarm comes off or is made, the stronger must it be if it has still to carry in enough for its winter requirements. In such a swarm the putting in of quite long combs is especially useful, because it has no longer the time to form a complete set of combs, and its comb remains too tender and too cold for the winter, as probably not more than one generation of young bees will be raised in it. If the store of empty combs is exhausted a supply may be obtained by clipping or entirely cutting out the parent stocks, which have either thrown off swarms or had swarms driven from them, three to four weeks after the departure of the old queen, before the young queen has begun to lay eggs. Stocks with old comb are best renewed in this way, or driven out of damaged or otherwise unserviceable hives into better ones. So a quantity of comb, both empty and full of honey, may be gained, which may very advantageously be used for feeding and furnishing young swarms, especially the later ones.

The combs that are already somewhat old we do not put into the brood-nest of the young stocks, but more towards the side from which later on they may be easily removed. If these contain much honey and pollen, as is generally the case in a favourable season, they must be previously put into the bees for them at least to use up the pollen and honey from the cut surfaces so that they may be fastened to bars. The feeding of young swarms, when the pasture is not good and the weather not favourable, is now of the greatest advantage, and cannot sufficiently be recommended. Comb-building makes the quickest progress at the beginning. Gradually the swarm builds less diligently as more bees get lost and the comb has gained in extent, and therefore can only be more thinly covered. If the swarm has at last quite left off building it will hardly be able to begin again till a larger number of young bees has come out, which in late swarms come too late, as the pasture has then been over some time ago. We, therefore, when the weather makes it necessary, assist the young stocks with food, which may be considerably diluted, so that they may build quite zealously as long as they will and can build. For later swarms we always put in fewer comb-beginnings. It is better for a swarm to build a few combs completely, or to extend them at least to a considerable length, than to begin too many and leave them all half finished. In the first case a complete set of combs may easily be made in autumn by adding full combs out of other wealthy stocks, whereas out of a number of combs that are only short it is not easy to prepare a warm winter nest for the bees.

July.

July is for most districts of Germany the decisive month. On its quality principally depends whether the year is to be a good, moderate, or bad bee year. It is the bees' special harvest month. The stock that is to bring profit to its owner must now gradually diminish breeding, and direct its activity principally to the increase of its stores of honey. This the bees generally do themselves, but not always. If July is damp and more favourable for brood than for carrying in honey, the bees continue to breed regularly, and perhaps still more extensively, and continue to rear drones. If the pasture then ceases suddenly the combs that have been occupied by brood remain later on empty, and the stock has to meet the winter with a scarcity of honey. The bee-keeper must try to prevent this as far as possible. His management and his whole activity are therefore determined this month by the prevailing weather. If this is dry and allows the bees to carry in honey in quantity he has nothing further to do than provide room for its storage by opening the honey-rooms for the bees, giving collaterals or supers to small hives, but best by taking away the obvious surplus so that the bees have nothing more to do with it and can direct all their powers to the collection of fresh stores. It is just when the bees are carrying in the most honey that the most can be taken away in a good honey season by their owner. By this means the industry of the bees is continually kept lively, whereas when they feel themselves to be in possession of a considerable surplus they fall off remarkably in carrying in, and do not fly any longer in proportion to their strength. If we can put in empty combs instead of the full ones taken away, we can repeat the taking of honey the sooner if the honey harvest continues abundant. If the bees carry in honey very abundantly, breeding diminishes of itself, because nearly all the cells are immediately filled, and the queen can deposit but few eggs. In damp weather, and a more scanty honey harvest, the bee-keeper himself must try to check breeding as much as possible, or at least make it impossible for the bees to extend it further. This is attained in upright hives by inserting a board horizontally which prevents the bees from prolonging the brood-combs downwards. It is better if the stocks, especially the new swarms, build up a small space quite close, and partially fill it with honey, than make long combs which in autumn are for the most part empty, because the honey carried in has been expended on comb and brood. But if we want not only to put a limit to breeding, but to prevent it altogether, the queen must either be caged or removed entirely if, on account of age or any defect, she is no longer worth wintering. Since the massacre of the drones not unfrequently takes place towards the end of this month, it is now time for the renewing of queens, that is, to provide for the removal of those that are two or three years old, if we have no reserve hives, which are cashiered in autumn, and from which queens may be given to the ones that are to be wintered. From many a stock, which has either swarmed voluntarily or from which a swarm has been driven, its young queen may have been lost on her marriage trip without it being immediately observed. When we cannot, or will not, investigate the interior, about six weeks after the departure of the old queen we take notice whether the parent-hives have young bees again playing in front, and later on whether they make preparation for banishing the drones. Parent stocks whose queens have become fertile generally do this earlier than those that still have their queens of the year before. If we want to put to rights those that show they are queenless, we give them speedily a good fertile queen when this is possible; but if an unfertile or otherwise useless one is present in a stock, it must of course be previously removed.

We must now effectually help young swarms, that may be too weak or too late, if we intend to winter them. This is done by putting in combs of brood and honey, but only in case we have hives that have a surplus of population and honey. Otherwise it is better to unite the young stocks in autumn with others, and to reserve their comb for future use. So long as the bees still carry in fairly they are more in a position to combine into one the combs put in to them with their previous comb, to fill up gaps, and to concentrate the stores of honey whereby a better and a safer wintering is made possible.

If pasture in the fields declines considerably, or has entirely ceased, the bees accustomed to carrying in are much disposed to intrude into other hives and rob them. We apply therefore, especially in hives that are weak or suspected of being queenless, the means previously mentioned for the prevention of robbing; especially not allowing hives that are undoubtedly queenless to stand any longer, but either putting them to rights or entirely cashiering them. There are, however, districts in which the honey-harvest from the blossom of the buckwheat is just now beginning. In that case, what has been said for June holds good for this month. We may still receive swarms or make them artificially, and they are still able to become very heavy and good when this flower—so exceedingly rich in honey on a poor, sandy soil and in a warm, moist air—extends still into August, or when the bees later on have the opportunity of working on the heather. In many districts and seasons the first swarms themselves throw off swarms again, which are called maiden swarms. Whether they are better accepted and set up, or given back again to their parent hive, consideration of the pasture in prospect and experience of other kind must teach.

August.

In most districts the entire close of the bee-pasture occurs in this month. The stocks continue to expel the drones. We take notice of it, and carefully have an eye upon those hives in which drones are tolerated. They are, at least, very much to be suspected of being queenless. To prevent robbing the bee-keeper must now double his vigilance, because the bees are very urgent. The taking of surplus stores of honey, the furnishing of stocks that have not brought in enough for their requirements, must be seen to in the cool hours of the morning, or, better still, of the evening, and in cool weather, making quick work of the business, leaving no honey-combs lying open uncovered nor any hives standing open. Avoid enticing other people's bees by the smell of honey; for although individuals may go home loaded with honey, they come again with strong reinforcements and attack the hives vigorously. The entrances should be narrowed to make it easier for the bees to defend themselves against the attacks of robbers. All crevices should be stopped up through which bees may penetrate, or through which the smell of honey may escape from the hives. When many bee-hives are kept in the place, and when the season is poor in honey, more precaution must be used, or the bee-keeper will suffer great loss through robber-bees.

But in heath districts a principal honey-harvest from the blossom of the heather begins toward the middle of this month. It is usual, when the heath is too distant for the bees to be able to fly to it from their stand, to remove the hives or to travel with them to it. This is the more to be advised to the bee-keeper to whom such an opportunity presents itself when the stocks have not been able to carry in any great store of honey from the summer harvest. To travel with hives already heavy is less remunerative; for one thing, because they increase less in weight, and, again, because their removal is more difficult. The danger of the comb collapsing and of the bees being suffocated is much greater than in light hives. In the hives described by the Author this danger does not occur. Before the journey the honey magazines may be emptied, and even the heaviest combs may be taken out of the brood-room and left at home. The hives to be removed to the heath will increase in weight in proportion as they are deprived of their stores, and empty comb is put in for them, and bees are added to them from other stocks. We shall find more honey in them at the close of the pasture if a stop has been put to breeding by caging the queen about three weeks previously. For hives that are to be wintered this would be less applicable, because in autumn there might be a deficiency of bees, especially of young and vigorous ones. In hives, however, that are to be cashiered this is very advisable. There is saved thereby the honey which otherwise would have been expended on brood; and the bees, especially if they have already sufficient comb, can carry in the more, because they have no need to be burdened with care of brood and always gain more empty cells for the deposit of honey.

September.

In heath districts the bees may, in favourable weather, still have some pasture till towards the middle of September; otherwise they can only gather honey when honey-dew has formed, which this month is not rare on pines. But this generally brings the stocks more loss than profit, because in this month little or no fresh brood is reared; but bees when they fly get lost in quantities by becoming entangled in the numerous spiders' webs, and many are caught and eaten by the hornets. The stocks are in this way much reduced; the honey that is carried is now in itself of bad quality, is no longer properly refined, remaining for the most part unsealed, and is often the cause of dysentery when a long and severe winter ensues.

When all the brood has come out—which is generally about the middle or toward the end of the month, varying with the district and the weather—the hives which we do not want to winter may be emptied of bees, either to break up the combs for the honey and wax or to preserve them for future use. In hives with moveable combs there is no need to wait this time, since the combs that still contain brood may be put into other hives. Uniting might be still more advantageously undertaken at the same time, if it is already evident which stocks are worth wintering and which must be cashiered. There will certainly be no desire to winter every one that we have standing in the apiary, but rather to unite whatever, from any reason, is not suited for wintering. Making artificial swarms is an invaluable art if only for this reason—that it gives occasion for the rearing of young vigorous queens. But it will be unprofitable and injurious for the bee-keeper who winters light and weak stocks. Those stocks that do not possess at least ten pounds of their own honey, mostly sealed, an adequate comb, a vigorous queen, and so many bees that in autumn they fully cover about five combs, it is better to unite than to leave standing. When the season has been a bad one for bees we must be more free to unite stocks; so that those that are left over may receive the requisite store of honey, and we may still keep a quantity of combs—some of them full and some empty—with which both the wintered stocks and the artificial swarms to be made from them next year may be assisted. A good artificial swarm after a month is much better than an old weak stock whose wintering has caused much trouble and expense. Whoever winters light and weak stocks burdens himself for the spring—if they live so long—with much trouble, and has from his bee-business more loss and vexation than profit and joy.

Although, after a favourable summer, all the stocks ought to have carried in beyond their own requirements, yet a part of them will not be able to be wintered because their queens are already too old and are no longer worth wintering. If a stock has an adequate winter store and a fine comb, neither too old nor too new, but has a queen that is too old, she is removed and replaced by a young, fertile queen from another stock, that is to be cashiered after the former has been left standing queenless for some time. If we want to Italianise an ordinary stock by introducing an Italian queen, the most favourable time for it is in this month, because the bees cannot form any queen-cells after the removal of their queen, therefore there will be no need to destroy any for the purpose of defending the introduced queen from all danger, and because we may keep the queen caged for a week without any disadvantage till the bees have become perfectly acquainted with her and have accepted her, and seeing that, if she were not caged, there would be no more breeding going on. Besides, we may now obtain fertile Italian queens more certainly and cheaply than in spring, inasmuch as every one cannot then make up his mind to rob a stock that has come through the winter, of its queen. The Author is certainly ready at any time of the year to send off Italian queens when transmission is possible, but at a different price according to the difference of the time of year. In order to meet unnecessary inquiries he will in future, when orders will have in some measure diminished, lay down a rule in this way that a fertile, genuine Italian queen costs in March, April, and May 12 shillings, and from June to October 9 shillings. Unusually beautiful specimens, which only occur as rarities, are of a higher price, and are even in autumn not sent off under 15 shillings. For a small swarm 3 shillings more is reckoned than for the queen alone.

Whoever has not provided for the increase of his stocks in the swarming-time must winter what he has in his apiary, and will make his honey-harvest only by cutting the combs or taking away the surplus, which may partly be done in this month. A part of the surplus may now be taken without injury from log hives, Lager skeps, and other similar hives that have to be cut, only we must not come too near to the proper winter nest of the bees, so as to make it too cold, and we must leave sufficient stores for the bees in case there is a long and severe winter and a cold spring. The possible surplus may indeed be taken away at the spring clip. In cutting honey we must take proper precautions against attracting strange bees and occasioning robbing. For, from want of pasture, the bees are now extremely disposed to rob.

Stocks which still tolerate drones are as a rule queenless; since they have mostly only a few old bees it is best to cut out the comb and cashier them. A large quantity of pollen is generally found in this comb. It may be specially preserved for putting in spring into young stocks that possess little or no store of pollen.

October.

If from want of time or because the bees continue to breed in fine weather, the uniting of stocks that are not to be wintered has not been already undertaken in the previous month, it may be done now just as well, or perhaps better. All the brood now is more sure to have come out, and, for this reason, and because of the coolness of the time of year, there is less fear of the empty comb being attacked by moths; and robbing, though by no means impossible, is not so much to be feared now since the bees, for the most part, are remaining quietly in their hives, and only play in front of the hives sometimes at noon. Frequent flights are just as injurious to them in late autumn as in early spring; by this means bees get lost which can no longer be replaced by young ones. Therefore, we must not disturb the stocks with repeated feeding; if what is wanting to them cannot be given in sealed honeycombs, and we have to give them liquid honey, it is best given in large supplies in the evening, when all flight is at an end. If, after an unusually bad summer, there should be a very large deficiency for winter requirements in the stocks that are to be wintered, let at least a part of their supply be given to them in pieces of candy, because too much liquid food given in autumn makes wintering dear and uncertain, inasmuch as the bees are cold in their quarters, consume more largely, and often suffer from dysentery. The candy, of which the best kind is the yellow, is put in for them over or near their winter quarters. If this is not possible on account of the construction of the hive, it may be put underneath. The comb must then be shortened so much that the bees can properly cover it. The food may be pushed under on an old firm comb, which, at the same time, keeps off the cold from below. In severe cold the bees can only use from the stores that are in their comb, which, therefore, must not be too scanty; but on warm days they come down and feed on the candy put underneath. The sugar may be given to straw skeps in a little box made of thin board put on over the open bung-hole, or the hive is reversed, and, after a kind of lattice has been made from thin boards, the sugar is laid upon this, so that the bees may very well ascend, but the pieces of sugar cannot fall down between the combs. Since the hive has to remain standing reversed in this way, the food should not be put on before the bees will have no further occasion for a flight. Boards are put over the food put in, and the remaining space is filled up with hay. Pieces of honeycomb, if we have any, may be put into the reversed hive, immediately on its comb, perhaps in the sides or corners not built up. In a short time the bees attach the pieces laid on, perhaps the same night, and the hive may be again set up. In this way the necessary winter provision may not only be supplied to the bees, but an incomplete set of combs may be completed and protected from draught, and a stock that otherwise would scarcely pass may be wintered. But the wintering of such stocks is only excusable when unusual losses have been suffered from foul-brood or other accidents, and there is a deficiency of good stocks to stand the winter. For to winter them is troublesome, costly, and extremely uncertain. The bees, although we may have kept them into the spring, may forsake their cold comb on some fine day, and fly off or cause damage and disorder in the apiary. When the bees have drawn more together in the cooler time, and no longer guard the entrance, wasps make themselves free of the hive, and steal, when they are numerous, many a drop of honey and kill many a bee, to which they are generally superior in battle. Their nests should be destroyed or the wasps should be caught in an open bottle partly filled with a sweet liquid. In the late autumn mice venture into the hives. Their entrance must be prevented by nails or a slide, without, however, preventing the free exit and flight of the bees.

November.

The days are becoming shorter and colder. The bees are drawing together into a closer cluster. For the most part they keep themselves in perfect quiet, so that not the least hum is to be noticed in the hive. Generally, however, there occur days in this month on which the bees play at noon, and are still able to cleanse themselves before the winter. Such a late flight is very serviceable to them, because they are then able to hold out longer in the spring without a flight. If a part of the stocks should play merrily on a genial day while the others are keeping quiet, these may be incited to play at the same time if they have not flown out for some time by knocking on the hives or breathing into the entrances. A late flight is especially very serviceable to the bees when they have been carrying honey late in the season from honey-dew or have been fed with liquid honey. To give such food to them now would be somewhat too late. The stocks, however, may always be provided with sealed honey-combs, solid honey, or pieces of candy, now as well as in the proper winter months. But this must be done in such a way that the bees are not too much disturbed and excited.

Whoever is accustomed to house his hives for the winter in a cellar or other place that may be suited for the purpose need not be in any haste about it. Sometimes it snows and freezes in the first half of this month, and later there come warm days, when the bees may still fly out and cleanse themselves. Towards the end of the month or at the beginning of the next, when the winter threatens to begin in earnest, the housing may take place. The bees must not be debarred from exit and entrance in their winter-quarters. Mice are kept off most easily and certainly by catching them with traps, otherwise they may be debarred from access into the hive by nails driven in across the entrances. The hives left standing on their usual stand must be furnished with protection from the cold in another way by filling up with straw the empty spaces there may be at the top and side. To put straw on the outside of the hives is not well, because of the mice; they hide themselves in it, and may at least disturb the stocks even if they are not able to gain an entrance into them.

December.

If the hives have not been prepared for the winter in the previous month it must be done now. Where there is danger of robbery the hives should be secured. The piles of press-like hives standing apart are pushed together and combined so that they cannot be moved apart and robbed. The double-hives are screwed to one another so that they cannot easily be carried off, the doors are fastened so as not to allow of being easily opened, or there is put over them an outer door, common to them all, that will easily lock up, or they are secured in some other way. But all dangers which threaten the hives from man, birds, severe frosts, and cutting winds, from snow and sunshine, they escape by being housed. It is, therefore, very advisable for every bee-keeper who wants to winter his hives safely and wants to multiply them quickly to make for himself winter quarters of this kind for his hives, if he does not already possess such a place in a quiet vault, or dry cellar, or the like. The cost of making will very likely be repaid to him in one year by a good and cheap wintering of his hives. How he may apply the same advantages to the bees in the Pavilion or in the separate hives with adjoining compartments as if they were housed or buried has been mentioned in the section where the Twin-stock and Pavilion are described. But there must be no delay before the winter in opening properly the slits always propolised by the bees, as these openings are meant to convey air to the bees of the same temperature as the ground. Winter may then bring what weather it may, cold or mild, dry or snowy, settled or changeable. Secluded from all changes the bees will live through it with small consumption and insignificant loss, and joyfully welcome the spring. The operations already mentioned begin then afresh for the bee-keeper, and there await him the old but ever new joys which the care of bees affords in such abundant measure.

INDEX.

ABBOTT, C. N., notes by, on,—large and small drones, 8 n.; the ability of bees to remove eggs, 11 n.; the time required for the evolution of the queen, 12 n.; the power of the queen to lay fertilised or unfertilised eggs, 17 n.; the utility of drones, 22, 23 n.; the ability of drones produced from a fertile worker's eggs to pair with queens, 24 n.; the building of drone-cells for storage purposes, 27 n.; the advantages of giving worker-foundation comb to queenless stocks, 28 n.; the removal of dried or mouldy pollen by bees in spring, 29 n.; the aim of British bee-keepers to produce honey rather than wax, 29 n., 61 n.; the colour of pea-flour pellets, 31 n.; caution against early and late breeding, 32 n.; the temptation to bees to take up water, and the evil attending it, 33 n.; comb-building by queenless stocks, ib.; the quietness of bees during mild temperature, 35 n.; the queen depositing eggs in royal cells built upon drone-comb, 37 n.; the time required for hatching supplementary queens, 38 n.; the position of the bee-cluster in shallow hives in cold weather, 40 n.; sunshine on hives in winter, 42 n.; the proximity of hives to railways to be avoided, ib.; Lager and Ständer hives, 47 n., 48 n.; the Pettigrew hives, 49 n.; the use of the crown-board, 57 n.; the advantages of comb-foundation, 60 n., 61 n., 140 n.; the practice of utilising scraps of old comb for guides, 61 n.; close-ended frames, 64 n.; the uselessness of bottom-rails of frames, 65 n.; the chamber above the brood-nest in twin-stocks, 77 n.; the cross-sticks in skeps, 115 n.; the indisposition of stocks long queenless to receive a queen, 161 n.; introduction of queens to after-swarms, ib.; the selection of drones by the queen, 166 n.; the introduction of queens to swarms, 175 n.; the advantages of the English frame-hive, 177 n.; the best position of the queen-cage in cold weather, 179 n.; the utility of condemned bees, 180 n.; the author's method of making artificial swarms, 183 n.; best means of Ligurianizing, 191 n.; the influence on drone-progeny by fertilisation, 195 n.; drone-traps, 197 n.; judgment required in making artificial swarms, 200 n.; the difficulty of regulating hives with immoveable combs, 204 n.; the facility of bees building downward, 205 n.; checking the production of brood in autumn, 218 n., 220 n.; raising drones by second swarms in their first year, 221 n.; the cells in the outer combs of a hive, 222 n.; bees removing eggs to queen-cells, 230 n.; administering barley-sugar to hives, 235 n.; the condensation of vapour in hives, 237 n.; the use of open-ended frames, ib.; dysentery not frequent in England, 255 n.; stimulative feeding, 256 n.; caution respecting using honey from affected stocks for feeding, 273 n.; the sources of contamination by foul-brood, 275 n.; the effect of dry weather on wax-moths, 284 n.; wax as a remunerative product of bee-keeping, 307; the solution of sugar or candy, 323 n.

Abbotts' longitudinal Combination-hives, 50 n., 114 n.

Abdomen, distension of, 247, 271.

Acacia, 308.

Activity of bees, 24, 25; aim of, 36-40.

After-swarms, 38, 154; time of, may be predicated, 155; their attachment to the young queen that has led them forth, 161; their joyful reception of a fertile queen, ib.; prevention of, 163; good stocks obtained from, 171.

Age of bees, 20; affected by the flowers they frequent, 20, 21; by the amount of work they perform, 21; of queen-bee, 18; of drones, 22 n.

Ailments of bees, 270-278.

Alder, 307.

American honey, caution respecting the purchase of, 277.

Anger of bees, 288; how subdued, 167, 288

Angular hives preferable to round, 47; moveable combs manageable in, ib.
Animals, food of, 30.
Ant-poison, its effect on bees, 267.
Ants enemies of bees, 285, 286; how hives may be guarded from, 286.
Aphis honey, or honey-dew, 308, 309, 332.
Apiaries, location of, 41, 318; should be protected from violent storms, 42; should be at a distance from tumbling noises, ib.; should not be near large surfaces of water, ib.; shallow ditches and brooks should be close to it, ib.; should be sheltered by birches and pines, 43; must be hedged round to prevent bees molesting men or cattle, ib.; convenience to the bee-keeper of possessing two, in making artificial swarms, 169, 227, and in cases of contamination from foul-brood, 274; removal of bees from old to new, 317, 318.
Apicultural implements, 287-304.
Apple, 277.
April, 321.
Artificial swarms, 165, 198, 324, 326; preferable to natural, 165; increase of stocks attainable by, ib.; requirements of, ib.; acquirement of bees for, 166, 167; prevention of the bees composing, returning to their original hives, 168; retention of, in the apiary, 169, 170; by means of the twin-stock, ib.; by small hives, 170; method of making them swarm naturally, 174; treatment of, 178; remarks on Dr. Dzierzon's method of making, 183 n.; judgment required in making, 200 n.; warning against, 333.
Aspen, or trembling poplar, 307.
Assafœtida, its effect on bees, 163.
Associations for the promotion of bee-culture directed to seek the use of honey for making honey-wine, 3, 306.
August, 331.
Autumn, life of bees in, 37; checking the production of brood in, 218 n.

Badger, the, in winter, 241.
Baldenstein, Herr von, the first to call attention to Italian bees, 193; result of his preference of natural swarming to artificial, ib.
Balm, smell of, enticing to bees, 156.
Barley-sugar, a substitute for honey, 234; in what manner administered, ib.; best position of, in different hives, 234, 235, 235 n.

Beans, 2.
Bears, their fondness for honey, 280.
Bees, pleasure in the study of the wonders of nature in a colony of, 1; valuable products from, ib.; service of, in the fertilisation of plants, 2; natural history of, 5; live in communities, ib.; are cold-blooded, ib.; heat necessary for their existence produced by their being united in swarms, ib.; sexual relations of, 6; various races of, 6, 7; sexes of, 8, 9; ability of, to remove eggs, 11 n.; duration of their lives, 20, 21; activity of, 24, 25, 36-40; food of, 30-34; organs of, by which chyle is secreted, 32; require both honey and pollen for their sustenance, 34; are more active when obliged to construct cells, ib.; their vital activity, 35, 36; use their legs and wings in winter to promote circulation, 35 n.; their renewed vitality on the return of spring, 36; the aim of their activity, 36-40; their accumulation of honey in the autumn, 39; positions of cluster of, in the hive in winter, 40, 40 n.; thrive best in temperate and tropical countries, 41; their domiciles when in a wild state, ib.; their domestication by man on account of their honey and wax, ib.; their dwellings, 41-43; methods of transferring, from old to new hives, 145, 318; loss of, in showers and storms, 201; their facility for building downward, 205 n.; stimulating, to the greatest industry, 219; variety of, 221; removing eggs to queen-cells, 230 n.; winter best on old comb, 233; their need of moisture, 236; their ability to withstand extreme cold, 241, 242; consumption by, of honey during the winter, ib.; burying, in winter, 242; effect on, of cutting winds in winter, 245-247; excitement of, caused by the exhaustion of their stores, 246; should not be incited to breed in winter, 247, 316; chilled, 248-250; effects on, of sudden cold, 251; cleaning, after winter, 254; activity of, corresponds to the condition of the weather, 255; gaps in their pasture to be supplied by artificial means, 269; ailments of, 270; enemies of, 280-286, 332; irascibility of, 288; pressing them to be avoided, 290; pasture for, 310; transporting them to better pasture, ib.; providing with fresh food in winter, 317;

enticing them to play round the entrances to hives, ib.; removal of, from old to new apiaries, 318; not to be disturbed by feeding, 336.
Bee-calendar, 313.
Bee-cap, or bee-hood, 290.
Bee-cluster, position of, in shallow hives in winter, 40, 40 n.
Bee-culture, different methods of, 139.
Bee-hives. See Hives.
Bee-houses, their inconveniences and disadvantages, 41, 42.
Bee-keeper, conditions required in a successful, 3; his chief anxiety in spring, 269; his attention directed to bee-pasture, 269, 307; to artificial swarming, 270; his work in the respective months of the year, 313-338.
Bee-keeping, pleasantness of its pursuit, 1, 338; styled by Ehrenfels the 'poetry of agriculture,' 1; material advantages derived from, ib.; contributes to the prosperity of a country, ib.; knowledge of the economy of bees necessary for the perfect conduct of, ib.; remunerative in districts extensively cultivated, 2; attention recently bestowed on, ib.; its importance acknowledged by governments, 3; its probable future importance, ib.; theory of, 5-36; practice of, 41-136.
Bee-louse, 283.
Bee-moths. See Wax-moth.
Bee-pasture, 307-310.
Bee-stand, 41.
Bee-stings, protection against, 290.
Bell-glasses as supers, 210; management of, 211.
Bienenzeitung, the organ of the Association of German bee-keepers, 16, 242, 244, 265, 266, 283, 301, 312, 321.
Bilberries, 216, 219, 308.
Birch-trees, juice from, much relished by bees in spring, 43; advantages of, near an apiary, ib.
Bird-cherry, 308.
Black bees erroneously considered to be a distinct kind of bees, 19; their colour, how caused, 19, 20; are generally robbers, 20.
Bogenstülper hive, 70, 71.
Bottom-rails of frames, 65; uselessness of, 65 n.
Box-hives, removing combs from, before travelling, 214; uniting stocks in, 227; the best mode of placing barley-sugar in, 234; how to make them retentive of heat, 239; preparation of, for wintering, 240.

Breeding, caution against early and late, 32 n., 313; effect of its being checked by ungenial weather, 194; young bees after the honey-harvest to be discouraged, 218; in winter to be discouraged, 316.
Brood, food of, 31; preparation of its food by nurse-bees, 32; increased amount of food required by, on the return of spring, 36; raised largely in wide and round hives, 49; at what temperature it may be destroyed, 61; is not to be placed in the sun, ib.; water required for the preparation of, 236; feeding for increased deposit of, 258.
Brood-cells, wax required for sealing, 33; effect of introduction of, among swarm-bees, 162, 163.
Brood-nest, frames not desirable in, 64; chamber above, in twin-stocks, 77 n.
Buckthorn, 308.
Buckwheat, 2, 21, 207, 213, 216, 235, 309, 325, 329, 330; curious phenomenon connected with bees when working on, 279; colour of honey from, 306.
Busch attributes queen depositing only drone-eggs to incomplete fertilisation, 15.

Candy, 203, 277, 335; insertion of, in winter, 227; as food for winter, 314; provision for the solution of, 322, 323 n.
Carbon an essential element in honey, 241.
Carniolan bee, its qualities, 7; difficulty in maintaining its purity, ib.
Cashiered stocks, 223, 275, 326, 332; bees of, to be allotted to other hives, 225.
Cellars, burying bees in, in winter, 242; advisability of making them adjuncts to apiaries, ib.; uses to which they might be put in summer, 242.
Cells of bees, construction of, 25; are inclined upward for the storage of honey, ib.; purposes of, 26-28; drone and worker, 26, 27; in the outer combs of a hive, 222 n.
Cheese, old strong smelling, its effect on bees, 164.
Chilled bees, how they may be revived, 248-250, 254.
Chyle, preparation of, in the body of bee, and its conveyance into the cells, 32; its nutritive power, ib.
Chyle-glands, 32.
Clamps for wintering bees, 242.
Clay, hives of, 46; should have a secur

foundation, ib.; inconveniences of, 46, 47.
Cleaning stocks, 265.
Cleansing flight, 239, 253, 316, 318; under what circumstances dispensed with, 36; inducing bees to, 243; how long bees can do without, 250; effect of hindrance of, 251; not to be encouraged in bees showing no need for, 316, 317.
Close-ended frames, 64 n.
Clovers, 2, 309.
Clipping combs in autumn, 334.
Cold, precautions against, 238, 239.
Collaterals, 136, 209, 210.
Colonies perishing from water-dearth, 33; roused to new life on the return of spring, 36; existing in crevices of rocks and walls would thrive in clay hives, ib.; in hollow trees and buildings, 51; requirements for the multiplication of, 165.
Combs, construction of, 25; passages left in, ib.; consist of double rows of hexagonal cells, 26; utilising, 61, 61 n.; preservation from moths, 61, 62; cleaning, 62; damaged and mouldy, 62; affected with foul brood, not to be used again, 63; value of, for new hives, 145; furnishing new hives with, 146; amount of, that bees can cover, 147; with brood, inducing bees to settle in, ib.; of cashiered stocks, preservation of, 199; shortening of, 201, 228, 327; benefits from extracting honey from, 207; used for brood-rearing gain in firmness, 215; securing when on a journey, ib.; preservation of, 226, 327; clean set of, free from drone-comb necessary for wintering, 233; mode of cleaning, 283; obtaining empty, for artificial swarms, 321; placing, in hives, ib.
Comb-bars, shape of, 54; width of, 55; number of, in hive, 55, 56; distances between, how maintained, 56, 57.
Comb-beginnings, 326; for later swarms, 327.
Comb-cutting in spring, 264-6; opponents of, 264, 267; objects of, 264, 265; advantages from, 265, 266; fittest time for, 266; to what extent it may be practised, 266-268; results of, 321; reasons for, ib.
Comb-foundation, advantages of giving worker, to queenless stocks, 28 n.; objections to, 60, 61.
Comb-guides, 212.
Comb-horse, the, 293, 294.

Comb-roller, 303.
Condemned bees, utility of, 180 n.
Condensation of the vapours from bees, 237, 237 n.
Convenience of handling in hives, 51.
Corn-flower, effect produced by, on the appearance of bees, 20, 21; yields honey, 308, 309, 325.
Cover-boards of box-hives, 237; best material for, 237.
Covering boards, 56, 57.
Cow-dung, dried, used for smoking bees, 289.
Cross-bars of frames, 65; advantages of, 66.
Cross-sticks in skeps, 115 n.
Crow-foot, 279.
Crown-board, 57 n.
Crystallised honey dissolvable by bees, 235.
Curd-cement, 58.
Cutting out the honey recommended, 207.
Cyprian bees, vii.

Dampness in hives, 246; not to be found in well-constructed ones, ib.; to what extent necessary in hives, 315.
Danish bee-keeper, a, employment by, of thick wire as sides of frames, 66.
Dead bees, removal of, 62, 63, 252.
Death's-head moth, 283.
December, 337.
Diseases of bees, 270.
Disturbances, injurious effects of, on bees in winter, 244.
Dividing-boards, use of, in log-hives, 69.
Division of stocks, when injurious, 199, 333; reasons assigned for, 217.
Door-hive, 72.
Dragon-flies, impregnation of, 14.
Driving, 171, 226, 326; purposes of, ib.; best time for, ib.; from Ständers, 171, 172, 176; process of, 172; discovering the queen in, 173; from box-hives, 174, 177; from straw skeps, 176; from Lager hives, ib.; from manifold hives, 177.
Drones, or male bees, have no stings, 8; description of, ib.; their cells, ib.; small, abnormal in a hive, ib.; large and small, 8 n.; do not survive the act of fertilisation, 13; how attracted by queens, 14; sole purpose of, to fertilise young queens, 21; are not intended to produce heat in the hive, 22; their utility asserted, 22 n., 23 n.; their short lives, 22; breeding of, re-commences in spring, 23; purity of, not affected by the mingling of races, ib.;

those produced by fertile workers are capable of pairing with queens, 24; necessity of the presence of, during the queen's matrimonial excursions, 166; selection of, by the queen, 166 n.; breeding early, 194, 320; method of treatment of superfluous,197; raising, by second swarms in their first year, 221 n.; tolerated late in the season by queenless stocks, 221, 334; massacre of, 327.

Drone-brood, presence of, contributes to the pacifying of queenless stocks,189; caution respecting the introduction of, ib.; presence of, a sign of a queenless stock, 230.

Drone-cells,8; dimensions of,26; seldom built during the first year, 27; purposes of, 27 n.; why built by queenless stocks, 28; when occupied, 37.

Drone-comb, 268; built by queenless stocks, 33 n.; how dealt with in adverse weather, 194; utilisation of, in the honey-season, 204; discouraging the building of, 323.

Drone-traps, 197, 297.
Drumming. See Driving.
Dryness of hives, 236.
Dwellings of bees. See Hives.
Dysentery, produced by the bees taking up water in cold weather, 33 n.; its effect on bees, 254; not frequent in English apiaries, 255 n.; causes of, 270, 271; its contagious nature, ib.; the queen not subject to, ib.

Dzierzon, Dr., honours conferred on him as a bee-keeper, 2, 3; his experience in bee-keeping, 4; opposition to his views as to the sexual relations of bees, 6; results from his introduction of the Italian bee into Germany, 6, 7, 193; his views as to large and small drones questioned, 8 n.; his views as to the inability of worker-bees to remove eggs from cell to cell questioned, 11 n.; proves that queens may be reared from larvæ as long as they remain unsealed, 12; opposition to his views on the fertilisation of the eggs of bees, 16; his views respecting the utility of drones controverted, 22 n., 23 n.; his statement that eggs are deposited in queen-cells not borne out by ocular testimony, 37 n.; his hives of straw and wood combined, 46; discontinues the use of magazine hives, 50; his use of twin-stocks, 98; his multiplication of Italian stocks by means of artificial swarming, 193; recommends twin-stocks and double Lager hives for wintering bees, 311.

Eggs, can worker-bees remove them from one cell to another? 11, 11 n.; power of the queen to deposit, in drone or worker cells, 15; of drones do not require fertilisation; 16; fertilisation and non-fertilisation of, 16, 17; laying, discontinued at the departure of the old queen, 38.

Egg-and-milk feeding, v, 258.
Egyptian bee, its qualities, 7.
Egyptians, travelling culture practised by, 216.
Ehrenfels, Baron von, his views as to the age attained by bees,21; considers wintering bees the masterpiece of the bee-keeper, 310.

Eight-fold hive, 134; preparation of, for wintering, 240.
Ekes, 209.
Elm, 308.
Empty combs, utilisation of, during the honey harvest, 219.
Enemies of bees, 280.
Entrances, not to be closed in winter with perforated zinc, 239; narrowing of, 245, 331.
Equalizing driven swarms, 174, 175.
Eyebright, 309.

Feather for brushing bees from combs, 295.
February, 316.
Feeding, 255, 259, 324; young swarms, 202; when stimulating, is dangerous, 256; when advantageous, 256, 257; speculative, 258, 261; with eggs and milk, ib.; with sealed honeycombs, 259; caution respecting, ib.; leads to robbing, ib.; in winter, 313, 314; cutting combs convenient for,322; time for, ib.; in autumn, 338; in box-hives and in skeps, ib.

Feeding-box, 299, 300.
Fertile queens, use of, in making artificial swarms, 178-180; how obtained, 188.
Fertile workers, 16; deposit their eggs irregularly, 19.
Fertilisation of young queens, 188; signs of, ib.; no effect produced by, on drone progeny, 195, 195 n.; late in the season, 221.
Field-radish, 309, 325.
Fighting, 225; prevention of, 167, 226.
Firs, 308.
First swarms, 153; their objection to a young unfertilised queen, 162.

Flight, inciting bees to, 336.
Flights, frequent, in autumn injurious,
Food, of bees, 30; consumption of, by weak stocks, 335.
Flour a substitute for pollen, 31.
Fly-catcher, 281.
Fly-wolf, a destroyer of bees, 282. 229.
Fork, its use in comb-cutting time, 294, 295.
Foul-brood, 271-278; nature of, 271; danger of, 271, 272; symptoms of, 272; mild and malignant, 273; causes of, 273, 274, 277; treatment of stocks infected with, 274; germs of, not destroyed by sulphuring, 276; cure of, by means of salicylic acid, 277, 278.
Fourfold hive, 134.
Frames of hives, 63; distance of, from walls and from each other, 64; their convenience for the storage of honey, ib.; description of, ib.; under part of, made of one piece, 64, 64 n.; lower cross-bar of, 65; disadvantages of, ib.; preference given to those without bottom-rail, ib.; with cross-bars, various uses of, 66; use of strips of tin as sides of, ib.; made of thick wire, ib.
Frame-hives, English, advantages of, 177 n.
Fruit-trees, 308.
Fungoid growth in stomachs of bees, 280.
Füssgangerei, a disease of bees, 279.

German bees, divided into honey and swarming bees, 7.
Gloves, use of, deprecated, 290.
Gravenhorst's Bogenstülper, its superiority to straw hives, 70; description of, 71; division-boards in, ib.
Grooves in hives for accommodation of comb-bars, 56, 57; advantages of a pair of, in the honey-room, 206; second pair of, to be inserted to secure combs, 216.
Ground-ivy, 308.
Guide-combs, 58; mode of attachment to combs, ib.

Hawthorn, 308.
Hazel-nut, 307.
Heat, retention of, in hive to be provided for, 45; escape of, from hive, 146; influence of, on bees, 287.
Heath-bee, its propensity to swarm, 7.
Heather, 213, 216, 309, 331.
Hilbert, Emil, his experiments in egg-and-milk feeding, 258; his experiments on the treatment to be observed for the cure of foul-brood, viii, 277.

Hives, position of cluster in, in winter, 40; proximity of, to railways to be avoided, 42 n.; materials of, 44; requisites of, ib.; walls of, 44, 45; retention of heat to be provided in, 45; respective advantages of straw and wood for, 46; of straw and wood combined, ib.; of clay, ib.; shape of, 47, 49; angular to be preferred to round, ib.; Lager, 47; Ständer, 48; width of, 48, 49; size of, 49, 50; to be constructed so that the honey may be easily taken, 49, 51; divisible, 50; convenience of handling, 51; side-door of, 52; different kinds of, 67-136; for queen-raising, 135; do not require to be made so large as formerly, through use of extractor, 136; furnishing new, 146; difficulty of regulating, with immoveable combs, 204 n.; dangers to which they are liable in travelling, 214; preparations of, for travelling, 214, 215; ease by which they can travel down a stream, 216; construction of, for retention of heat in winter, 238; when well constructed do not require outer packing in winter, 240-244; cleaning in spring, 253, 255; treatment of, when infected with foul-brood, 275; improvements in, 312; change of position of, 317; preparations of, for swarming and for swarms, 326, 327; securing, for the winter, 337.
Honey and wax, production of, contributes to the prosperity of a country, 1; considered valuable products by the ancients, ib.; demand for, at the present time, 1-3; methods of obtaining, 139; are the objects of bee-keeping, 218; refining, 303; separation of, 304; further use and manufacture of, 305
Honey, its probable future employment in the manufacture of wine, 3; cells built for storage of, inclined upwards, 26; is non-nitrogenous, 31, 258; its utility in maintaining the process of respiration and generating heat, 31; with pollen required to keep up the strength of bees, 34; also for the production of wax, ib.; amount of, required to yield one pound of wax, ib.; consumption of, in winter, 35; amount of, required by the bees to prevent death, 35; time for making surplus, 39; bees' eager desire after, ib.; large returns derived from the use of spacious hives, 49; not pro-

cured so well in round and wide hives as in narrower and long ones, 49; to get, for the table, 210; how the production of, may be increased, 218; one pound of, per month consumed by bees during winter, 234; given as food, 259; caution respecting using, from affected hives, 273 n.; extraction or straining of, 303; separation of, from wax, 304; increase in consumption of, 306; its uses in preserving and sweetening fruits, &c., ib.; for making mead and wine, ib.: prospect for a market for, ib.; introduction of, to hives in winter, 314; extracting, from hives, 320; to be extracted before travelling, 332.

Honey-combs, mode of emptying, 60.
Honey-dew, 309, 332.
Honey-extractor, v, 49, 60, 136, 207, 301.
Honey-harvest, prevention of breeding during the, 329; enlarging the honey-room in, ib.; end of the, ib.
Honey-magazine, 326.
Honey-production may be increased by the removal of the queen, 221; by the introduction of a race of bees more productive of honey, ib.
Honey-room, emptying of the, 206-210; passages to, in winter, 314.
Honey-stocks, management of, 203-206; facility for their travelling, 213.
Honey-slinger. See Honey-extractor.
Honey-wine, promotion of the manufacture of, 306.
Hook for pulling out comb-bars, 295, 296.
Horn, a disease of bees, 279.
Hornets are destructive of bees, 282.
Horse-beans exude honey, 309.
Horse-chestnut, 368.
Housing hives, 337.
Hruschka, Major von, his invention of the honey-slinger, 301.
Huber considers the appearance of drone-eggs only in hives due to queen's retarded impregnation, 15.
Hubler, Mr., of Altenberg, on the use of puff-ball, 192; on the destruction of the germs of foul-brood by sulphuring, 276.
Humble-bee, 166.
Hybrids, industry of, 189.
Hymenoptera, 5.

Impregnation of queen-bees, 13; takes place at a distance from the hive, ib.; drones do not survive the act of, ib.; signs of, 14.

Interchangeability of comb, 57, 58.
Italian bees, marks distinguishing them from the German and English bee, 6, 20; their advantages known to Virgil, 6; their superiority to the common bee, ib.; increased knowledge of the sexual relation of bees resulting from their introduction, 7, 19; decide the duration of the life of bees, 20; propagation of, 189, 193; require less trouble than black, 222; their industry and productiveness of honey, 227; rarely sting, 290.
Italian drones required to keep stocks pure, 193; breeding, 194.
Italian queens more easily discovered than black in a hive, 10; facts in the natural history of the bee proved by their introduction, 15, 20, 21; their crossing with black drones does not affect their male progeny, 23; their introduction to after-swarms, 161; their introduction to stocks, 190, 191; their fertility, 222; their price, 334.
Italianising stocks, 191, 192; favourable time for, 333.

January, 313.
Jelly, royal, 11; pollen a necessary constituent in, ib.
July, 328.
June, 325.

Kalinski, Mr. J., recommends Phacelia congesta as a honey plant, 310.
Klotzstock or Klotzbeute hive, 67.
Knauff, Mr., his method of uniting, 228.
Knife for loosening combs, 295; for unsealing combs, 303.
Köhler, Pastor, on spring comb-cutting, 265.
Kolbe, Professor, discovers salicylic acid to be a cure for foul brood, 277.
Krüger on Dr. Dzierzon's improvement in twinstocks, 311.

Lager hives, 47, 67; are easy to transport, ib.; inconvenience of, in winter 48; description of, 48 n.; recommendation respecting, ib.; length of comb-bars in, 56; introduction of queens into, 189; taking honey from, 208; insertion of barley-sugar in, 235; empty space in, how to be filled in winter, 239; shortening combs in, 265, 321.
Lager-hives, double, 95-104; with four

compartments, 105, 108; suitable for travelling, 213, 217; recommended for wintering bees in, 310, 311.

Larvæ of honey-bee, development of, 10-12; royal, 10, 11; age of, necessary for the production of queens, 11, 12.

Ligurianizing with small swarms, 190; best method of, 191 n.

Lime-trees, 207, 216, 308.

Log-hives, side-door to, 52; description of, 67; continuance of the employment of, not desirable, 69; brood-combs taken from, 144; transferring bees from, 145; making artificial swarms from, 170; signs of the presence of queens in, 201; cutting honey-combs from, 207; the best place in, to put barley-sugar in, 234; not necessary that they should be wrapped up for the retention of heat, 241; brushing rime from, in winter, 246.

Lüneburg Heath, transport of hives to, 216.

Lupins, 309.

Magazine, or divisible hives, 50; making artificial swarms in, 168.

Maiden swarms, 330.

Manifold hives, 67, 128; with three compartments, 129; not suitable for travelling, 213; treatment of, in winter, 240, 241; changing position of, 317.

Maple, 308.

March, 319.

Martens, their fondness for honey, 280.

May, 324.

May-flies, impregnation of, 14.

Mead, 306.

Mice, enemies to bees, 238, 280, 315, 336, 337.

Micropyle, 14.

Middle wall of the comb to be preserved, 59, 60.

Milk-feeding, v, 258.

Milk-glands, 32.

Moisture, how it should be obtained, 236; effect of, in hives with open-ended frames, 237; condensation of, 322.

Moss suitable for filling up empty spaces in hives in winter, 239; for conveying to bees candy or crystallized honey, 239.

Mountain-ash, 279.

Moveable, or division-board, 208.

Moveable-comb hives, advantages possessed by, in artificial swarming, 199.

Moveable combs, advantages of, 53, 60, 220; adaptation of, in log-hives, 68; ease with which stocks are united in hives with, 227.

Moveable pieces of board (comb-bars), 53.

Mustard, wild, 2.

Nadir boxes, 210.

Natural history of bees, 40.

Natural swarms, suggested mode of producing artificially, 153; inconveniences of waiting for, 164.

New hives, occupation of, 143, 144.

Nitrogenous foods, 30, 31.

November, 336.

Nurse-bees, young bees employed as, 20; their preparation of food for the brood, 32.

Nutt's collateral hives, 50 n.

Oatmeal, 194, 257.

Observation-hive, 136-138.

October, 334.

Oettle's machine for making hives, 73.

Old hives, changing for new, 144.

Open-ended frames, 237 n.

Ovaries of worker-bees undeveloped, 9, 19; of queens examined, 14.

Oviduct, 16.

Palm-willow, 307.

Parent stock, effect of making artificial stocks from, 170.

Partition of superfluous space in hives, 50; boards for, ib.

Pasture to be taken advantage of, 219.

Pauper swarms, 152.

Pavilion hive, 123-128; not suitable for travelling, 213; preparation of, for winter, 240, 338.

Pea-flour, a substitute for pollen, 31; colour of the pellets of, 31 n.

Pettigrew hives, 49 n.

Phacelia congesta, 310.

Phylloxera, ravages of, 3.

Pincers for removal of comb-bars, 57.

Pines, afford a pleasant shelter to an apiary, 43; honey from, liable to cause dysentery in winter, 207, 235, 308; honey from, 308.

Pipe for smoking bees, 288.

Pliers for taking out and putting in combs, 294.

Poisoning robber bees, 263.

Pollen, or bee-bread, its nitrogenous qualities, 11; only deposited in worker-cells, 28; exceptions to this rule, 28 n.; when dried or mouldy is removed by bees in spring, 29 n.; is eagerly gathered by bees in spring, 31, 36; sub-

stitutes for, 31, 257; uses to which it is put by bees, 34; is extensively procured from birch-trees, 43; preservation of, 62.
Pollen-gathering, 320, 321.
Potatoes, preservation of, in winter, 242.
Preserving fruits, honey used for, 306.
Press hives, 90; uniting stocks in, 227.
Preuss, Dr., his investigations into the nature of foul-brood, 277.
Propolis, 39, 57.
Puff-ball, 192.

Queens, natural history of, 9, 10; points of resemblance between them and worker-bees, 10; generally only one in a hive, ib.; development of, 11; may be produced from larvæ as long as they remain unsealed, 12; time required for the evolution of, 12 n.; fertilisation of, 13, 188; duration of their lives, 14, 18; their power of adapting eggs to the respective cells, 15, 17, 17 n., 18; cause of their production of drone-eggs only, 15, 16; their desire for oviposition stimulated by bees, 17 n.; their egg-laying powers, 18; time of departure of, from the hive, 38; length of their cell-life, 38, 38 n.; prevented from returning to the hive when swarming, 157; their matrimonial flight, 160; battles between, 162; impregnation of, takes place in the air, 165; indifference of, in the selection of drones, 166; capturing, during driving, 173; taking, from log-hives, 174; from Lager hives with two moveable doors, 177; examination of, after having left queen-cells, 186; introduction of young, 187; when old can be used as decoys, 189; danger of introduction of, into stocks previously possessing old queens, 192; propagation of, 195; removal of, during honey-gathering, 220, 329; signs of the possession of good, 229; sound and vigorous, requisite for wintering, 233; unfertile, 269; not subject to dysentery, 271; their fertility to be ascertained, 323; value of fertile, 325; loss of, in their matrimonial flights, 325; removing of old, ib.; caging, 332.
Queen-cages, 158, 179, 179 n., 296.
Queen-cells, two classes of, 10; construction of, 11; position of, in hives, 12; colour of the wax of, 30; temperature of the hive when they are built, 37; eggs deposited in, built on drone-comb, 37, 37 n.; formation of, 183; time required for maturing, 184; their insertion in artificial swarms, ib.; use of surplus, 184, 185; cutting out, 185; introduction of, 185, 186; securing, in boxes, 187; danger of sending, by post, ib.
Queenlessness, when continued, disposes stocks to receive a queen, 161, 187; signs of, 229, 233, 257, 258, 268, 269, 319.
Queenless stocks, advantages in giving worker-foundation to, 28 n.; their tendency to build drone-comb, 33 n.; are subject to being robbed, 39; tolerate the existence of drones, 221; how dealt with, 330.
Queen-raising, hives for, 135; compartments in hives suitable for, ib.
Queen-wasps, 166.

Railway, hives travelling on the, 216.
Rain-water preferred by bees, 43.
Rape, 2, 216, 219, 268, 279, 308.
Rational bee-culture, 142.
Red clover, 309.
Redstart, 281.
Refining wax, 207.
Reinforcement of stocks, 225.
Rime on entrances of hives, 245, 246.
Robbers known by their black colour, 20, 261; may be utilised for the formation of artificial swarms, 263.
Robbing, of queenless and weak stocks, 39, 219; Italian bees addicted to, ib.; inclination of bees to, in autumn, 39; disposition of bees to, 259-264; in early spring, 259; evil effects of, 260; prevention of, 261, 330.
Rye, 257.
Rye-flour, 194.

Sainfoin, 309.
Salicylic acid, a preservative against foul-brood, v, viii, 277; administration of, 278.
Schmidt, Mr., of Ingolstadt, his hives and bee-appliances, 112; his apparatus for honey-slinging, 302.
Scholz, Pastor, recommends clay hives, 46; recommends placing bees in pits in winter, 242; on the results of comb-cutting, 321.
Schönfeld, Pastor, his investigations respecting foul-brood, 277.
Scooping vessel, 299.
Scouts sent out to discover new homes for swarms, 24, 25, 327.
Screw, the, for opening doors, 300.

Scriptures, honey and wax employed in, to denote the fertility of a country, 1.
Sealing combs, 219.
Sections, vii.
Separableness of bee-hives, disadvantages of, 50.
Separate hives, 90.
September, 332.
Shrew-mice, precautions against, 238, 315.
Side-door of hives, 52.
Siebold, Theodor von, his examination of drone and worker eggs as to their fertilisation, 17; his lecture on the organs of bees by which chyle is secreted, 32.
Sieve, its use in bee-culture, 290, 291.
Silesia, weather in, 195.
Single hives, making artificial swarms from, 170.
Sixfold press hive, 130; in pairs, 132; not suitable for travelling, 213; treatment of, in preparing for winter, 240.
Size of hives dependent on abundance of bee-pasture and mode of management, 49.
Skeps, reversed when travelling, 215; uniting with, 226; in what part of, barley-sugar should be placed, 234; covering up, in winter, 241.
Smoke, its effect on bees, 167, 288.
Smoker, consisting of a bellows and tin capsule, 288, 289.
Snow, effect of, on bees, 248.
Speculative feeding, injurious effects of, 319.
Spermatheca, 17.
Spermatozoa in worker-eggs, 16, 17.
Spiders, enemies of bees, 286.
Spring, life of bees in, 38.
Spring-cleaning, 252-255, 314.
Spring-feeding, 255; to be continued without interruption, 269.
Ständer hives, description of, 47 n., 109-122; comb-bars in, 56; divisions or storeys in, ib., 67; their superiority as hives for raising brood, 108; in some respects superior to Lager hives, 109; with one compartment, 109-113; principle of, similar to that of Abbotts' Combination hive, 114 n.; the use of bars in, 115; with two compartments, 117-119; with four compartments, 119; divisions in, 149; driving bees from, 171, 172, 176; introduction of queens into, 189; treatment of artificial swarms in, during the first year, 203; treatment of honey stocks in, 205; more suitable for rearing bees than producing honey, 209; taking honey from, ib.; treatment of empty room in, in winter, 239.
Stewarton hive, 50 n., 56 n.
Sting-poison, 261.
Stings, females of bees, wasps, and hornets, provided with, 8; are absent in male bees, ib.; protection against, 290.
Stock, number of bees in a, 233.
Stocks, continued queenlessness disposes them to receive a queen, 161; selection of, to winter, 235.
Stohr, Director, recommends clay hives, 46.
Storing stocks, necessary for wintering, 233.
Stork, 281.
Stosch, Count, mode recommended by him of placing hives in winter, 243
Straw as a material for hives, 46.
Straw hives, their antiquity, 69; their adaptation for raising brood rather than yielding honey, ib.; placing supers on, 70; how brood-combs can be taken from, 69, 144; how to obtain swarms with young Italian queens from, 198.
Suffocation of bees, how prevented in travelling, 214.
Sugared milk, 258.
Sulphuring bees, 139, 220.
Summers, moist, favourable for the production of brood, 208, 209.
Sunshine on hives in winter, 42 n., 315.
Supers, 136, 209.
Supplementary cells, 10, 11, 37.
Swallows, 281.
Swarming, signs of, 37, 38.
Swarm method, 139, 140.
Swarm, the first, 153; making a large, 175.
Swarming-time, the work of the bee-keeper in, 325, 326.
Swarms, number of bees in, 21; their desire to possess a new home, 24; send out scouts, 24, 25; their settlement in the hive, 25, 156; their construction of combs and cells, 25, 26; more abundant from small hives, 49; origin and treatment of different kinds of, 151, 152; securing the, 158; mode of dealing with two flown together, 161; fertile queens acceptable to, ib.; mode of division of, with two queens, 162; how dealt with when settled in unlikely places, 163; that have once flown away, their liability to go off again, 164; prevention of, ib.;

introduction of queens to, 175 n.; attention necessary to, for wintering, 200; to be kept strong, ib.: feeding, 328; attention to young, 330.
Swarm-bees, 221.
Swarm-cells, 10, 11, 37.
Swarm-net, or sack, 158, 298.
Sweet water gained in washing wax, useful as food for swarms, 207.
Sweet beer-wort, 269.
Sycamore, 308.
Syringe, uses of the, 300.
Syrup, 259.

Temperature in the hive, higher required in extended comb-making, 33; when queen-cells are built, 37; required for the bees to fly out, 39.
Thirst of bees, 31-33, 42, 43, 194, 236, 257, 322.
Thorstock (door-hive), 72; its construction and dimensions, 73.
Three compartments, the hive of, 129
Tin-pan for melting wax, 300.
Titmice, precautions against, 238; disturbance caused by, to bees, 281.
Toads eat bees, 281.
Tobacco, 2; flowers of, yield honey, 2, 309; smoke, its effect on bees, 288.
Touchwood, 226, 288, 289; soaked in a solution of saltpetre, its effect on bees, 163, 289.
Transport hive, facility of making artificial swarms in, 168; uniting bees in, 227; use of, for carrying swarms, 291; various sizes of, ib.; manufacture of, 292; used for hanging up combs, 292, 293.
Travelling bee-culture, hives suitable for, 136; its advantages, 213-218, 331; practised by the Egyptians, 216.
Travelling by night preferable, 215.
Tree, the best time for felling a, for the extraction of a swarm, 164.
Trifolium hybridum, 309; incarnatum, ib.
Triple hive, 129, 130.
Tube, insertion of, to prevent robbing, 263.
Tuft, a disease of bees, 279.
Turnips, 308.
'Tüt' swarms, 152, 154; why so called, 38.
Twinstock hives, prizes awarded to, at Dresden and Stuttgart, 76; their extensive use, ib.; construction and dimensions of, 76, 95; doors to, 80; entrances to, 81; moveable board, or dummy, in, 82, 83; pile of, 92, 93; advantages of, 97-104; natural swarms from, 153; convenience of, in dividing double swarms, 162; favourable for making the artificial swarms remain in the same apiary as the natural, 169; taking honey from, 208; suitable for travelling, 213, 217; ventilation of, on journeys, 215; securing combs in, when travelling, 216; ventilation of, in winter, 219, 241; uniting with, 227; mode of placing, in cellars in winter, 242, 243; convenience of, in winter, 252; recommendation of, for wintering bees, 311; preparation of, for winter, 338.

Unfertilised queens produce drone-eggs only, 16.
Uniting bees, 217, 225, 332; stocks should be at some distance from each other, 227; November the best time for, ib.; in spring, 319.
Uniting spirit, 226.

Vegetables, preservation of, in winter by burying, 242.
Ventilating doors, 215.
Ventilation of hives in order to stimulate bees, 219.
Vertigo, 278.
Vetches exude honey, 2, 309.
Virgil, his description of the Italian bee in the Georgics, 6; refers to the queen-bee as a king, 8.
Vitality of bees in winter, 35.
Vogel, Mr., his opinion of the Egyptian bee, 7.
Volatile oil, effect of the application of an offensive, on bees, 164.

Warmth, a chief necessity for bees and brood, 44; retention of, in bee-hives, 45; how best obtained in winter, 315.
Wasps, 283; destruction of, 336.
Water is indispensable to bees in preparing food for brood, 31-33, 194, 322; places where it is procured by bees, 33; temptation to bees to take up, and the evils attending it, 33 n.; should be supplied to bees, 42, 43, 257.
Water-dearth, colonies perishing from, 33.
Wax, no substitute for, discovered by chemistry, 2; production of, one of the principal aims of bee-keeping, 29; is a secretion of the bees, ib.; scales of, 30; colour of, ib.; old and new, used by bees in lengthening

cells, 30; difference of opinion among bee-keepers as to what it is secreted from, ib.; is a non-nitrogenous substance, 34; is produced from honey and pollen, ib.; number of pounds of honey required to yield one pound of, ib.; production of, 61; when evaporated by being put on burning charcoal, its effect on bees, 163; tin pan for melting, 300; separation of, from honey, 304; refining of, ib.; suggestions for the greater production of, 304, 305, 307; uses of, 305.

Wax-comb, burnt, used for smoking bees, 288, 289.

Wax-dust, preservation of, 252.

Wax-making, discontinued by the interruption of breeding, 33; cessation of, in autumn, 39.

Wax-moth, precaution against, 26, 225, 239, 283, 327; species of, 283, 284; injury to combs by grubs of, 284; effect of a dry atmosphere on, 284 n.; how to get rid of, 285.

Weak stocks, in danger of being robbed, 39; not to be retained during the winter, 223; treatment of, in winter, 248; strengthening of, 322, 323; assisted by the insertion of sealed brood, 202.

Weather, destructive influences of, on bees, 286; in winter, ib.; in spring, 287; in summer, ib.; alleviations of its effects, ib.

Weiser, or Weisel, the ancient German name for the queen, 8.

Wheat-flour, a substitute for pollen, 194, 257.

White clover, 207, 216, 308, 325.

Whortle-berry, 308.

Width of hives, 48, 49.

Willows, 307.

Wine, honey as an ingredient in, 3.

Winter, life of bees in, 40; preparations for, 223, 238, 337; strong stocks only to be kept during, 223; requisites of a stock which is to stand the, 233; absence of disturbing noises in, 244; suitable temperature for, 245; alternations of cold and heat in, ib.; perfect rest required in, 316; management of weak stocks in, 330.

Wintering bees, hives recommended for, 310, 311; in spare rooms, 311.

Wintering hives, 337.

Wire-pins for regulating distances between frames, 64.

Wire, use of, for sides of frames, 66, 67.

Wood as a material for hives, 46, 47; placed parallel with the length of the hive renders comb secure, 115, 216.

Woodbury, T. W., xi.

Woodpeckers, precautions against, 238, 281; are destructive of bees, 281.

Worker-bees not to be described as neuters, 9; are undeveloped females, 9, 19; egg-laying, appear in exceptional cases, ib.; why so called, 19; supposed at one time to produce all the drones in the hive, ib.; distinctive duties performed by the young and old, 20; duration of their lives, 20, 21; number of, in a hive, 21.

Worker-brood to be limited during honey-gathering, 220.

Worker-cells, 8; dimensions of, 26; their suitability for fixing a measure, 27.

Worker-eggs, rate at which they are deposited, 37.

Young bees to be preferred to old, 193.

Zeidel, or depriving method, 141.
Zinc slide, 239.

www.ingramcontent.com/pod-product-compliance
Lightning Source LLC
Chambersburg PA
CBHW062211220526
45471CB00009B/3161